Categorical
Longitudinal Data

Categorical Longitudinal Data

Log-Linear Panel, Trend, and Cohort Analysis

Jacques A.
HAGENAARS

SAGE PUBLICATIONS
The International Professional Publishers
Newbury Park   London   New Delhi

For information address:

SAGE Publications, Inc.
2111 West Hillcrest Drive
Newbury Park, California 91320

SAGE Publications Ltd.
28 Banner Street
London EC1Y 8QE
England

SAGE Publications India Pvt. Ltd.
M-32 Market
Greater Kailash I
New Delhi 110 048 India

Printed in the United States of America

Library of Congress Cataloging-in-Publication Data

Main entry under title:

Hagenaars, Jacques A.
    Categorical longitudinal data : log-linear panel,
trend, and cohort analysis / Jacques A. Hagenaars.
        p.   cm.
    Includes bibliographical references.
    ISBN 0-8039-2957-9
    1. Multivariate analysis. 2. Longitudinal method. I. Title.
QA278.H33   1990
519.5'35 — dc20   90-8254
                                                                      CIP

FIRST PRINTING, 1990

Sage Production Editor: Susan McElroy

Contents

Preface   8

Acknowledgments   11

1. **The Analysis of Social Change**   13
   1.1 Availability of Social Change Data   13
   1.2 The Nature of Social Change Data   14
   1.3 Research Designs for Establishing Social Change   17

2. **Log-Linear Analysis of Categorical Data**   23
   2.1 Introduction   23
   2.2 Basic Concepts and Notation   26
   2.3 The Saturated Frequency Model   33
   2.4 Unsaturated Frequency Models   46
   2.5 Testing and Fitting Unsaturated Models   56
   2.6 The Collapsibility Theorem   68
   2.7 The Effect Model   70
       2.7.1 The Modified Multiple Regression Approach:
             Effect Models with a Dichotomous Dependent Variable   71
       2.7.2 The Modified Multiple Regression Approach:
             Effect Models with a Polytomous Dependent Variable   75
       2.7.3 The Modified Path Analysis Approach   77
   2.8 Ordinal Categorical Variables   82
   2.9 Some Problems: Small Sample Size and
       Polytomous Variables   87

3. **Latent Class Analysis and Log-Linear Models with
   Latent Variables**   93
   3.1 Introduction   93
   3.2 Latent Class Analysis: Log-Linear Models with
       One Latent Variable   95
       3.2.1 The Basic Model   95
       3.2.2 Obtaining the Maximum Likelihood Estimates:
             The EM- Algorithm   103

3.2.3 Restricted Models 108
3.2.4 Identifiability: Testing and Fitting 111
3.3 Association between Latent and External Variables:
Latent Class Scores and Quasi-Latent Variables 113
3.4 Models with Two or More Latent Variables 119
3.4.1 Saturated Models for Latent Variables 119
3.4.2 Unsaturated and Modified Path Analysis Models for
Latent Variables: A Modified LISREL Approach 123
3.5 Local Dependence Models: Direct Effects between Indicators 126
3.6 Simultaneous Analyses in Several Groups 127
3.7 Causal Models with Latent Variables 135
3.8 Ordinal Latent Variables 143

4. Panel Analysis: Answering Typical Panel Questions 147
4.1 Introduction 147
4.2 Changes in One Characteristic: Log-Linear Analysis
of a Turnover Table 148
4.2.1 Introduction 148
4.2.2 (Quasi-)Independence 152
4.2.3 (Quasi-)Symmetry and Marginal Homogeneity 156
4.2.4 A Turnover Table with Ordered Categories 162
4.2.5 "Difference" Scores 164
4.3 Analyzing Tables with More Than Two Variables 165
4.3.1 Comparing Changes in Different Subgroups 166
4.3.2 Comparing Changes in Related Characteristics 169
4.3.3 Testing Changes in Association 176
4.3.4 Comparing Changes in Different Periods 180
4.4 Latent Variable Models 181
4.4.1 Introduction: Change and Unreliability 181
4.4.2 Analyzing a Turnover Table by Means of
Latent Variable Models 183
4.4.3 Analyzing Tables with More Than Two Manifest Variables
by Means of Latent Variable Models:
Latent (Quasi-)Symmetry 193

5. Panel Analysis: Investigating Causal Hypotheses 203
5.1 Introduction: The Panel Design as
a Quasi-Experimental Design 203
5.2 The One-Group Pretest-Posttest Design 203
5.3 The Nonequivalent Control Group Design 215
5.4 Modified Path Models 233
5.5 Typical Panel Problems 248
5.5.1 Nonresponse 249

    5.5.2 Test-Retest Effects and Correlated Errors:
           Local Dependence Latent Class Models   263

6. Trend Analysis   **271**
  6.1 Introduction   271
  6.2 The Trend Design   272
  6.3 Net Changes in One Characteristic   274
    6.3.1 Time as an Independent Variable   274
    6.3.2 Net Changes in Subgroups:
           Introduction of Additional Variables   285
  6.4 Net Changes in More Than One Characteristic   289
  6.5 Interrupted Time-Series Design   300
  6.6 Models with Latent Variables   306

7. Cohort Analysis   **314**
  7.1 Introduction   314
  7.2 Clarifying the Key Concepts Age, Period, and Cohort   317
  7.3 Alternative Designs   320
  7.4 The Identification Problem   326
  7.5 Log-Linear Analysis of a Cohort Table   332
    7.5.1 Introduction: Obtaining the Data and Clarifying
           the Meaning of Age, Period, and Cohort   332
    7.5.2 The Age × Period Design   337
    7.5.3 The Age × Cohort Design   340
    7.5.4 The Period × Cohort Design   343
    7.5.5 The Three-Factor Design: Age × Period × Cohort   345
  7.6 Extending the Cohort Table   354
    7.6.1 Introducing Additional Variables   354
    7.6.2 An Example: Introducing Gender   355

8. A Summary View   **360**

References   **363**

Author Index   **385**

Subject Index   **390**

About the Author   **398**

Preface

The main impetus for writing this book arose from the firm conviction that our insight into processes of social change can be greatly enhanced by making more extensive and more appropriate use of social survey data. This includes data not only from newly designed studies, but also from past surveys which are becoming more readily accessible to the social researcher as a result of the efforts of several Data Archives.

However, when using these survey data, one has to reckon with the fact that they are very often of a qualitative categorical nature and are generally available for only a few discrete, irregularly spaced moments. Textbooks on longitudinal analysis rarely deal with this kind of data. This book, on the other hand, focuses completely on the analysis of categorical data obtained at a few discrete points in time and thus supplements the standard textbooks.

Because a large variety of important substantive problems involving categorical data can be dealt with within the log-linear framework, the log-linear model occupies a central position in this book.

Special attention is paid to log-linear models with latent variables. Since most of the observed (survey) data social scientists are interested in is more or less unreliable, techniques are needed that make it possible to assess the degree of unreliability and to correct for it. Latent variable models can perform these functions.

This book has been written for social scientists who are interested in the systematic empirical investigation of social change. The exposition of the various techniques has been integrated into general methodological discussions about the potentialities and weaknesses of various research designs. Extensive use has been made of "real world" examples to show the potentialities of the techniques for answering genuine research questions. Advanced knowledge of (social sciences) statistics is not required; knowledge of log-linear modeling at the introductory level is helpful, but not necessary.

After a short introductory chapter (*Chapter 1*) on the types of analyses of social change discussed in this book, an attempt has been made in *Chapter 2* to supply the reader with enough information about log-linear modeling to be able to read the remaining chapters. Extensive references to all important subjects have been provided. Readers already familiar with log-linear models may just skim through this chapter to become familiar with the (Goodman) notation and to get an idea of the points that are considered to be especially important for this study. Readers without any knowledge of log-linear modeling have to read this chapter very carefully. As teaching experiences have shown, this chapter can be understood by those with no more than an intermediate level of statistical knowledge and no knowledge at all about log-linear modeling. But as is unavoidable in a monograph of this kind, the treatment of log-linear modeling in Chapter 2 is somewhat concise.

The next chapter on log-linear modeling with latent variables, *Chapter 3*, is less condensed. Because unreliability constitutes such an enormous problem in the study of change and general treatments of unreliability in categorical data are lacking, a rather extensive and comprehensive overview is presented in this chapter. It starts with the standard latent class analysis model many readers may be familiar with. In line with Goodman's and Haberman's work, this standard model is formulated as a log-linear model with latent variables. It is then shown that this formulation offers the same possibilities for the analysis of "qualitative" longitudinal data as factor analysis, LISREL, etc. do for "quantitative" data.

The core of this book is formed by Chapters 4 and 5 on Panel Analysis, and Chapters 6 and 7 on Trend and Cohort Analysis respectively.

In *Chapter 4* it is shown how panel data can be used to answer a large number of research questions having to do with the nature of both gross individual and net aggregate change. The analyses introduced in the first sections of this chapter implicitly assume that the observed change is identical to real change. In Section 4.4 this assumption is abandoned and log-linear models with latent variables are used to distinguish between real, true change and apparent change caused by unreliable measurements.

Chapter 5 deals with the ways in which the panel design can be regarded as a quasi-experimental design to detect causal relations. The methods developed in Chapter 4 are applied within a causal context. The usefulness of the causal modeling approach for attacking two sets of typical panel problems, namely, those caused by nonresponse and by reinterview effects, is demonstrated in the last section of this chapter.

The chapter on trend analysis, *Chapter 6*, begins with a comparison between the panel and the trend design (Sections 6.1, 6.2). Section 6.3 contains a discussion of how net changes in a particular categorical characteristic can be followed over time, both for the total group and for several subgroups. In Section 6.4 the study of net changes in just one characteristic is extended to the simultaneous analysis of changes in two or more characteristics. In Section 6.5 the trend design is treated as a quasi-experimental design, the interrupted time-series design. As in the panel design, unreliability causes problems in the trend design with regard to separating true and apparent change. Again these problems can be attacked successfully by log-linear models with latent variables. This is the subject of the last section of this chapter.

Chapter 7 deals with cohort analysis. Because the cohort design consists of a series of panel or trend studies, no new analysis techniques need to be introduced. Instead this chapter focuses on the main problem of cohort analysis: how to describe and explain changes in terms of Generational, Life-Course, and Period effects.

At the end of the book in *Chapter 8*, a summary of the main arguments is given, along with a presentation of several computer programs with which the techniques discussed in the previous chapters can be routinely applied.

Acknowledgments

Most of the analyses in this book were carried out when writing my dissertation. Helpful comments were then received from Philip Stouthard, Emmanuel Bijnen, Cees de Graaf, and Jo Segers, all associated with the Department of Sociological Methods and Research of Tilburg University. Several other friends and (ex)colleagues of this department contributed to this study. At many times, there was an intensive cooperation with Ruud Luijkx. Martin van der Walle and Frans Hummelman wrote several computer programs, most noteworthy LCAG and DESMAT. Marcel Croon and Ton Heinen provided, perhaps without being aware of it, the necessary intellectual stimuli to carry on with the book.

The original Dutch manuscript was first translated by Eric Williams and then edited by Marianne Sanders with amazing skill and ingenuity. I treasure the best of memories of our cooperation. The Netherlands Organization for Scientific Research (NWO) made the translation financially possible (Grant number p40-50). I am also very grateful for the additional financial support received from the Faculty of Social Sciences at Tilburg University.

On the road from the original Dutch manuscript to the final English version were many obstacles. Mitch Allen and later on C. Deborah Laughton from Sage Publications have been very helpful in removing them. Masako Ishii-Kunz from University of California reviewed the whole manuscript. Her comments and suggestions have very much improved the book. The same is true for the reactions of J. Scott Long to Chapters 1 and 2.

Inclusion in these Acknowledgments gives to all those mentioned the credit for the positive elements of this book; at the same time, they remain free — regrettably — from its shortcomings: The author is solely responsible for them.

1. The Analysis of Social Change

1.1 Availability of Social Change Data

Valid insights into the nature of past and future social changes can be gained only from theories that have been based on the systematic analyses of empirical data. However, historians and social scientists do not always have at their disposal the data necessary to make these systematic analyses.

As Thomas (1973) put it in the Foreword of his monumental study *Religion and the decline of magic; Studies in popular beliefs in sixteenth- and seventeenth-century England*:

> I particularly regret not having been able to offer more of those exact statistical data upon which the precise analysis of historical change must so often depend. Unfortunately, the sources seldom permit such computation, although it is to be hoped that the information contained in the largely unpublished judicial records of the time will one day be systematically quantified. My visits to these widely scattered archives have been less frequent and less systematic than I should have liked. In my attempt to sketch the main outlines of the subject I have only too often had to fall back on the historian's traditional method of presentation by example and counter-example. Although this technique has some advantages, the computer has made it the intellectual equivalent of the bow and arrow in a nuclear age. But one cannot use the computer unless one has suitable material with which to supply it, and at present there seems to be no genuine scientific method of measuring changes in the thinking of past generations. As a result, there are many points in my argument at which the reader can be given no statistical evidence on which to accept or reject the impressions I have formed after my reading in contemporary sources. (p. x)

This lengthy quotation eloquently summarizes the causes and consequences of the lack of suitable data. On some subjects no data at all are available; on others, the data are not readily accessible. As a consequence, the validity and reliability of the conclusions drawn by the investigator cannot be evaluated adequately.

Students of the more recent past are in a much better position to obtain the data they need. The official bookkeeping of modern societies covers more ground than ever, and does so in a more tractable form. Direct information on the thoughts and sentiments of people is provided in particular by large-scale social surveys whose data are being made increasingly available to the scientific community by well organized Data Archives such as the Archives of the Inter-university Consortium for Political and Social Research housed at the University of Michigan, the Roper Center for Public Opinion Research at the University of Connecticut, the Zentralarchiv für Empirische Sozialforschung at the University of Cologne, and the Steinmetz Archives in Amsterdam.

Investigators can therefore use not only the surveys they design and carry out themselves, if necessary repeated over time to measure social change, but they can also make use of surveys conducted by others in the past. By using surveys from different periods, many processes of social change can be closely followed.

Carrying out secondary analyses of already existing (survey) data is not without serious difficulties (Hakim, 1982; Hyman, 1972; Jacob, 1984; Kiecolt & Nathan, 1985). But social scientists and historians have never had this "genuine scientific method of measuring changes in the thinking of past generations" (Thomas, 1973: x). The aim of this book is to provide the tools appropriate for exploiting the richness of these surveys.

1.2 The Nature of Social Change Data

The kind of social change data an investigator will collect or look for in archives is of course determined by the substantive questions that are being asked and by the theoretical notions an investigator entertains. From a statistical point of view, in order to choose the most appropriate analysis techniques to answer these research questions, it is important whether the data refer to "quantitative" continuous characteristics, such as income measured at interval/ratio level, or whether the data are of a "qualitative" categorical nature, for example, marital status or party preference, resulting from measurements at nominal or ordinal level (disregarding "continuous" ordinal level data as well as discrete interval level data).

Most textbooks on longitudinal analysis deal almost exclusively with the analysis of interval level data, proposing techniques that require assumptions which have little or no plausibility in the case of categorical data.[1] Nevertheless, many of the changes social scientists want to study pertain to changes in discrete categorical characteristics such as family composition, occupation, types of political activism, etc. At other times one may be interested in characteristics that in principle can be measured at interval level (as many attitudes can), but the only data available are the result of coarse measurement of the underlying phenomenon using just two or three broad categories.

Log-linear techniques, most ardently propagated among social scientists by Goodman (see References), are excellently suited to analyze categorical data. And, as this study will show, a large variety of questions about social change that involve categorical data can be answered within the framework of the log-linear model.[2]

Thus, while the theoretical interests of the researcher is the main determinant of which hypotheses about social change will be investigated and which analysis techniques have to be chosen, the nature of the available data itself must also be considered.

In this respect, it is also very important to know how these data have been collected, particularly whether the data have been obtained through continuous or through discrete observation over time. Assuming that in principle the same cases have been followed over time, continuous observation implies that, at each and every moment of the relevant period, the characteristics of the cases are known. With discrete observations, occurring at certain fixed intervals, the characteristics of each case are known only at particular moments.

For example, in a political study one may be interested in the changes in party preference between 1970 and 1985. Continuous observations will then lead to knowing the party preference for each and every moment between 1970 and 1985 for each case. Discrete observations will provide information about the party preference of each case only for those particular moments at which the surveys have been done.

Of course, continuous observations are the ideal and with them the researcher can use very powerful analysis techniques to investigate completely the processes of social change. Particularly relevant in this respect are techniques such as event-history analysis, in which the probability that an individual undergoes a change at any point of time in a particular characteristic, for example, occupation, is studied as a function of the previous changes in the characteristic concerned and the values of exogenous variables like age and gender (Allison, 1984; Tuma & Hannan, 1984). However, with some exceptions, like data on educational or occupational careers, data resulting from continuous observations are rarely available. Because of this, only discrete observations will be dealt with in this study.

The restriction to discrete observations does not necessarily imply restricting attention to underlying processes of change that are also discrete in nature. Coleman was one of the first to study continuously ongoing processes of attitudinal change with the aid of just a few discrete observations over time (Coleman, 1964). To be able to do this, rather strict assumptions about the nature of the ongoing process have to be made. In a certain sense one has to estimate the values of the missing in-between observations. However, there is usually no way of telling whether the assumptions made about the underlying processes are valid. So unless the number of discrete observations approaches a situation of continuous observations, or unless there are very strong theoretical reasons to believe that the changes follow a certain pattern, the results of analyses postulating continuous change based on discrete observations may be very misleading.

On the other hand, adapting the nature of the underlying processes of change to the discrete "spacing" of the observations also does not seem to be a good idea. First, most social change theories would not predict that changes occur only at a few discrete points in time. And, second, even if changes did occur at certain discrete points, it would then be necessary to choose the moments of observation in such a way that all these discrete changes will be detected. There is no way of determining whether this is the case.

In this study, a more or less pragmatic attitude toward the solution of this dilemma has been assumed. Time will be treated as a discrete variable, not with the implication that social change occurs only at certain discrete moments but rather from the viewpoint that the consequences of the possible continuous processes of change will be investigated at certain discrete points in time.[3]

In short, the techniques discussed in this book are relevant as far as the substantive questions that are being asked or the nature of the available data necessitate dealing with changes in *categorical* characteristics whose consequences are analyzed at *discrete points in time*.

1.3 Research Designs for Establishing Social Change

The approach to the analysis of social change advocated is to make use of several cross-sectional surveys repeated over time, since the *one-shot cross-sectional survey* by itself, being conducted at just one point in time, is not suited for the study of social change. Although it is used to this purpose, this is always problematic.

In principle, within the one-shot survey, respondents can provide information not only about their present behavior, attitudes and beliefs, but also about their past. By comparing data from the present and the recollected past, changes can be detected. However, given the fallible and selective nature of human memory, it is clear that this information about the past will be reliable and valid only under very specific circumstances.

Within a particular one-shot survey one can also compare older people with younger people and project the characteristics, for example, the political preferences, of the older ones into the past and those of the younger ones into the future. But these cross-sectional age differences can only be used to tell us something about the past and the future and therefore something about social change when two conditions are satisfied: first, that there is justification to interpret the differences between age groups at one particular point in time entirely as differences between generations and, second, that the generational characteristics are stable over time.

But generational characteristics need not remain the same over time, and cross-sectional age group differences must also be interpreted in terms of life-course differences, as differences between older and younger people and not entirely as differences between generations. Also, whereas (stable) generational differences can easily be related to social change, cross-sectional life-course differences as such tell us nothing about social change. As the one-shot survey provides no way of distinguishing between generational and life-course effects, statements about social change based on cross-sectional age group differences are of doubtful validity. More will be said about this in the discussion of the cohort design in Chapter 7.

A further way in which the one-shot survey is used for the study of social change is by investigating the relations between certain variables at one point in time and explaining past, and especially, future social developments with the aid of these cross-sectional relations. For example, at one particular moment in time one finds a relationship between education and unemployment: The better educated have lower unemployment rates. From this it is concluded that in the future unemployment will decline, as the trend is toward increasing education.

This line of reasoning is correct only if the changes in education and in unemployment over time are related to each other in the same way as the cross-sectional differences in these variables. Or, as it is sometimes stated: The system of variables under study must be in (a dynamic) equilibrium. This of course is an assumption which is hard to justify either theoretically or empirically. (For an extensive discussion, see Carlsson, 1972; Schoenberg, 1977).

So, although one-shot surveys may give some hints as to the way processes of social change develop, in general they do not provide reliable and valid evidence. What is needed is direct evidence from measurements that have followed the process over time — in other words: surveys repeated over time.

There are several designs which involve repeated surveys. In a study using a *panel design*, data on the same topics are collected from the same people at two or more (discrete) points in time (t, $t + 1$, etc). It can be established how the individual score on a particular variable, for example, religion, has changed, that is, whether or not the score on religion at time $t + 1$ is different from the score at time t and, if so, what the difference is. Cases can be identified as nonchangers or changers, and this last group can be divided into subgroups according to the directions the change has taken. Then the differences between these (sub)groups in terms of other characteristics (e.g., age, education) can be investigated to determine the causes and consequences of the changes.

A basic requirement for panel analysis is that the measurements at each period of observation, or wave, provide reliable information about the true scores at that moment. Otherwise, observed (lack of) change is not necessarily indicative of true underlying (lack of) change. The disentanglement of "apparent" and "true" change is one of the most complicated issues in panel analysis.

Panel studies are plagued by other problems as well. A panel tends to wear out. With each new wave, fewer and fewer people are inclined to cooperate. It is not unusual to lose half of the original sample in the second wave. This nonresponse phenomenon should be taken explicitly into account.

Another problem is the reinterview effect (test-retest effect). Particularly when the time span between the successive waves is rather small, it is likely that what happened at a previous interview will in some way influence the answers at a later interview. Although, in general, less of a problem than the nonresponse phenomenon, reinterview effects, when they are anticipated, also have to be dealt with.

Although the basic feature of the panel design is the possibility of detecting individual change, that is, gross change, it appears from many reports of panel investigations that a major part of the analyses does not concern individual change, but involves only aggregate, net change, trends. Panel analysis is then used to answer questions like: Is the total number of supporters of a particular political issue at time $t + 1$ larger or smaller than at time t? Is this trend to be seen more clearly among younger than among older people? These kinds of questions can also be answered by so-called *trend studies*, repeated surveys which do not involve the same cases.

As in the panel design, trend design data are gathered at different points in time from the same population, but, in contrast to the panel design, not from the same people. At each moment of observation a new sample is drawn from the (same) population under study.

While panel studies can be used to investigate both individual and aggregate change, trend studies provide information about the aggregate net change only. They will tell you, for example, how much the total number of supporters of each of the various political parties has declined or increased, but not how this net change came about. One does not know whether the increasing support for the Republican Party is just the result of some Democrats going over to the other side and all Republicans being loyal to their party, or the result of defections from both sides, but more heavily from the Democratic side. Although in principle the trend design is less powerful than the panel design for studying social change, there are many occasions when the substantive questions refer just to aggregate and not to individual change.

In general, the availability of trend data is much greater than of panel data. Panel studies have to be designed as such, trend studies do not. The numerous surveys and polls present in the archives often contain information on the same kind of topics. And although many of them were never intended to provide information about social change, these "unrelated" one-shot surveys can be rearranged into a trend design. Because of this greater availability, trend data will often encompass a much larger time span than panel data.

Studies of long-term social change inevitably have to confront the issues of generation succession and generation replacement. Inevitably, long-term social change has to be understood within the framework of successive generations in each of which people grow up and eventually die in an ever changing historical context, continuously being confronted with new generations.

In a long-term election panel study one may start with a group of people that are representative of the electorate at that moment. But as time passes people from the original electorate that were (partly) present in the panel die and new people not represented in the original panel become eligible to vote. For the study of electoral changes at the societal level, the original panel members provide less and less relevant information. The panel has to be supplemented by new (panel) groups selected from the new generations.

In trend studies a new sample is taken at each period from the same population, for example, the Dutch population. But a trend study in which a sample representative of the Dutch population in 1960 is compared with a sample representative of the Dutch population in 1985 cannot be regarded as consisting of two samples from the same population in a statistical sense. The Dutch population of 1960 does not contain the same people as the Dutch population of 1985. Between 1960 and 1985 some generations have died out and new generations have been born. To understand fully the changes that have taken place, it is almost always necessary to take this process of generation replacement explicitly into account.

This generational aspect gets full attention in the *cohort design*.[4] In a cohort study a number of generations are followed over time, over their life-course. If in each particular generation the same people are investigated, a cohort study amounts to a series of panel studies; if in each generation at each period of observation a new sample is drawn, a cohort study consists of a series of trend studies.

With the cohort design, social changes are studied from three viewpoints, namely, generation, age, and period. The several generations that are included in the cohort study may systematically differ from each other, accounting for several observed social changes. Even just differences in size between the cohorts may have an enormous social impact. For example, large birth cohorts may first lead to overcrowded schools and later on to overcrowded labor markets; at old age they may impose a real "burden" to be borne by the next and smaller generations.

As each generation is followed over its life-course, the consequences of growing up and becoming older can be made visible. It becomes clear in what way the life-course differences are the same or different for the various generations, whether, for example, being old in one generation means the same as in other generations.

Finally, by following the generations over their life-course, they are by definition also being followed over time. Therefore, the influences of the various events that take place in the course of time can be investigated. It can be seen whether or not a certain event (e.g., a war, an economic depression) influences all generations in the same way. After all, each generation experiences the same event at a different stage of its life-course.

Cohort analysis enables the researcher to unravel the complex, often interacting, influences of the three closely related variables, generation, age, and period, on the processes of social change.

With these three forms that repeated surveys may assume, the panel, the trend, and the cohort design, social investigators have powerful instruments at their disposal to get to the core of many processes of social change.

Notes

1. Some authors of books on longitudinal research indicated in the title that they would pay attention only to variables measured at interval level, for example, Kessler and Greenberg (1981). Other titles promised more general surveys, but still dealt almost exclusively with quantitative variables, for example, Goldstein (1979), Hsiao (1986), Nesselroade and Baltes (1979), and Visser (1985). The literature on time series analysis, even when explicitly intended for use by social scientists, also deals only with quantitative data (e.g., Gottman, 1981; McCleary & Hay, 1980). However, the tide seems to be turning. Plewis (1985) devoted three chapters to the analysis of categorical data, and Van der Heijden (1987) presented an overview of the possibilities of correspondence analysis for the analysis of categorical longitudinal data.

2. More in particular, log-linear modeling using the maximum likelihood principle for the estimation of the parameters and testing the fit of the model will be employed. Categorical data can be analyzed by means of log-linear models using estimation methods other than the maximum likelihood method, or by means of models which are based on principles totally different from those underlying the log-linear models. Examples are Davis's d-system (Davis, 1975a, 1978), prediction analysis (Hildebrand, Laing, & Rosenthal, 1977), information theoretical approaches (Gokhale & Kullback, 1978; Krippendorff, 1986), weighted least squares and minimum chi-square procedures (Grizzle, Starmer, & Koch, 1969; Landis & Koch, 1979; Reynolds, 1977; Swafford, 1980), correspondence and homogeneity analysis (Caillez and Pagès, 1976; De Leeuw, 1973; Gifi, 1981; Van der Heijden, 1987). Bishop, Fienberg, & Holland (1975, ch.10) discuss several of these approaches; Goodman (1987) points to similarities of correspondence analysis and log-linear modeling. However, so far the approach chosen in this study

is the most versatile and can handle the most diverse problems within one consistent framework.

3. Consequently differential equations will not play a role in this study. Difference equations implicitly play a role as far as characteristics measured at time t are related to characteristics at time $t + 1$. Explicit systems of difference equations that are embodied in "dynamic models" (Harder, 1973; Huckfeldt, Kohfeld, & Likens, 1982) will not be dealt with. For an interesting discussion of continuous versus discrete processes of change in relation to ways of analyzing social change data, see Kohfeld and Salert (1982).

4. For the time being, the terms cohort and generation are treated as synonyms. In Chapter 7 the two will be distinguished from each other.

2. Log-Linear Analysis of Categorical Data

2.1 Introduction

The starting point for the analysis of the relations among categorical variables is a multivariate frequency table. In most studies done before the 1970s such tables were analyzed by means of the *elaboration procedure* (Hyman, 1955; Lazarsfeld, 1955; Rosenberg, 1968). In the early 1970s, however, social scientists became familiar with loglinear modeling, particularly through the work of Goodman (1972a, 1972b, 1973a), and embraced it wholeheartedly. The fact that the log-linear model became popular so quickly has its roots in the serious shortcomings of the elaboration procedure and in the lack of alternatives.

The standard elaboration procedure starts with a two dimensional frequency table in which the relation between two variables, education and marital status, for example, is represented. Then three and higher dimensional tables are inspected to see if and in what way the original relationship changes when other variables are taken into account and are held constant. It is then possible to discover whether or not the relation between education and marital status is different for older than for younger people or if the original relationship disappears when it is investigated separately among Catholics, among Protestants, etc. Lazarsfeld (1955) presented a general scheme to describe the outcomes of the elaboration procedure in causal terms, distinguishing among "spuriousness," "interpretation," and "interaction."

The results of the elaboration procedure soon become unmanageable. If one wishes to investigate the relation between two variables, holding just three trichotomous variables constant results in as many as $3^3 = 27$ subgroups within which the original relationship has to be examined. Moreover, many of these subgroups will have a small number of cases, resulting in large sampling fluctuations. All this calls for a procedure that is more formalized than the standard elaboration procedure.

Partly because of ignorance of its potentialities and partly because of the shortcomings of the regression model for the analysis of categorical data, most researchers do not consider *regression analysis* to be a real alternative. The standard regression model, in which linear relations are assumed between variables measured at interval level, and which requires the supposition that there are no interaction effects, as well as a number of strict assumptions about the error term that has to justify the use of ordinary least squares estimation, is just not applicable to categorical data measured at nominal or ordinal level.

Variants of this standard model sometimes offer solutions. These variants make use of multivariate procedures, of dummy variables, of product terms, of generalized least squares estimation, etc. (see Bock, 1975; Holm, 1979; Johnston, 1972). However, researchers are often not familiar with these procedures. The literature in this field is not very accessible and the majority of these procedures are not included in standard computer programs. Furthermore, and more important, the results of multivariate procedures, for example, in terms of generalized variances, are often difficult to interpret.

But even disregarding these difficulties, there are other major problems with the regression analysis approach, one of which is the potential occurrence of "impossible" predicted values for the dependent variable. For example, predicted values for a dichotomous dependent variable with categories coded 0 and 1 may well be outside the range 0 - 1. Moreover, in regression analysis, statistical tests are usually based on the assumption of underlying normal distributions, at least of the error terms. This is practically always a very unrealistic assumption for categorical variables.

Essentially, the language of regression analysis was not developed for the analysis of categorical data. It can be adapted in many ways, but it will never be a natural way of talking about categorical data problems. Log-linear models, on the other hand, do provide a natural way of dealing with multivariate frequency tables.

Before going into the log-linear model itself, some basic preliminary concepts, such as maximum likelihood estimation, odds and odds ratios, along with some basic notations will be introduced in Section 2.2.

An important distinction is that between saturated and unsaturated log-linear models. In a saturated model no restrictions are imposed upon the relations between the variables, whereas in an unsaturated model the relations between the variables are restrained in some way, mostly by assuming that one or more (interaction) effects between variables are absent. The saturated log-linear model will be discussed in Section 2.3; Section 2.4 deals with unsaturated models.

The validity of the restrictions imposed by an unsaturated model can be tested empirically against the observed data. How to test and fit unsaturated models is the topic of Section 2.5.

In Section 2.6 an important aspect of log-linear analysis is discussed, namely, the collapsibility of the multivariate frequency table. Say we have three variables, A, B, and C, denoting religious affiliation, political preference, and gender, respectively, and a three dimensional frequency table ABC. If this table is collapsed by summing the frequencies over Gender (C), the result is a two dimensional table AB (Religious Affiliation × Political Preference). By using the collapsibility theorem, it can be established under what conditions the relation between Religious Affiliation (A) and Political Preference (B) is the same in the ABC as in the AB table, that is, with and without holding Gender (C) constant. Insight into the collapsibility theorem is necessary to understand and carry out many of the log-linear analyses in this book.

In the log-linear models discussed in Sections 2.2 through 2.6 no distinction is made between independent and dependent variables, between causes and effects. Such models in which all variables are treated as belonging to the same causal level are called frequency models. Log-linear models that do make the distinction between dependent and independent variables can be found in Section 2.7 under the label "effect model" or "logit model."

In principle, both log-linear frequency and effect models pertain to categorical nominal level data. However, sometimes we have ordinal level categorical data. Log-linear models can be specified in such a way that the ordinal character of the data is taken into account. This is discussed very briefly in Section 2.8.

This chapter is concluded with a short discussion of some common problems in log-linear analysis that are caused by small sample size and by the presence of polytomous variables.

This chapter does not include a complete survey of all the potentialities and limitations of log-linear analysis.[1] The discussion will focus primarily on those subjects which are most relevant for practical research and for what will be dealt with in the core Chapters 4-7.

2.2 Basic Concepts and Notation

The estimation and testing procedures in this book are all based on the *maximum likelihood principle*. This is a statistical principle for finding "good" parameter estimates. Hays (1981) described this principle in a concise but intuitively very clear way as follows:

> In effect this principle says that when faced with several parameter values, any of which might be the true one for a population, the best "bet" is that parameter value which *would* have made the sample actually obtained have the highest prior probability. When in doubt, place your bet on that parameter value which would have made the obtained result most likely. (p. 182)

This general intuitive sense of the maximum likelihood principle is sufficient to follow the main arguments on maximum likelihood estimation in the next sections.[2]

To obtain maximum likelihood estimates one has to compute the likelihoods of the sample results given particular "estimated" population values, and for this, one needs to know the sampling distribution of the data. In regression and factor analysis, it is usually assumed that the data follow a (multivariate) normal distribution. In log-linear modeling, a *multinomial* or *product-multinomial sampling distribution* is postulated.

The latter distributions occur when the population can be divided into K exhaustive and mutually exclusive categories, formed by (all combinations of) all categories of all variables that are part of the log-linear analysis. If a simple random sample with the sample size N fixed in advance is drawn from this population, the data follow a multinomial distribution. If the sample is a stratified sample in which a simple random sample per stratum is drawn with a fixed number of cases per stratum, we have a product-multinomial distribution.

In terms of log-linear analysis, the difference between having a product-multinomial distribution and having just the multinomial distribution means that the fact that the frequency distribution of the stratifying variables is fixed and not random should be taken into account. How this should be done will be explained in Section 2.4.

Categorical survey data of the kind discussed in this study easily meet the conditions necessary for the (product-)multinomial distribution to apply. Thus, maximum likelihood estimates for the parameters of the log-linear model can be obtained without imposing restrictions on the data that are unlikely to be true.[3]

To interpret log-linear parameters it is necessary to be familiar with the geometric mean and with the concepts of odds and odds ratios. Moreover, there has to be agreement on some basic notation. To illustrate these concepts and the notation, the data in Table 2.1 will be used in this section and throughout this chapter. These are fictitious data. It is the only place in the book where fictitious data are used as it was necessary for didactic reasons to have an example with a certain known structure.

Table 2.1, although it has only four variables, contains a lot of not easily detected information about the relations among Age (A), Religious Membership (B), Political Preference (C), and Voting Behavior (D). Political scientists may test and formulate several hypotheses from data like these, for example, about the sources of social support of the political democratic system. This complicated table will be analyzed in a stepwise fashion; at the end of this chapter its (causal) structure will be perfectly clear.

The *notation* that will be used throughout this book is, with some slight modifications, the notation Goodman used in most of his publications (Goodman, 1972-1987; Knoke & Burke, 1980). Observed frequencies are indicated by f; superscripts refer to variables and subscripts to categories of these variables. For example, the first observed cell frequency in Table 2.1, which equals 15, is denoted by f_{1111}^{ABCD}. (For typographical reasons the superscripts are sometimes omitted, e.g., f_{ijkl} instead of f_{ijkl}^{ABCD}.) The observed proportion p of individuals belonging to category i of variable A, j of variable B, k of variable C, and l of variable D is correspondingly denoted as p_{ijkl}^{ABCD}. Thus, p_{1111}^{ABCD} in Table 2.1 is equal to $15/750 = .02$. If N is the total sample size, then:

$$f_{ijkl}^{ABCD} = Np_{ijkl}^{ABCD} \qquad (2.1)$$

Probabilities in the population are denoted by π, so that π_{ijkl}^{ABCD} may be described as the probability that in the population a randomly selected element belongs to $A = i$, $B = j$, $C = k$, and $D = l$. Analogous to equation 2.1, the expected frequency F can be defined:

$$F_{ijkl}^{ABCD} = N\pi_{ijkl}^{ABCD} \qquad (2.2)$$

F represents the frequencies that would have been found if the sample were an exact reflection of the population without sampling fluctuations.

From a purely statistical point of view, the essence of table analysis can be stated as getting estimates of the population parameters π with

TABLE 2.1   Relations Among Four Variables

A. Age	B. Religious Membership	C. Political Preference	D. Voting Behavior			
			1.Yes	2.No	Total	% Yes
1.Young	1. Member	1. Left Wing	15	16	31	48.4
		2. Right Wing	30	5	35	85.7
		3. Christian Democratic	50	27	77	64.9
	2. Nonmember	1. Left Wing	18	50	68	26.5
		2. Right Wing	51	14	65	78.5
		3. Christian Democratic	31	29	60	51.7
1.Old	1. Member	1. Left Wing	35	9	44	79.5
		2. Right Wing	58	3	61	95.1
		3. Christian Democratic	127	20	147	86.4
	2. Nonmember	1. Left Wing	21	47	68	30.9
		2. Right Wing	40	4	44	90.9
		3. Christian Democratic	30	20	50	60.0
		Total	506	244	750	67.5

SOURCE: Fictitious data from Hagenaars and Heinen (1980, Table 2)

the aid of the observed proportion p departing from certain hypotheses about the population, for example, statistical independence between certain variables. In terms of frequencies, F is estimated using f and certain hypotheses.

Denoting the maximum likelihood estimator of π as $\hat{\pi}$, and of F as $\hat{F}$, then, analogous to the above equations, $\hat{F}$ can be defined in terms of $\hat{\pi}$:

$$\hat{F}_{ijkl}^{ABCD} = N\hat{\pi}_{ijkl}^{ABCD} \tag{2.3}$$

The frequencies F_{ijkl}^{ABCD}, $\hat{F}_{ijkl}^{ABCD}$ and f_{ijkl}^{ABCD} are cell frequencies of the full cross-classification of the variables A, B, C, and D in table $ABCD$. From this table, several *marginal tables* may be obtained. A marginal table results from summing over one or more variables of the original table. Table 2.2 is such a marginal table resulting from collapsing the full

TABLE 2.2   Age (A) and Religious Membership (B)

	B. Religious Membership		
A. Age	1. Member	2. Nonmember	Total
1. Young	143 (42.6%)	193 (57.4%)	336 (100%)
2. Old	252 (60.9%)	162 (39.1%)	414 (100%)
Total	395 (52.7%)	355 (47.3%)	750 (100%)

SOURCE: Table 2.1

ABCD Table 2.1 by summing the frequencies in Table 2.1 over Political
Preference (*C*) and Voting Behavior (*D*):

$$f_{ij}^{AB} = f_{ij++}^{ABCD} = \sum_{k=1}^{K} \sum_{l=1}^{L} f_{ijkl}^{ABCD} \qquad (2.4)$$

where summation is denoted by the usual summation sign Σ or equivalently by a plus (+) subscript indicating summation over that index.

With the assumption that Membership of a Religious Denomination
is indicative of being religious, the marginal table *AB* (Table 2.2) can be
used to learn whether older people are indeed more religious than
younger people as some generational as well as life-course theories
suggest. To answer this question, it is useful to work with *odds*, in this
case, the odds of being a member of a religious denomination rather than
a nonmember.

From the last row in Table 2.2 we may conclude that the probability
that an individual randomly selected from the total sample is a member
of a religious denomination is .527 and the probability that this individual is not a member is .473. Expressing the inequality of these two
probabilities by taking their ratio, we conclude that the odds of being a
member rather than a nonmember are .527/.473 = (395/750)/(355/750)
= 395/355 = 1.113 and the odds of being a nonmember instead of a
member are .473/.527 = 1/1.113 = 0.898. In the total sample, the probability of being a member is 1.113 times greater than being a nonmember.

If the two probabilities had been equal, the odds would have been 1.
Odds of being a member larger than 1 would have pointed to a numerical
preponderance of members, $+\infty$ being the upper limit if all people were
members. Odds of being a member smaller than 1 would have indicated
a preponderance of nonmembers, 0 being the lower limit if nobody was
a member.

Because these odds are computed on the basis of a marginal distribution of Table 2.2, they are sometimes called *marginal odds*. More important in terms of our research question are the *conditional odds*. These are like marginal odds, but then computed for subgroups, within categories of some other variable(s). In Table 2.2 the odds of being a member can be calculated separately for the younger and the older people. The ratio member/nonmember for the younger age group is $143/193 = .426/.574 = .741$ and for the older age group is $252/162 = .609/.391 = 1.556$. Among the younger people, members form a minority; among the older people, they constitute a majority. Older people are more religious than younger people.

The more the two conditional odds of being a member deviate from each other, the stronger the association is between age and religion. A measure of this deviation is the ratio of the two conditional odds, the *odds ratio*. In this case: $.741/1.556 = (143/193)/(252/162) = .476 (\approx .5)$. The odds of a younger person being a member of a religious denomination are only half those of an older person.

The odds ratio is equal to a well-known measure of association for 2×2 tables, the cross-product ratio α (Bishop et al., 1975, Section 11.2.2):

$$\alpha = \frac{f_{11}}{f_{12}} \bigg/ \frac{f_{21}}{f_{22}} = \frac{f_{11}}{f_{21}} \bigg/ \frac{f_{12}}{f_{22}} = \frac{f_{11}f_{22}}{f_{12}f_{21}} \qquad (2.5)$$

The first part at the right-hand side of equation 2.5 represents the way the odds ratio has been calculated above. From the equalities in equation 2.5 it follows that the same value .476 would have been obtained if we had set out from the conditional odds of being young rather than old among members and nonmembers. We might also conclude from $\alpha = .476$ that the odds young/old are twice as large among nonmembers than among members. In this sense, α, or the odds ratio, is a symmetric measure of association.

If the two conditional odds had been equal (and $\alpha = 1$), then the relative distribution of religious membership would have been the same among the older and the younger people. In other words: Age and membership would have been statistically independent of each other. Statistical independence implies $\alpha = 1$, just as $\alpha = 1$ implies statistical independence.

Another advantageous feature of α and odds ratios is that their value does not change when all cell frequencies within a row or a column are multiplied by a constant. Let us assume that the sample which led to Table 2.2 constituted a simple random sample. If we had drawn a stratified sample instead, with age as the stratifying variable, selecting 600 younger people and only 200 older people, the value of α would still have been the same as the one obtained above. After all, the conditional proportional distributions of membership among young and old people would not have changed (disregarding sampling fluctuations). In this sense, odds ratios are independent of the marginal distributions of the variables.

The line from odds to odds ratios can be extended to higher order odds ratios pertaining to the relationships among three or more variables. Assume that the relation between age and membership of a religious denomination is studied within the categories of the variable Gender (*C*), giving rise to the frequencies f_{ijk}^{ABC}. The extent to which the relation age-religion is different for men (*C* = 1) and women (*C* = 2) can be expressed as the extent to which the odds ratio age/religion among the men deviates from the odds ratio among the women. This *higher-order odds ratio* α' can be expressed as follows:

$$\alpha' = \frac{f_{111}\,f_{221}}{f_{121}f_{211}} \bigg/ \frac{f_{112}\,f_{222}}{f_{122}f_{212}} = \frac{f_{111}\,f_{221}\,f_{122}\,f_{212}}{f_{121}\,f_{211}\,f_{112}\,f_{222}} \tag{2.6}$$

As appears from equation 2.6, this higher order odds ratio is also symmetric: Whichever two-variable relationship is taken as a starting point, the same value for α' is always found. Therefore, α' in equation 2.6 also expresses the extent to which the odds ratio gender (*C*)/religion (*B*) differs among younger (*A* = 1) and older (*A* = 2) people: $[(f_{111}f_{122})/(f_{112}f_{121})]/[(f_{211}f_{222})/(f_{212}f_{221})]$; and the extent to which the odds ratio age/gender differs among members and nonmembers.

To understand log-linear modeling it is not only necessary to know what marginal and conditional odds and odds ratios are, but one should also be familiar with the concept of *partial odds*. In line with the parametrization of the log-linear model used throughout this book (see the next section) partial odds are defined as average conditional odds. For Table 2.2, the partial odds of being a member indicate what the odds of being a member rather than a nonmember are on average among the old and the young. (A numerical example will be presented below.)

The meaning of partial and marginal odds should not be confused. Marginal odds member/nonmember greater than 1 logically indicate that there are more members than nonmembers in the sample. Partial odds greater than 1 do not necessarily have this implication. To understand this we have to examine the way partial odds are computed.

Partial odds are calculated as the *geometric mean* of the conditional odds. The meaning of this measure of central tendency can be made clear by comparing it with the more common arithmetic mean.

The arithmetic mean is defined as $\overline{X} = (\Sigma_i\, X_i)/N$. The sum of the deviations from X equals zero: $\Sigma_i\, (X_i - \overline{X}) = 0$. This means that if we have a series X-values, let us say the ages of a number of people, and if we guessed that each person is $\overline{X}$ years old, in total we would overestimate the ages as much as we would underestimate them, that is, if the deviations between our guesses and the true value are expressed as the differences $X - \overline{X}$. In this sense, we would be right "on average."

The formula for the geometric means is as follows, where Π with a subscript is the product sign:

$$\overline{X}_{\text{geom}} = \left(\prod_{i=1}^{N} X_i \right)^{1/N} = \sqrt[N]{X_1 X_2 \ldots X_N} \qquad (2.7)$$

The geometric mean is commonly used for the calculation of average rates of change and appears, for example, in the compound-interest formula (Yamane, 1973). The following example is frequently used to explain the meaning of the geometric mean as an average. The value of a horse is $100. Two people may guess its value; the one who comes closest to its true value gets the horse. Person A guesses $10; person B $1000. Who gets the horse? In terms of differences between the guesses and the true value, person A wins of course. But the owner (an early amateur of log-linear modeling) does not know whom to give the horse to. Person A underestimated the horse's value by a factor of 10; person B overestimated the value by a factor of 10.

If the geometric mean were applied to this problem, you would see from equation 2.7 that the geometric mean of 10 and 1000 is 100. The geometric mean is defined such that the product of the deviations from X equals one: $\Pi_i\, (X_i / \overline{X}_{\text{geom}}) = 1$. This implies that if we have a series X-values, let us say again the ages of a number of people, and if we guessed that each person is $\overline{X}_{\text{geom}}$ years old, in total we would overestimate the ages as much as we would underestimate them, that is, if the deviation between our guess and the true value is expressed as a ratio $X/\overline{X}_{\text{geom}}$. In this sense we would be right "on average."

The common arithmetic mean is the logical choice as a measure of central tendency if one works with sums and differences, that is, with an additive model; the geometric mean is the natural choice if one works with products and divisions, with odds and odds ratios, that is, with a multiplicative model.

There is an interesting connection between the geometric mean and the arithmetic mean in terms of logarithms. Denoting the natural logarithm to the base e as ln and making use of the well-known rules: $\ln ab = \ln a + \ln b$; $\ln a/b = \ln a - \ln b$; $\ln a^b = b\ln a$, it can be shown that the logarithm of the geometric mean equals the arithmetic mean of the logarithms of the values:

$$\ln \overline{X}_{geom} = \ln \left(\prod_{i=1}^{N} X_i \right)^{1/N} = \frac{1}{N} \sum_{i=1}^{N} \ln X_i \qquad (2.8)$$

Returning to our problem about partial and marginal odds, we will use as an example the variables in Table 2.2, but with different cell entries. Let us say that there are 11 people who are young, 1 of whom is a member of a religious denomination, and 10 who are not. Furthermore, there are 100 older people, 60 of whom are members, and 40 who are not. In total there are 61 members and 50 nonmembers. The marginal odds member/nonmember are $61/50 = 1.22$. The conditional odds for the younger people are $1/10 = .10$ and for the older people $60/40 = 1.50$. The partial odds member/nonmember are computed as the geometric mean of the two conditional odds: $\sqrt{(.10)(1.50)} = .258$. On average, within categories of age, members are outnumbered by nonmembers, but because of the small number of younger people the marginal odds are in favor of the members.

It may sound like a gross simplification, but it is essentially true: Once familiar with the notation and thoroughly acquainted with the above concepts, the reasoning behind log-linear modeling and the interpretation of the parameters of the log-linear model will present no special difficulties.

2.3 The Saturated Frequency Model

The log-linear models introduced in this section are in a sense the most general variants of log-linear modeling. They are saturated: No

TABLE 2.3 Relations Among Three Dichotomous Variables

A. Age	B. Religious Membership	C. Voting Behavior [a]		Total	% Yes	Odds Yes/No
		1. Yes	2. No			
1. Young	1. Member	95	48	143	66.4	1.979
	2. Nonmember	100	93	193	51.8	1.075
2. Old	1. Member	220	32	252	87.3	6.875
	2. Nonmember	91	71	162	56.2	1.282
	Total	506	245	750	67.5	2.065

SOURCE: Table 2.1
[a] Voting Behavior has been denoted by D in Table 2.1.

a priori restrictions are imposed on the data. They are frequency models: No assumptions are made about the causal structure of the data.

The data in Table 2.3 will be used as an example. This table again results from collapsing Table 2.1, now over Political Preference. Note that, for practical purposes, in this section Voting Behavior has been labeled C instead of D.

In the previous section it was seen that, in the total sample, age and religious membership are associated. Table 2.3 shows us, among other things, how age and religious membership are related to voting: Do older and religious people vote more readily than younger and nonreligious people? Is the relation between religion and voting stronger among the old than among the young?

The relationships in a three dimensional frequency table can be fully described by the following log-linear model, rendered in its multiplicative form with parameters denoted by τ:

$$F_{ijk}^{ABC} = \eta \tau_i^A \tau_j^B \tau_k^C \tau_{ij}^{AB} \tau_{ik}^{AC} \tau_{jk}^{BC} \tau_{ijk}^{ABC} \qquad (2.9)$$

The frequencies F in equation 2.9 are described as the *product* of a number of parameters. Therefore the underlying model is called a *multiplicative model*. According to equation 2.9, the sizes of the cell frequencies F depend on a constant factor η; the sizes of the cell frequencies also depend on which categories of the single variables A, B, and C they pertain to, as indicated by τ_i^A, τ_j^B, and τ_k^C; moreover, as indicated by

τ_{ij}^{AB}, τ_{ik}^{AC}, and τ_{jk}^{BC}, they depend on which categories of the "joint" variables AB, AC, and BC they belong to; finally, according to τ_{ijk}^{ABC}, they depend on which categories of the "joint" variable ABC they pertain to. By analogy with analysis of variance, one could speak of an overall effect (η) on F, of three main or one-variable effects (τ_i^A, τ_j^B, and τ_k^C), three interaction or two-variable effects (τ_{ij}^{AB}, τ_{ik}^{AC}, and τ_{jk}^{BC}), and one second order interaction or three-variable effect (τ_{ijk}^{ABC}).

Within the context of frequency models, labeling the τ parameters "effects" or "effect parameters" does not have any causal connotation. All that is implied by the term "effect" is that the sizes of the cell frequencies depend on, that is, are a mathematical function of, the sizes of the τ parameters.

Before it is possible to discuss the meaning of the parameters in equation 2.9 in depth by relating them to odds and odds ratios in Table 2.3, some formal aspects of the log-linear model should be dealt with.

As will be clear in later sections and chapters, for some purposes it is more convenient to work with the logarithmic transformation of equation 2.9:

$$G_{ijk}^{ABC} = \theta + \lambda_i^A + \lambda_j^B + \lambda_k^C + \lambda_{ij}^{AB} + \lambda_{ik}^{AC} + \lambda_{jk}^{BC} + \lambda_{ijk}^{ABC} \qquad (2.10)$$

where

$$G_{ijk}^{ABC} = \ln F_{ijk}^{ABC} \quad \theta = \ln \eta \quad \lambda_i^A = \ln \tau_i^A \quad \lambda_{ij}^{AB} = \ln \tau_{ij}^{AB} \quad \lambda_{ijk}^{ABC} = \ln \tau_{ijk}^{ABC} \text{ etc.}$$

Equation 2.10 represents an *additive model*. It is this logarithmic, additive form that has given the log-linear model its name.

Equations 2.9 and 2.10 each contain too many parameters to be identifiable. Given the values of the expected frequencies F, there does not exist a unique solution for the τ parameters: There are already as many τ_{ijk}^{ABC} parameters (or λ_{ijk}^{ABC} parameters) as there are frequencies F (or G). However, we are not interested in the effects of the categories per se. The question is whether older people are more religious than younger people, not what the effect of being old per se and the effect of being young per se is on religiosity.

The fact that effects are defined by comparison provides a general solution to the *identifiability problem*, which can be worked out in several directions. For example, if A in equations 2.9 and 2.10 refers to the dichotomous variable Age in Table 2.3, one way to reduce the number of parameters is to set the effects pertaining to $A = 1$, being young, equal to an arbitrary constant and to see to what extent the effects pertaining to $A = 2$, being old, deviate from this arbitrary value. Usually, the arbitrary value chosen is the one of "no effect." This implies setting certain τ parameters equal to one or certain λ parameters equal to zero, very much like what is done in regression analysis when dummy variables coded 0, 1 are used.

In log-linear modeling a somewhat different solution is usually chosen, namely, the one that is used in analysis of variance or in regression analysis when effect coding $(-1, +1)$ is employed. Log-linear effects are usually expressed as deviations from the average effect. In the case of the dichotomous variable Age, being old enhances the frequencies F to the same extent that being young diminishes them.

Dummy coding and *effect coding* are arbitrary parametrizations of essentially the same model with the same (estimated) expected frequencies. The parameters of the one approach can simply be transformed into the parameters of the other approach. Both parametrizations lead to the same substantive conclusions about the relationships in the multivariate frequency table. Nevertheless, the values of the parameters themselves and their interpretations depend on the parametrization used. Interpreting parameter values obtained with the one parametrization as if they were obtained by the other will generally lead to the wrong substantive conclusions. Readers not familiar with these kinds of parametrizations are strongly urged to study the relevant literature. Kerlinger and Pedhazur (1973) and O'Grady and Medoff (1988) discussed these and other parametrizations within the (familiar) context of regression analysis and analysis of variance; Alba (1987), Kaufman and Schervish (1986, 1987), and Long (1984) presented excellent discussions of how to arrive at correct substantive interpretations of the log-linear parameters in terms of (estimated) expected frequencies, odds, and odds ratios using either dummy or effect coding.

Because most applications of log-linear modeling use effect coding, this is what will be done in this book. Consequently, the interpretations of the parameters offered below only apply when effect coding has been applied.

Effect coding implies that the log-linear λ parameters over any subscript sum to zero and the τ parameters of the multiplicative model multiplied over any subscript equal one. In equation 2.11 this is indicated for some of the parameters of the equations 2.9 and 2.10.

$$\prod_{i=1}^{I} \tau_i^A = 1 \quad \prod_{i=1}^{I} \tau_{ij}^{AB} = \prod_{j=1}^{J} \tau_{ij}^{AB} = 1 \quad \prod_{i=1}^{I} \tau_{ijk}^{ABC} = \prod_{j=1}^{J} \tau_{ijk}^{ABC} = \prod_{k=1}^{K} \tau_{ijk}^{ABC} = 1$$

$$(2.11)$$

$$\sum_{i=1}^{I} \lambda_i^A = 0 \quad \sum_{i=1}^{I} \lambda_{ij}^{AB} = \sum_{j=1}^{J} \lambda_{ij}^{AB} = 0 \quad \sum_{i=1}^{I} \lambda_{ijk}^{ABC} = \sum_{j=1}^{J} \lambda_{ijk}^{ABC} = \sum_{k=1}^{K} \lambda_{ijk}^{ABC} = 0$$

In the case of dichotomous variables, the constraints in equations 2.11 lead to:

$$\tau_2^A = 1/\tau_1^A \quad \tau_{12}^{AB} = \tau_{21}^{AB} = 1/\tau_{22}^{AB} = 1/\tau_{11}^{AB}$$

$$\tau_{112}^{ABC} = \tau_{121}^{ABC} = \tau_{211}^{ABC} = \tau_{222}^{ABC} = 1/\tau_{221}^{ABC} = 1/\tau_{212}^{ABC} = 1/\tau_{122}^{ABC} = 1/\tau_{111}^{ABC}$$

$$\lambda_2^A = -\lambda_1^A \quad \lambda_{12}^{AB} = \lambda_{21}^{AB} = -\lambda_{22}^{AB} = -\lambda_{11}^{AB}$$

$$(2.12)$$

$$\lambda_{112}^{ABC} = \lambda_{121}^{ABC} = \lambda_{211}^{ABC} = \lambda_{222}^{ABC} = -\lambda_{221}^{ABC} = -\lambda_{212}^{ABC} = -\lambda_{122}^{ABC} = -\lambda_{111}^{ABC}$$

With these restrictions, the equations 2.9 and 2.10 are exactly identified. If the three variables A, B, and C have I, J, and K categories respectively, there are $I \times J \times K$ cell frequencies. The number of independent τ_i^A or λ_i^A parameters is $I - 1$; there are $J - 1$ independent τ_j^B (λ_j^B) parameters and $(K - 1)$ independent τ_k^C (λ_k^C) parameters. The number of two-variable parameters AB is $(I - 1)(J - 1)$; the effects AC number $(I - 1)(K - 1)$; and the effects BC number $(J - 1)(K - 1)$. The three-variable effect τ_{ijk}^{ABC} has $(I - 1)(J - 1)(K - 1)$ independent parameters. Together with the overall effect, this sums to exactly $I \times J \times K$ independent parameters and all parameters can be calculated knowing F or G.

To see how the effects are actually calculated and as an aid in the discussion of the meaning of the parameters, formulas for some of the effects appearing in equations 2.9 and 2.10 are given, first for the general case of three polytomous variables, and then for the case of three dichotomous variables. The formulas for the effects not presented can easily be derived from this by setting up analogous formulas and using the kind of restrictions appearing in equations 2.11, 2.12.

$$\eta = \left(\prod_i \prod_j \prod_k F_{ijk} \right)^{1/IJK} \qquad \theta = \frac{1}{IJK} \sum_i \sum_j \sum_k G_{ijk}$$

$$\tau_i^A = \frac{\left(\prod_j \prod_k F_{ijk} \right)^{1/JK}}{\eta} \qquad \lambda_i^A = \frac{1}{JK} \sum_j \sum_k G_{ijk} - \theta$$

$$\tau_{ij}^{AB} = \frac{\left(\prod_k F_{ijk} \right)^{1/K}}{\eta \tau_i^A \tau_j^B} \qquad \lambda_{ij}^{AB} = \frac{1}{K} \sum_k G_{ijk} - \theta - \lambda_i^A - \lambda_j^B$$

$$\tau_{ijk}^{ABC} = \frac{F_{ijk}}{\eta \tau_i^A \tau_j^B \tau_k^C \tau_{ij}^{AB} \tau_{ik}^{AC} \tau_{jk}^{BC}} \qquad \lambda_{ijk}^{ABC} = G_{ijk} - \theta - \lambda_i^A - \lambda_j^B - \lambda_k^C$$

$$- \lambda_{ij}^{AB} - \lambda_{ik}^{AC} - \lambda_{jk}^{BC}$$

(2.13)

For dichotomous variables this becomes

$$\eta = \left(\prod_{i=1}^2 \prod_{j=1}^2 \prod_{k=1}^2 F_{ijk} \right)^{1/8} \qquad \theta = \frac{1}{8} \sum_{i=1}^2 \sum_{j=1}^2 \sum_{k=1}^2 G_{ijk}$$

$$\tau_1^A = \left(\prod_{j=1}^2 \prod_{k=1}^2 \frac{F_{1jk}}{F_{2jk}} \right)^{1/8} \qquad \lambda_1^A = \frac{1}{8} \sum_{j=1}^2 \sum_{k=1}^2 (G_{1jk} - G_{2jk})$$

$$\tau_{11}^{AB} = \left(\prod_{k=1}^2 \frac{F_{11k} F_{22k}}{F_{12k} F_{21k}} \right)^{1/8} \qquad \lambda_{11}^{AB} = \frac{1}{8} \sum_{k=1}^2 [(G_{11k} + G_{22k}) $$

$$- (G_{12k} + G_{21k})]$$

(2.14)

$$\tau_{111}^{ABC} = \left(\prod_{k=1}^2 \frac{F_{111} F_{221}}{F_{121} F_{211}} \cdot \frac{F_{122} F_{212}}{F_{112} F_{222}} \right)^{1/8}$$

$$\lambda_{111}^{ABC} = \frac{1}{8} [(G_{111} + G_{221} + G_{122} + G_{212}) - (G_{121} + G_{211} + G_{112} + G_{222})]$$

As they stand, formulas 2.13 and 2.14 are of little practical use: the expected frequencies F (G) are not known. It is necessary to find (the maximum likelihood) estimators $\hat{F}$ for F, $\hat{G}$ for G, respectively. In the saturated model, the maximum likelihood estimators for F (G) are the observed frequencies $f(g)$:

$$\hat{F}^{ABC}_{ijk} = f^{ABC}_{ijk} \qquad \hat{G}^{ABC}_{ijk} = g^{ABC}_{ijk} \qquad (2.15)$$

where

$$\hat{G}^{ABC}_{ijk} = \ln\hat{F}^{ABC}_{ijk} \qquad \ln g^{ABC}_{ijk} = \ln f^{ABC}_{ijk}$$

By using the observed frequencies $f(g)$ in equations 2.13, 2.14 instead of F (G), estimated values for the effect parameters in the saturated model can be obtained. These estimated effect parameters will be indicated as $\hat{\eta}$, $\hat{\tau}$, etc.

Table 2.4 contains the estimated values of the parameters of the saturated model for the data on the age-religion-voting relationships (Table 2.3) and for the data on the age-religion relation (Table 2.2). The main focus is on the parameters in the column "Table 2.3"; the other parameters are used for comparison.

Although the overall and the single variable effect parameters are only rarely of interest, they will nevertheless be discussed, both for the sake of completeness and to get a better feeling of the systematics behind log-linear analysis.

As can be seen from formulas 2.13 and 2.14, the *overall effects* η and θ reflect the mean level of all cell frequencies. The geometric mean of the cell frequencies in Table 2.3 is 81.408 and the arithmetic mean of the logarithms of the frequencies is 4.399 (Table 2.4). The overall effect is merely a reflection of the sample size. If the sample size for Table 2.3 had been doubled, all cell frequencies would have been twice as large and, accordingly, $\hat{\eta}$ twice its present value.

The *one-variable effects* reflect the fact that cell frequencies are expected to be larger or smaller depending on the distribution of the single variables. If there are more old than young people, the cell frequencies pertaining to the category old are generally larger than those pertaining to the category young.

The one-variable effects turn out to be partial odds. As can be seen from formulas 2.13, the one-variable effect of A (τ^A_i) is determined in two steps. First, the geometric mean of all cell frequencies pertaining to $A = i$ is computed. Second, a calculation is made to determine how much

TABLE 2.4 The Saturated Model for Table 2.3 and Table 2.2

			Table 2.3			Table 2.2		
Effect of:			$\hat{\tau}$	$\hat{\lambda}$	$\hat{\lambda}/\hat{s}_\lambda$	$\hat{\tau}$	$\hat{\lambda}$	$\hat{\lambda}/\hat{s}_\lambda$
"Overall"			81.408	4.399		183.211	5.211	
A.	Age	1 = young	0.991	−0.009	−0.208	0.907	−0.098	2.648
B.	Rel. Memb.	1 = member	0.925	−0.078	−1.865	1.036	0.035	0.946
C.	Voting	1 = yes	1.443	0.366	8.721			
AB		11	0.905	−0.100	−2.376	0.831	−0.185	−5.000
AC		11	0.837	−0.178	−4.227			
BC		11	1.331	0.286	6.812			
ABC		111	0.875	−0.134	−3.182			
$\hat{s}_\lambda$				0.042			0.037	

(how many times) this average frequency deviates from what was expected on the grounds of the overall effect. Using the partial odds τ_i^A, the average probability of being in category $A = i$ is compared with the overall probability of being in any one of the cells of the table.

For the data in Table 2.3, $\hat{\tau}_1^A$, which denotes the estimated partial odds of being young, is almost 1 (0.991 − Table 2.4). This means that, on average, within the four combined categories of the variables Religious Membership and Voting Behavior, the frequencies of the cells pertaining to the category young almost equal the average overall cell frequency.

The parameter $\hat{\tau}_2^A$ indicates how much larger or smaller than the overall effect is the mean frequency of the cells $A = 2$ (being old). Because Age (A) is a dichotomy and because of the choice of the identifying restrictions 2.11, 2.12, $\hat{\tau}_2^A = 1/\hat{\tau}_1^A = 1/0.991 = 1.009$.

The ratio $\hat{\tau}_1^A/\hat{\tau}_2^A$ then tells us how many times larger or smaller the mean frequency of the cells $A = 1$ is than the mean frequency of the cells $A = 2$. In other words: $\hat{\tau}_1^A/\hat{\tau}_2^A$ are the estimated partial odds of being young rather than old, calculated as the geometric mean of the conditional odds f_{1jk}/f_{2jk}. In this case, $\hat{\tau}_1^A/\hat{\tau}_2^A = (\hat{\tau}_1^A)^2 = 0.982$, that is, on average there are almost as many young people as old.

In principle, the values of the one-variable effects will change when additional variables are introduced. A striking example of this are the estimates for τ_1^B in Table 2.4. For table AB (Table 2.2), the estimated odds of being a member rather than a nonmember $(\hat{\tau}_1^B)^2$ are computed while holding age constant, that is, as the average of the two conditional odds f_{11}^{AB}/f_{12}^{AB} and f_{21}^{AB}/f_{22}^{AB}. This results in the value $1.036^2 = 1.073$. However,

$(\hat{\tau}_1^B)^2$ for table ABC (Table 2.3) is computed holding age and voting constant, that is, as the average of the four conditional odds of being a member (for younger voters, younger nonvoters, older voters, and older nonvoters, respectively), amounting to 0.856. According to the first estimate, based on table AB, there are on average more members than nonmembers, whereas the reverse is true according to the second estimate obtained from table ABC.

In the same way as marginal and partial odds may differ from each other, as explained in Section 2.2, partial odds based on tables of varying dimensionality may give different results. When exactly this occurs can be derived from the collapsibility theorem presented in Section 2.6.

Not all one-variable effects $\hat{\tau}$ presented in Table 2.4 have been discussed. But their meanings are easily derived from the above. The *log-linear parameters* $\hat{\lambda}$ have essentially the same interpretation as the multiplicative parameters $\hat{\tau}$, be it that the average is taken as an arithmetic mean of logarithms of cell frequencies and that deviations (from the overall effect or from each other) are expressed as differences instead of ratios.

One-variable effects are seldom of interest because they tell us nothing about the relations between the variables, and it is relationships and multivariate distributions, not one-variable (partial) distributions, which are the focus of most studies.

The sizes of the cell frequencies in a table can be expected to vary systematically not only with the overall sample size and with the average distributions of single variables, but also with the average bivariate distributions. If Age and Religious Membership are negatively associated (younger people are relatively less religious than older people), the cells young/nonmember and old/member are more heavily populated and the cells young/member and old/nonmember less heavily populated than one would expect on the basis of the overall sample size and the average distributions of age and membership as such.

Let us first look at the *two-variable effect* parameter τ_{ij}^{AB}. This parameter pertains to the average relationship between A and B within categories of the remaining variables. For a table ABC, τ_{ij}^{AB} measures the strength of the partial association between A and B, holding C constant.

As equation 2.13 shows, the computation of τ_{ij}^{AB} proceeds analogously to the calculation of the one-variable effects. First a geometric mean is computed to determine the average size of cell frequencies $AB = (i, j)$. Second, it is determined how much (how many times) this average frequency deviates from what was expected on the basis of the lower order effects η, τ_i^A, and τ_j^B. For the three dimensional Table 2.3, $\hat{\tau}_{ij}^{AB}$, which expresses the partial relation of Age (A) and Religious Membership (B) holding Voting Behavior (C) constant, is 0.905 (Table 2.4). On average, the combination young/member $(AB = 1\ 1)$ among voters and nonvoters occurs 0.905 times less often than expected on the basis of the overall effect $\hat{\eta}$ and the one-variable effects $\hat{\tau}_i^A$ and $\hat{\tau}_j^B$.

Because all variables in Table 2.3 are dichotomous and as a result of the choice of the identifying restrictions 2.11, 2.12 $\tau_{12}^{AB} = \tau_{21}^{AB} = 1/\tau_{11}^{AB} = 1/\tau_{22}^{AB}$. This implies that the combination old/nonmember $(AB = 22)$ on average also occurs 0.905 times less often than expected. But the combinations young/nonmember and old/member occur $1/0.905 = 1.105$ times more often than expected on the basis of the lower order effects. The partial relation between age and religion is such that being older and being more religious tend to go together.

If there are at least three variables, the two-variable effects are partial effects and can be written as average *conditional effects*. Conditional effects are just ordinary effects computed within subgroups. Let us indicate the two conditional effects for the relation age-religion (AB) within the two categories of voting (C) as: $\hat{\tau}_{ij}^{AB/C}1$ and $\hat{\tau}_{ij}^{AB/C}2$. The parameter $\hat{\tau}_{ij}^{AB}$ dealt with so far is nothing but the geometric mean of these two conditional effects:

$$\hat{\tau}_{ij}^{AB} = \sqrt[2]{\hat{\tau}_{ij}^{AB/C}1\, \hat{\tau}_{ij}^{AB/C}2}$$

$$\hat{\lambda}_{ij}^{AB} = \frac{1}{2}(\hat{\lambda}_{ij}^{AB/C}1 + \hat{\lambda}_{ij}^{AB/C}1) \tag{2.16}$$

In general, for any table with three or more variables with any number of categories, the partial effect AB equals the geometric mean of all corresponding conditional effects AB.

The conditional effects AB and accordingly the partial effects AB are closely related to the odds ratios and the cross-products α within the relevant subtables AB and have all the advantageous properties of the latter. This is most easily seen when all variables are dichotomous, as in Table 2.3. From equation 2.14 it is easily deduced that $(\tau_{11k}^{AB/C})^4$ equals the odds ratio α in the subtable AB for $C = k$ and that $(\hat{\tau}_{11}^{AB})^4 = 0.905^4 = 0.671$ is the geometric mean of two conditional odds ratios.

If the variables A, B, and C are polytomous, a number of 2×2 tables can be defined within an $I \times J$ subtable AB by selecting in several ways rows i and i' ($i' > i$) and columns j and j' ($j' > j$). For each of these 2×2 tables, the relationship between the conditional two-variable effects and the cross-product ratio is as follows: $\alpha = (\tau_{ij\ k}^{AB/C} \tau_{i'j'\ k}^{AB/C}) / (\tau_{ij'\ k}^{AB/C} \tau_{i'j\ k}^{AB/C})$.

As with one-variable effects, two-variable effects generally take on different values when additional variables are introduced. For example, $\hat{\tau}_{11}^{AB}$ for the AB Table 2.2 equals 0.831, whereas τ_{11}^{AB} for the ABC Table 2.3 is 0.905. Holding voting constant has weakened the (negative) association between age and religious membership, bringing its value closer to one. Exactly when two-variable effects change after introducing additional variables is explained in Section 2.6.

There are two more two-variable effects for Table 2.3. For most researchers these would be the most important effects because they indicate what the direct partial relation is between religion and voting and between age and voting. It can be inferred from $\hat{\tau}_{11}^{AC} = 0.837$ that if religion is held constant, younger people are less inclined to vote than older people. The direct relation of religion and voting when age is held constant is rather strong, stronger than the relation age-voting: $\hat{\tau}_{11}^{BC} = 1.331$ and $\hat{\tau}_{21}^{BC} = 0.751$. Members of a religious denomination are generally much more inclined to vote than nonmembers.

There is still one effect for Table 2.3 that has not been discussed: the *three-variable effects* τ_{ijk}^{ABC}. These concern the question of whether or not there is interaction in Table 2.3. There are three ways to formulate this question, which, however, will appear to amount to the same. Is the relation religion-voting different among young and old people? Is the relation age-voting different among members and nonmembers? Is the relation age-religion different among voters and nonvoters?

The parameter τ_{ijk}^{ABC} indicates, as equation 2.13 shows, to what extent the frequency of the cell $ABC = (i, j, k)$ deviates from what is expected on the basis of all two and lower order effects. While, according to equation 2.16, τ_{ij}^{AB} tells us what the average conditional two-variable effects AB are, τ_{ijk}^{ABC} tells us how much the conditional two-variable effects deviate from each other, or rather how much each conditional effect deviates from the average partial effect τ_{ij}^{AB}. As such, τ_{ijk}^{ABC} is directly related to the higher order association coefficient α' (equation 2.6). τ_{ijk}^{ABC} and α' are both symmetric measures of interaction: The same value is always obtained regardless of which two-variable relationship is taken as the starting point:

$$\tau^{ABC}_{ijk} = \tau^{AB/C}_{ijk} / \tau^{AB}_{ij} = \tau^{AC/B}_{ikj} / \tau^{AC}_{ik} = \tau^{BC/A}_{jki} / \tau^{BC}_{jk}$$

$$\lambda^{ABC}_{ijk} = \lambda^{AB/C}_{ijk} - \lambda^{AB}_{ij} = \lambda^{AC/B}_{ikj} - \lambda^{AC}_{ik} = \lambda^{BC/A}_{jki} - \lambda^{BC}_{jk}$$

(2.17)

If all τ^{ABC}_{ijk} parameters are equal to one, then all conditional two-variable relationships are the same and according to the log-linear model, there are no (higher order) interaction effects.[4]

For the data in Table 2.3, $\hat{\tau}^{ABC}_{111}$ is 0.875. This implies that the conditional relation between religion and voting among younger people, as indicated by $\tau^{BC/A}_{111}$, is 0.875 times weaker than the average relation, which was $\hat{\tau}^{BC}_{11} = 1.331$. So for the younger people the conditional relation is $\hat{\tau}^{BC/A}_{111} = 1.165$. Because $\tau^{ABC}_{211} = 1/\tau^{ABC}_{111}$, it follows that for the older people the relation religion-voting is $1/.875 = 1.143$ times stronger than average, resulting in 1.521. The direction of the relation is not different: Regardless of age, members are more inclined to vote than nonmembers, but the relation between voting and religion is much stronger among the old than among the young. The bonds between religious membership and voting are stronger for the older generations.

Through the same kind of reasoning we come to the conclusion that the relation between age and voting is in the same direction, but much stronger among members ($0.875 \times 0.837 = 0.732$) than among nonmembers ($1.143 \times 0.837 = .957$). And comparable conclusions may be drawn for the relation between age and religion among voters and nonvoters.

The analysis of the relations among age, religious membership and voting as they were present in the three dimensional Table 2.3 is now complete. Extensions to more variables follow easily. If (and only if) the identifying restrictions in equation 2.11 are chosen and effect coding is used, each partial effect τ can be calculated as the geometric mean of the corresponding conditional effects, and each higher order effect can be calculated in terms of the extent to which the conditional effects deviate from the partial effect. General formulas for calculating all effects of the saturated model for any number of polytomous variables can be found in Bishop et al. (1975) and Haberman (1978).

When interpreting the *magnitudes of the effects*, it is good to keep some anchor points in mind. If a τ parameter exercises no influence, its value equals one; for a λ parameter the corresponding value is zero. The maximum positive effect is $+\infty$ for both τ and λ. The maximum "negative" value is zero for τ and $-\infty$ for λ.

Thus, the values for λ are symmetrically situated around the value zero. Owing to this, the strength of positive and negative λ effects can be directly compared. The τ values are asymmetrically situated around one, and positive and negative effects are not directly comparable. For instance, it cannot immediately be seen that the relation between religion and voting, indicated by $\hat{\tau}_{11}^{BC} = 1.331$, is stronger than the relation between age and voting, indicated by $\hat{\tau}_{11}^{AC} = 0.837$. First the inverse of one of these coefficients, for example, $1/0.837 = 1.195$, has to be computed and compared with the other to see that this is true. A disadvantage of λ is that it has to be interpreted in terms of logarithms of frequencies. The τ parameters can be interpreted more simply in terms of (ratios between) frequencies and probabilities.

The maximum values indicated above are only limits which are approached but cannot be reached. As can be seen in equations 2.13 and 2.14 the maximum values are "reached" when one or more cell frequencies are zero. But in that case not all effects can be determined, as dividing by zero and taking the logarithm of zero are nondefined operations.

It is important to notice that, in general, in both the saturated and the unsaturated model none of the estimated expected frequencies $\hat{F}$ is allowed to be zero. What to do when this nevertheless happens is discussed at the end of this chapter.

Table 2.4 contains some parameters, the *standardized effects*, which have not yet been dealt with. Large sample estimates of the variance of the $\hat{\lambda}$ parameters in a saturated model with dichotomous variables may be obtained by means of the following formula (Goodman, 1972b):

$$\hat{s}_{\lambda}^2 = \frac{1}{16^2} \sum_{i=1}^{2} \sum_{j=1}^{2} \sum_{k=1}^{2} \sum_{l=1}^{2} \frac{1}{f_{ijkl}} \qquad (2.18)$$

Formulas for the variances of the estimated effects in a saturated model for polytomous variables are somewhat more complicated and are presented by Bishop et al. (1975) and Haberman (1978). Given that a particular log-linear effect equals zero in the population, the standardized effect (asymptotically) follows the standard normal distribution and may thus be tested for significance. All two- and three-variable effects in Table 2.4 are significant.

These standardized effects may be used to set up a more parsimonious model, a model with fewer effects. This topic will be dealt with in Section 2.5.

There has been some misunderstanding of the meaning of the standardized parameters. Where the λ coefficients were compared with the unstandardized regression coefficients b, the standardized effects were analogously equated with the standardized β weights. However, the standardized λ effects are not standardized effects in the sense the β weights are, but they are test statistics. If a parallel is drawn, $\hat{\lambda}/\hat{s}_\lambda$ should be equated with the t-values, the statistics for testing the significance of β.

2.4 Unsaturated Frequency Models

In the saturated model the observed data are described and reproduced exactly in terms of a number of effects. All possible effects are present without a priori restrictions apart from the identifying restrictions 2.11 being imposed on any of these effects. Unsaturated models, on the other hand, express the investigator's expectations concerning the data in the form of a priori restrictions upon the effect parameters.

Taking the data in Table 2.3 as an example, a researcher might be of the opinion that the relation between religion and voting in the population is the same among older and younger people. Translated into log-linear terms, this would imply an unsaturated model in which the three-variable effects are absent, that is, a model equation like equations 2.9 and 2.10 but with the restriction that $\tau_{ijk}^{ABC} = 1$, $\lambda_{ijk}^{ABC} = 0$ respectively. Moreover, it could be assumed that age and religion are equally strongly related to voting, leading to the additional restriction $\tau_{ik}^{AC} = \tau_{ik}^{BC}$, or that age has no direct relation with voting but only indirectly through religion, leading to the restriction $\tau_{ik}^{AC} = 1$.

Such a priori expectations of the investigator can be tested empirically by setting up the relevant log-linear model, computing the estimated expected frequencies $\hat{F}$ for this model and comparing $\hat{F}$ with the observed frequencies f. If the model need not be rejected, the hypotheses are corroborated and it then makes sense to estimate the effect parameters of the model. If the model has to be rejected, the expectations were wrong and the researcher will look for a model that does fit the observed data, mainly guided by theoretical notions, but also making use of several statistics which make clear to what extent and at which points particular models deviate from reality.

This is a global description of how log-linear modeling proceeds. Each of these points will be elaborated upon below making a distinction between hierarchical and nonhierarchical models.

A log-linear model is called *hierarchical* if when any τ parameter is set to 1, all effects of the same and higher order are set to 1, that is, all parameters in which all the letters of the superscript of this τ parameter occur. For example, if in a model for a four dimensional table *ABCD*, the parameter τ_{ij}^{AB} is set to 1, the hierarchy principle implies that τ_{ijk}^{ABC}, τ_{ijl}^{ABD}, and τ_{ijkl}^{ABCD}, each containing the superscript *AB*, are also set to one.

The hierarchy principle may also be formulated the other way around: If a particular τ parameter is not set to 1, all parameters of the same or lower order are not set to one, that is, all those parameters whose superscript consists of only one or more letters of the superscript of this τ parameter. For example, when τ_{ijk}^{ABC} is not set to 1, then neither are τ_i^A, τ_j^B, τ_k^C, τ_{ij}^{AB}, τ_{ik}^{AC}, and τ_{jk}^{BC}.

Models that do no follow the hierarchy restriction are called non-hierarchical. The distinction hierarchical/nonhierarchical is, among other things, important for the way in which the maximum likelihood estimates can be obtained.

The simplest example of a hierarchical model is the independence model for a 2 × 2 table. Let us depart from the hypothesis that there is no association between religion and age in the population. Relatively there are as many members of a religious denomination among the young as among the old. Table 2.2 provides the observed frequencies for testing this hypothesis. (We have seen in the previous sections that contrary to what the hypothesis states, age and religious membership are significantly associated with each other in Table 2.2, but let us disregard this information for the moment.)

Independence in Table 2.2 implies $\alpha = 1$, or equivalently $\tau_{ij}^{AB} = 1$. This leads to the following unsaturated model:

$$F_{ij}^{AB} = \eta\tau_i^A\tau_j^B$$

$$G_{ij}^{AB} = \theta + \lambda_i^A + \lambda_j^B$$
(2.19)

To test the empirical validity of this model and to be able to compute the maximum likelihood estimates of the parameters, the maximum likelihood estimates of the expected frequencies F are needed. We need an estimate of how Table 2.2 would have looked if the frequencies mirrored exactly the probabilities in the population where, according to the hypothesis, age and religion are independent of each other. Unlike the saturated model, in an unsaturated model it no longer applies that the observed frequencies f are the maximum likelihood estimators of F.

The estimated expected frequencies for the independence model in equation 2.19 have the following, familiar form:

$$\hat{F}_{ij}^{AB} = f_{i+}^{AB} f_{+j}^{AB} / N \qquad (2.20)$$

To test the independence hypothesis, $\hat{F}$ has to be compared with the observed f either through the usual *Pearson-χ^2* or by the perhaps somewhat less familiar *(log)likelihood ratio chi-square L^2*:

$$\text{Pearson-}\chi^2 = \sum_i \sum_j [(f_{ij}^{AB} - \hat{F}_{ij}^{AB})^2 / \hat{F}_{ij}^{AB}] \qquad (2.21)$$

$$\text{(log)likelihood ratio chi-square } L^2 = 2 \sum_i \sum_j [f_{ij}^{AB} \ln(f_{ij}^{AB} / \hat{F}_{ij}^{AB})]$$

If the model is valid for the population, both statistics approach asymptotically the theoretical χ^2-distribution. The number of *degrees of freedom* (df) can be found as follows. First, determine the number of parameters to be independently estimated, that is, count the number of parameters that have to be computed for a particular model and subtract from this count the number of identifying restrictions (equation 2.11). Next, to obtain df, subtract the number of independent parameters from the number of cell frequencies. Or put differently: The number of degrees of freedom equals the number of independent a priori restrictions imposed on the parameters of the saturated log-linear model to obtain the model to be tested (not counting the identifying restrictions).

Equation 2.19 implies the restriction $\tau_{ij}^{AB} = 1$. As age and religious membership in Table 2.4 are dichotomies, the number of τ_{ij}^{AB} parameters to be independently estimated is $(2 - 1)(2 - 1) = 1$, hence $\tau_{ij}^{AB} = 1$ means imposing one restriction, and consequently, df = 1.

The test statistics for Table 2.4 are: Pearson-χ^2 = 24.9 and L^2 = 25.1, which with df = 1 gives $p < .001$. The independence hypothesis for Table 2.4 has to be rejected; the hypothesis that $\tau_{ij}^{AB} = 1$ in the population cannot be maintained. Age and religion are related to each other.

If the model did not have to be rejected, it would have made sense to use the estimated expected frequencies to estimate the parameters in model 2.19 by replacing the expected frequencies F by $\hat{F}$ in the formulas 2.13 and 2.14. In this case, L^2 and Pearson-χ^2 have almost exactly the same numerical value. For large samples L^2 and Pearson-χ^2 can be expected to have the same value when the model is true, because they both approach the same χ^2-distribution. When the sample size is small or the data are sparsely distributed in the table, they have different distributions, different from the χ^2-distribution and different from each other.

Usually both statistics will be presented, as different values may indicate that something is wrong with the approximation of the χ^2-distribution, and similar values may be conceived as a signal that they both are χ^2-distributed.

There has been much debate on how well, in the case of small samples and a large number of cells, both test statistics approximate the χ^2-distribution, if the null hypothesis is true. Fienberg (1980, Appendix IV), who based his views on Larntz (1978) and Margolin and Light (1974), concluded that the approximation of the theoretical χ^2-distribution by Pearson-χ^2 is fairly good, even in the case of minimal expected cell frequencies of one. But in the case of small cell frequencies, L^2 often underestimates the probability of a type I error and yields too high a value. Similar conclusions were drawn by Haberman (1978, Section 5.3), Koehler (1986), Lawal (1984), and Milligan (1980).

Very definite results cannot be given. The above conclusions are mainly based on computer simulations which have been able to take into account only a very limited number of different kinds of tables and hypotheses. Moreover, different studies sometimes use different criteria to judge the adequacy of Pearson-χ^2 and L^2, which makes it difficult to compare their results. An excellent summary can be found in Read and Cressie (1988: 133-153).

In the following chapters we will rely somewhat more on L^2 than on Pearson-χ^2, in spite of the above evidence regarding the superiority of Pearson-χ^2 as an approximation of the theoretical χ^2-distribution.

In a certain sense, L^2 is the logical choice. Application of the principle of maximum likelihood for calculating $\hat{F}$ implies that the differences between the estimated and the observed frequencies are minimized in terms of the test statistic L^2. Minimizing in terms of Pearson-χ^2 may lead to somewhat different estimates for the expected frequencies F. The test statistic L^2 belongs to the estimation principle that has been chosen. Moreover, L^2 can be partitioned into a number of components which can each be attributed to different effects, submodels, or subgroups. This is particularly important when looking for a well fitting, parsimonious model. With Pearson-χ^2, such partitioning is possible only by approximation.

The estimated frequencies of hierarchical models have an interesting relation to the observed frequencies. The independence model contains the single variable effects τ_i^A and τ_j^B. This implies that the marginal observed frequencies of A and B are reproduced exactly using the independence model. The estimated frequencies of the independence model in equation 2.20 are such that $\hat{F}_{i+}^{AB} = f_{i+}^{AB}$ and $\hat{F}_{+j}^{AB} = f_{+j}^{AB}$. The inclusion of a particular parameter in a hierarchical model always implies that the observed marginal frequency corresponding with the superscript of this parameter is reproduced exactly by the model.

Unsaturated hierarchical models may be interpreted as models in which one proceeds on the fact that the information contained in certain marginal frequencies in combination with the hypotheses (the a priori restrictions) is sufficient to describe what is going on in the total table. The *exactly reproduced marginals* are the *sufficient statistics* for the estimation of the parameters of a particular model (Bishop et al., 1975: 64; Mood, Graybill, & Boes, 1974: 300).

Therefore hierarchical unsaturated models can be uniquely represented by these reproduced marginal frequencies. The independence model 2.19 may then be denoted as $\{A, B\}$; the saturated model for a two dimensional table as $\{AB\}$. In the case of three variables, model $\{AB, AC, BC\}$ refers to a hierarchical model in which only the three-variable effect τ_{ijk}^{ABC} has been set to one.

This reproducing of marginal frequencies is of special importance in the case of a *stratified sample*, because the frequency distribution of the stratifying variables is fixed by design and is not random. This fixed character should be taken into account by reproducing the frequency distribution of the stratifying variables exactly and including in the model at least those parameters that will take care of this. When, with regard to a multidimensional table, the joint variable AB (e.g., Age-Gender) has been used as the stratifying variable, then any model for these data must reproduce exactly the observed frequencies f_{ij}^{AB} and must contain at least the parameters τ_i^A, τ_j^B, τ_{ij}^{AB} (and η). Because the AB distribution is fixed by design, all these "corresponding" parameters are also fixed and it does not make sense to calculate the standard errors of the estimates of these parameters.

Nonhierarchical models do not have this "reproduction property." This can be easily demonstrated through a simple model for Table 2.2. It is not the kind of model that an investigator would normally apply to this table, but this particular model will be very important in the next chapter(s).

According to Table 2.2, the conditional probability that a young person is a member of a religious denomination (.426) appears to be practically equal to the conditional probability that an older person is not a member (.391). One could entertain the hypothesis that these conditional probabilities in the population are exactly equal to each other. Though this may not be immediately clear, this hypothesis can be rendered through the following log-linear model:

$$F_{ij}^{AB} = \eta \tau_i^A \tau_{ij}^{AB}$$

$$G_{ij}^{AB} = \theta + \lambda_i^A + \lambda_{ij}^{AB} \tag{2.22}$$

This model is a nonhierarchical model because it does not contain the parameter τ_j^B although the higher order parameter τ_{ij}^{AB} is present. The estimated expected frequencies can be found as follows (Fienberg, 1978; Magidson, Swan, & Berk, 1981):

$$\hat{F}_{11}^{AB} = f_{1+}^{AB}(f_{11}^{AB} + f_{22}^{AB})/N$$

$$\hat{F}_{12}^{AB} = f_{1+}^{AB}(f_{12}^{AB} + f_{21}^{AB})/N$$

$$\hat{F}_{21}^{AB} = f_{2+}^{AB}(f_{12}^{AB} + f_{21}^{AB})/N \tag{2.23}$$

$$\hat{F}_{22}^{AB} = f_{2+}^{AB}(f_{11}^{AB} + f_{22}^{AB})/N$$

Using some simple calculations, it can be demonstrated that the estimated expected frequencies in model 2.23 satisfy the hypothesis of equal conditional probabilities (proportions): $\hat{F}_{11}^{AB}/\hat{F}_{1+}^{AB} = \hat{F}_{22}^{AB}/\hat{F}_{2+}^{AB}$.

Both test statistics L^2 and Pearson-χ^2 are equal to 0.90, which with one degree of freedom is not significant: $p = 0.34$. The difference between the two conditional probabilities .426 and .391 is not statistically significant.

From equation 2.23 it can also be learned that the marginal frequency of A is reproduced exactly: $\hat{F}_{i+}^{AB} = f_{i+}^{AB}$. However, in spite of the occurrence of the effect τ_{ij}^{AB}, this nonhierarchical model does not reproduce the observed frequency f_{ij}^{AB}.

It is possible to provide direct formulas, that is, closed form expressions to calculate the estimated expected frequencies $\hat{F}$ for both of the simple unsaturated models described above, the independence model (equation 2.19) and the nonhierarchical model (equation 2.22). If such direct estimates of F exist, there are also closed form expressions for the computation of the variances of the effects (Bishop et al., 1975; Fienberg, 1980).

However, there are many unsaturated models for which closed form expressions for $\hat{F}$ or the variances of the estimates do not exist. In those cases the estimated expected frequencies have to be found through an iterative procedure. There are several iterative procedures available, two of which, iterative proportional fitting and Newton-Raphson, will be discussed further.

The standard computer programs for log-linear analysis use exclusively iterative procedures, because these procedures work for all models, including those for which direct estimates can be found.

The most frequently used iterative procedure, suited for hierarchical models, is the *iterative proportional fitting* procedure (IPF) (Bishop et al., 1975; Fienberg, 1980). As some elementary knowledge of how IPF works is necessary to understand the next chapter, its principles will be described very briefly for the case of three variables A, B, and C and the "no three-variable interaction" model $\{AB, AC, BC\}$.

To find the estimated expected frequencies $\hat{F}$, the procedure starts with initial values for $\hat{F}_{ijk}^{ABC}$, denoted as $\hat{F}_{ijk}^{ABC}(0)$. Starting values may be chosen at will, albeit with the restriction that effects which are not part of the model may not occur in the initial estimates either. In this case all sets of initial values $\hat{F}_{ijk}^{ABC}(0)$ can be used which will satisfy $\tau_{ijk}^{ABC} = 1$. Usually all starting values are set to one:

$$\hat{F}_{ijk}^{ABC}(0) = 1 \qquad \text{for all } i, j, k \qquad (2.24)$$

As a matter of course, the marginal estimated expected frequencies $\hat{F}_{ij+}^{ABC}(0)$ are not equal to the observed marginal frequencies f_{ij}^{AB} as the hierarchical model $\{AB, AC, BC\}$ dictates, nor are the observed distributions of the joint variables AC and BC reproduced exactly. The first step of the first cycle is to adjust $\hat{F}(0)$ in such a way that the observed AB distribution is reproduced, resulting in new estimates $\hat{F}(1)$. The estimates $\hat{F}(1)$ are then adjusted at the second step of the first cycle to reproduce the observed distribution of the joint variable AC, resulting in the estimates $\hat{F}(2)$, which are finally adjusted at the third step to reproduce the observed *BC* distribution.

$$\hat{F}_{ijk}^{ABC}(1) = \frac{\hat{F}_{ijk}^{ABC}(0)}{\hat{F}_{ij+}^{ABC}(0)}\, f_{ij+}^{ABC}$$

$$\hat{F}_{ijk}^{ABC}(2) = \frac{\hat{F}_{ijk}^{ABC}(1)}{\hat{F}_{i+k}^{ABC}(1)}\, f_{i+k}^{ABC} \qquad (2.25)$$

$$\hat{F}_{ijk}^{ABC}(3) = \frac{\hat{F}_{ijk}^{ABC}(2)}{\hat{F}_{+jk}^{ABC}(2)}\, f_{+jk}^{ABC}$$

This ends the first cycle. With these three steps, in which the estimates $\hat{F}(0)$ have been successively adjusted to reproduce certain observed marginals, the initial estimates are improved by making use of the model's sufficient statistics, in this case: f_{ij}^{AB}, f_{ik}^{AC}, and f_{jk}^{BC}. At the end of the first cycle, the estimates obtained are better, closer to the final maximum likelihood estimates, than at the beginning of the cycle. If the test statistic L^2 was computed at this stage, it would be smaller than if computed at the beginning of the cycle.

If direct estimates of the expected frequencies of a particular model exist, the estimates found via the steps in the first iteration cycle are the wanted maximum likelihood estimates. Otherwise, as in the model dealt with here, some steps within the first cycle may partly nullify the results of previous steps. For example, after the first step in cycle 2.25 the estimates $\hat{F}(1)$ reproduce the observed *AB* distribution. But this is not necessarily true for $\hat{F}(2)$. It is then necessary to start a second cycle, having again three steps as in cycle 2.25, but now using $\hat{F}(3)$ as starting values instead of $\hat{F}(0)$. These cycles are repeated until the outcomes converge, that is, until the outcomes of the various cycles differ from each other less than some arbitrary small number, say .01. The estimates finally obtained are the desired maximum likelihood estimates which reproduce, as required, particular observed marginal distributions.

The *Newton-Raphson procedure* will not be discussed in the same manner as IPF, as this is outside the scope of this book. Haberman (1978) provided the essential details. However, both procedures will be compared on some practical aspects which are of importance when carrying out log-linear analyses. Extensive comparisons of Newton-Raphson and IPF are to be found in Bishop et al. (1975, Section 3.4) and Haberman (1978, Section 3.4).

An advantage of IPF is that if direct estimates exist, IPF will have to perform only two cycles to discover that the final estimates at the end of the second cycle are the same as those at the end of the first cycle. The Newton-Raphson procedure generally needs more cycles. If no direct estimates exist, IPF normally uses more iterations than Newton-Raphson, but as each cycle consists of very simple operations, the total computing time required by IPF is often less than that of Newton-Raphson, certainly for large multivariate tables.

On the other hand, compared to Newton-Raphson procedures, IPF has two main drawbacks. One disadvantage of IPF is that it does not provide the variances of the parameter estimates, whereas Newton-Raphson gives the standard errors as a by-product of its calculations. Because of this, programs such as ECTA or SPSS^{x}' HILOGLINEAR which use IPF compute the standard errors for all (unsaturated) models as the standard errors of the corresponding effects in the saturated model. The latter may be considered as the upper limits of the actual standard errors. This leads to conservative significance tests in favor of the null hypothesis. Programs such as FREQ (Haberman, 1979, Appendix A.1) and SPSS^{x}' LOGLINEAR which use the Newton-Raphson procedure provide the correct estimates of the standard errors.

Perhaps the main disadvantage of IPF is that it can be used only for hierarchical models. This makes programs using IPF very user friendly: All the user has to do is indicate which observed marginal frequencies have to be reproduced exactly. But this is also a serious restriction as appears from the fact that Bishop et al. (1975), Duncan (1975a), Magidson et al. (1981), and many others discussed tricks to transform a non-hierarchical model into a hierarchical model whose parameters can be estimated through IPF. Though very inventive, these tricks lack the simplicity and lucidity of directly handling nonhierarchical models, as is possible with the Newton-Raphson procedure.[5]

At the same time, this causes programs using Newton-Raphson to be less user friendly. The input to be provided by the user is more complicated. The user has to know how to work with design matrices and needs a program for setting up complicated design matrices.[6]

Design matrices are routinely discussed in treatments of regression analysis or analysis of variance. Until rather recently they were seldom dealt with in the log-linear literature. Because understanding design matrices is necessary to carry out many of the longitudinal analyses presented in this book, a very brief explanation will be presented. An extensive exposition is provided by Evers and Namboodiri (1978); Haberman (1978, 1979) gives numerous applications.

As an example, the saturated log-linear model for two trichotomous variables is taken. This model is represented in its additive logarithmic form, because in this form the function of design matrices can be clarified:

$$G_{ij}^{AB} = \theta + \lambda_i^A + \lambda_j^B + \lambda_{ij}^{AB} \qquad (2.26)$$

In terms of matrix algebra, this model is expressed as:

$$G = X\lambda \qquad (2.27)$$

where X is the design matrix.

Written out, equation 2.27 looks as follows:

$$
\begin{bmatrix} G_{11} \\ G_{12} \\ G_{13} \\ G_{21} \\ G_{22} \\ G_{23} \\ G_{31} \\ G_{32} \\ G_{33} \end{bmatrix}
=
\begin{bmatrix}
1 & 1 & 0 & 1 & 0 & 1 & 0 & 0 & 0 \\
1 & 1 & 0 & 0 & 1 & 0 & 1 & 0 & 0 \\
1 & 1 & 0 & -1 & -1 & -1 & -1 & 0 & 0 \\
1 & 0 & 1 & 1 & 0 & 0 & 0 & 1 & 0 \\
1 & 0 & 1 & 0 & 1 & 0 & 0 & 0 & 1 \\
1 & 0 & 1 & -1 & -1 & 0 & 0 & -1 & -1 \\
1 & -1 & -1 & 1 & 0 & -1 & 0 & -1 & 0 \\
1 & -1 & -1 & 0 & 1 & 0 & -1 & 0 & -1 \\
1 & -1 & -1 & -1 & -1 & 1 & 1 & 1 & 1
\end{bmatrix}
\cdot
\begin{bmatrix} \theta \\ \lambda_1^A \\ \lambda_2^A \\ \lambda_1^B \\ \lambda_2^B \\ \lambda_{11}^{AB} \\ \lambda_{12}^{AB} \\ \lambda_{21}^{AB} \\ \lambda_{22}^{AB} \end{bmatrix}
\qquad (2.28)
$$

$$G \quad = \qquad\qquad X \qquad\qquad\qquad \lambda$$

The design matrix determines how the (logarithms of the) expected frequencies are linked to log-linear effects. For example, the equation for G_{11} can be found by multiplying the first row of X by the effect vector λ:

$$G_{11} = (1)\theta + (1)\lambda_1^A + (0)\lambda_2^A + (1)\lambda_1^B + (0)\lambda_2^B + (1)\lambda_{11}^{AB} + (0)\lambda_{12}^{AB}$$

$$+ (0)\lambda_{21}^{AB} + (0)\lambda_{22}^{AB}$$

$$G_{11} = \theta + \lambda_1^A + \lambda_1^B + \lambda_{11}^{AB}$$

And the equation for G_{13} is (remembering that the log-linear effects over each subscript sum to zero):

$$G_{13} = (1)\theta + (1)\lambda_1^A + (0)\lambda_2^A + (-1)\lambda_1^B + (-1)\lambda_2^B + (-1)\lambda_{11}^{AB}$$

$$+ (-1)\lambda_{12}^{AB} + (0)\lambda_{21}^{AB} + (0)\lambda_{22}^{AB}$$

$$G_{13} = \theta + \lambda_1^A - (\lambda_1^B + \lambda_2^B) - (\lambda_{11}^{AB} + \lambda_{12}^{AB}) = \theta + \lambda_1^A + \lambda_3^B + \lambda_{13}^{AB}$$

By properly manipulating the design matrix, effects can be left out or set equal to each other; linear or parabolic effects can be created; other parametrizations, for example, dummy coding instead of effect coding, can be chosen. When one knows how to handle design matrices, all kinds of different hierarchical and nonhierarchical models can easily be defined and their parameters estimated.

At the same time, the more flexible computer programs become and the easier it becomes to employ a large variety of models, the more difficult it will be to choose the "right" model. The next section presents some guidelines on how to proceed in this matter.

2.5 Testing and Fitting Unsaturated Models

The object of most model selection procedures is to find the true model, that is, the population model that generated the observed data. In this respect, a particular selected model may be wrong for two reasons. It may contain too many parameters, that is, parameters that are not part of the population model—the problem of overfitting—and it may contain too few parameters, that is, parameters that occur in the population model are excluded from the selected model—the problem of underfitting.

The major part of the literature on model selection is devoted to the question of how to avoid the errors of under- and overfitting. Just a few highlights of these discussions will be presented in this section. More extensive and excellent expositions are provided by Long (1988; albeit in the context of LISREL models and generally using a slightly different definition of over- and underfitting).

It cannot be emphasized enough that theoretical notions should play a major role in model selection. Even in exploratory research where only statistics seem to determine which model has to be chosen, theoretical considerations are at least as important. From a purely statistical point of view, usually hundreds of different models may be formulated for a particular set of data, many of which will fit the data equally well. Which models will actually be considered and which of the well-fitting models will ultimately be selected as the right one(s) has to a large extent to be determined by theoretical interests and interpretations. The importance of theoretical guidelines becomes even greater when one realizes that the many goodness-of-fit statistics will not always give the same answers and that, moreover, we often rely on the outcomes of statistical testing procedures whose use cannot be properly justified. Below it is mainly statistics which will be of concern, but that discussion will make clear how overriding is the importance of good substantive ideas.

First a brief outline will be presented of the testing procedures that may be applied when the investigator has very definite a priori ideas about the relationships among the variables in the population. After this textbook case of model selection, the opposite case will be discussed in which a researcher has none or just very vague notions about the relations among the variables. Elements from these two extreme cases will then be used to show how to conduct what might be called a theory-driven data exploration.

From these discussions it will become clear that the standard χ^2-test statistics and testing procedures may have limited use. This has given rise to the development of alternative descriptive measures of goodness-of-fit and to a different approach toward model selection. These are the topics of the last part of this section.

Model selection seems to present no special difficulties when the investigator has explicit a priori ideas about the relationships in the population. If the theoretical expectations lead to one particular unsaturated model, one simply uses the standard χ^2-statistics (equation 2.21) to test the empirical validity of the postulated model.

If rival theories exist leading to not one but a few models and one wants to know which one is supported by the data, *conditional L^2-tests* can be used, provided that these rival theories can be translated into models which are hierarchically related to each other, that is, that one model can be obtained just by imposing restrictions on the parameters of the other model.

Let us take as an example three unsaturated models for a table ABC:

model 1: $\{A, B, C\}$ $F_{ijk} = \eta \tau_i^A \tau_j^B \tau_k^C$

model 2: $\{AB, C\}$ $F_{ijk} = \eta \tau_i^A \tau_j^B \tau_k^C \tau_{ij}^{AB}$ (2.29)

model 3: $\{AB, AC, BC\}$ $F_{ijk} = \eta \tau_i^A \tau_j^B \tau_k^C \tau_{ij}^{AB} \tau_{ik}^{AC} \tau_{jk}^{BC}$

In terms of the variables in Table 2.3, the first model implies that the variables Age (A), Religious Membership (B), and Voting Behavior (C) are all independent of each other. The second model indicates that age and religion are related to each other, but voting does not depend on either of them. In the third model, all three variables have direct relationships, but the three-variable interactions are absent.

The three models in example 2.29 are hierarchically nested in the sense that model 2 contains only a subset of the parameters of model 3, and model 1 contains only effects which are also in model 2. Model 1 is a *restricted* model with respect to the *unrestricted* model 2, and model 2 in its turn is a *restricted* model with respect to the *unrestricted* model 3.

If models are hierarchically nested in this manner, they can be tested against each other by computing the usual L^2-statistics for the restricted model and the unrestricted model, denoted as L_r^2, L_u^2 respectively, and subtracting them from each other to obtain the conditional test statistic $L_{r/u}^2$:

$$L_{r/u}^2 = L_r^2 - L_u^2 = 2 \sum_i \sum_j [f_{ij} \ln(\hat{F}_{ij-u} / \hat{F}_{ij-r})]$$

$$\text{df}_{r/u} = \text{df}_r - \text{df}_u$$

(2.30)

where the subscript r refers to the restricted and the subscript u to the unrestricted model.[7]

If the restricted model is valid for the population, $L_{r/u}^2$ asymptotically approaches a χ^2-distribution with $\text{df}_{r/u}$ degrees of freedom, given that the unrestricted model is approximately true. Using the conditional test statistic $L_{r/u}^2$, one tests the null hypothesis that in the population the restricted model is true against the alternative hypothesis that the unrestricted model holds. This is different from the unconditional test in which L^2 tests the same null hypothesis that the restricted model is true, but against the alternative that the saturated model is true. If a conditional test is employed, rejecting the null hypothesis implies accepting the unrestricted model as true; if an unconditional test is used, rejecting the restricted model means accepting the saturated model as true.

It could be said that $L_{r/u}^2$ tests the hypothesis that the expected frequencies of the restricted and the unrestricted models are the same: $F_r = F_u$, and that the differences between $\hat{F}_r$ and $\hat{F}_u$ are just the result of sampling fluctuations. In this way $L_{r/u}^2$ provides a test of the significance of the extra restrictions that are imposed in the restricted model relative to the unrestricted model.

Assuming that model 3 in equation 2.29 is approximately valid for the population, $L_{2/3}^2$ is used as a test of the hypothesis that $\tau_{ik}^{AC} = \tau_{jk}^{BC} = 1$ in the population (Fenech & Westfall, 1988). Thus, given that there is no three-variable interaction among age (A), religion (B), and voting (C) in the population, but that otherwise all lower order effects may exist, $L_{2/3}^2$ can be used to test simultaneously whether the direct effects of age and religion on voting are significant.

Tests of such a hypothesis within a more restricted model have more power than a test of the same hypothesis within a less restricted model (Goodman, 1981a). In general, conditional tests are to be preferred to unconditional tests, as the (conditional) test of a particular restricted model against a more unrestricted but unsaturated alternative has more power than the (unconditional) test of this same restricted model against the saturated alternative.

Another advantage of conditional L^2-tests is that the approximation of the theoretical χ^2-distribution under rather general conditions can be expected to be satisfactory, even in circumstances in which the approximation for the nonconditional tests is problematic, as in the case of sparse tables (Haberman, 1978, Chapter 5).

Closely related to conditional testing is the *partitioning* of L^2. The test statistic L^2 for model 1 in example 2.29 can be written as:

$$L_1^2 = (L_1^2 - L_2^2) + (L_2^2 - L_3^2) + L_3^2 = L_{1/2}^2 + L_{2/3}^2 + L_3^2 \qquad (2.31)$$

The extent to which the estimated expected frequencies of model 1, the most restricted model, deviate from the observed frequencies is indicated by L_1^2. This discrepancy can be partitioned into three components $L_{1/2}^2$, $L_{2/3}^2$, and L_3^2. The test statistic $L_{1/2}^2 = L_1^2 - L_2^2$ indicates to what extent this discrepancy can be significantly reduced by the addition of τ_{ij}^{AB} to model 1. $L_{2/3}^2 = L_2^2 - L_3^2$ represents the extent to which the addition of the parameters τ_{ik}^{AC} and τ_{jk}^{BC} then further reduces this discrepancy. L_3^2 can be used to check whether it is necessary to add the three-variable effect τ_{ijk}^{ABC} in order to reproduce the observed frequencies satisfactorily.

In the above discussion it was assumed that the investigator has explicit a priori ideas about reality which lead to the formulation of one or just a few unsaturated models. This is the textbook situation in which statistical testing theory can be applied to see whether or not the hypotheses are acceptable in the light of the data.

However, even in this case, model selection may turn out to be problematic. The standard χ^2-tests are overall tests by which only particular kinds of theoretical expectations are tested. One tests whether or not particular effects are absent in the population, whether or not certain effects are equal to each other or to a particular constant, and whether or not the relations among variables have a specific, for example, linear, form. If the model need not be rejected, the investigator's hypotheses are confirmed as far as they pertain to these kinds of restrictions.

However, once the estimated expected frequencies of an accepted model have been obtained and the effect parameters estimated, the investigator might still discover discrepancies between theory and data. It might have been expected, but not included in the model tested, that certain relations among variables are positive and others negative, or that particular relationships are stronger than others: Several of these expectations may turn out not to be true. Moreover, inspection of the standardized effects may lead to the conclusion that some of the effects that were supposed to be present are not significant. Such outcomes may still lead to rejection of the postulated theory and to the exploration of different models (see below).

It should also be kept in mind that, from a practical point of view, the statistical testing procedures work well only when the sample size is neither too large nor too small.

A very *large sample* may cause all effects in the saturated model to be significant, even very small ones which have no substantive meaning, and may lead to the rejection of all unsaturated models. This problem (closely related to the problem of overfitting) can be avoided by choosing models not only on the basis of their statistical significance, but also on account of the sizes of the effects. If a particular unsaturated model has to be rejected on the basis of the test outcome, but adding parameters does not change the estimates of the parameters that were already in the model and the added parameters themselves are small, this particular model should be accepted (see Wheaton, 1987: 123-125). Otherwise, it should be rejected in agreement with the test outcome.

Small samples cause more difficulties. In addition to the fact that with small samples Pearson-χ^2 and L^2 cannot be expected to follow the χ^2-distribution, there is the problem that tests based on small samples have little *power*. This is exactly the opposite of the large sample case: Even large effects that do exist in the population will be labeled as insignificant (the problem of underfitting). If the danger of falsely accepting the null hypothesis (type II error) is large and consequently the power is low, it may be better not to choose between the null hypothesis and the alternative hypothesis, but in line with Hays's recommendations to "suspend judgment" (Hays, 1981: 244).[8]

The situation in which the researcher has definite hypotheses is decidedly not without difficulties; situations in which no clear a priori hypotheses exist are even more complicated. Given the lack of precise hypotheses, the investigator explores the data in search of a model that can be given a meaningfully theoretical interpretation and that fits the data with as few parameters as possible. In principle, the fit of the model which is finally selected cannot be improved significantly by adding parameters, but will be worsened when some of its parameters are deleted.

Besides theoretical meaningfulness, the leading principle of exploratory search procedures is the *parsimony principle*, Occam's razor. Given a certain level of accuracy, a less complex explanation of the data is to be preferred above a more complex one and a model with fewer parameters should be preferred to a model with more parameters (other things being equal). As Popper (1959) has argued, simple models are not to be preferred for aesthetic reasons, nor for their elegance, but because simple models provide us with more information about the state of the world, as they exclude more possible states of the world than do complex models. Simple models have a higher chance of being rejected; they have a higher degree of falsifiability (Mulaik et al., 1989; Popper, 1959, Chapter VII). Because of the importance of the concept of falsifiability in science, parsimony is a desirable property and not just an aesthetically pleasing feature of a model.

Many authors, Aitkin (1980), Benedetti and Brown (1978), Bishop et al. (1975: 165-175), Goodman (1971a, 1973c), among others, have given extensive guidelines on how to undertake such data explorations from a statistical point of view. These procedures boil down to mixed forms of forward and backward selection.

In the case of *forward selection*, the data exploration is started with a very parsimonious model, that is, a model in which all variables are independent of each other. When this model has to be rejected on the basis of the test statistics, one tries by successive addition of parameters to find a model that has as few parameters as possible, but has not to be rejected according to the (conditional) tests. In searching for this well-fitting parsimonious model, use is made of several statistics, especially the conditional test statistic $L^2_{r/u}$ and the residual cell frequencies $(f_{ij} - \hat{F}_{ij})$.

Although it requires some practical experience, a careful study of the *residual cell frequencies*, especially the pattern of + and –, can give us hints as to where and why a particular model fails to fit the data and which parameters may be fruitfully entered into the model. Several examples of this will be given in later chapters.

Mostly the residuals ($f_{ij} - \hat{F}_{ij}$) are not employed, because the sizes of these residuals are too dependent on the sizes of the cell frequencies themselves. The variant occurring most often is the *standardized residual* ($f_{ij} - \hat{F}_{ij}$)/($\hat{F}_{ij}^{1/2}$). Squaring these quantities and summing them over all cells results in Pearson-χ^2. The standardized residuals show to what extent each particular cell contributes to the size of Pearson-χ^2. Haberman (1978: 78-79) presented *adjusted residuals*, residuals divided by their respective estimated standard deviations. If the model is valid in the population, adjusted residuals approximate the standard normal distribution, in which the approximation is probably adequate if each expected cell frequency is at least 25.

Backward selection procedures work the other way around: one starts with the saturated model and, by leaving out certain parameters, tries to find a more parsimonious model that still fits the data according to the χ^2-tests. To determine which effects may be deleted, usually the standardized effects $\hat{\lambda}/\hat{s}_\lambda$ are employed. Because deleting from a particular model all insignificant standardized effects at the same time may lead to a model that does not fit the data, (conditional) L^2-tests in the end determine which model is acceptable and which parameters from the saturated model can be eliminated.

The main problem with these purely statistical search procedures is that the outcomes may depend on what particular forward or backward procedure or combinations thereof has been used. As was remarked above, usually a rather large number of models can be found which fit the data about equally well. The statistical search procedure chosen arrives at one of these models, but a different search procedure might have resulted in a different choice and there is no way of telling within this statistical framework which one represents the true population model. The dangers of over- and underfitting loom large.

Our main tool for finding the true population model and diminishing the dangers of over- and underfitting is to derive our models from social theories that have been corroborated at several points by solid empirical evidence. Luckily, researchers are seldom or never totally deprived of such theoretical notions when carrying out an investigation. With this in mind, the following set of guidelines based on the above discussion and on the literature on model selection within the context of LISREL models may prove useful (Kaplan, 1988; Luijben, 1989; MacCallum, 1986; Saris, Satorra, & Sörbom, 1987; Suyapa, Silvia, & MacCallum, 1988). At every step of the search process, use should be made of whatever theoretical insights are available.

To get powerful tests and to avoid problems of underfitting, the sample size should be (very) large. Ideally, the investigator will determine what size of the effect parameters the tests are able to detect given the sample size (see note 8).

If at a particular step, the search procedure may be continued in different but all theoretically meaningful directions, it is best to try them all. If different theoretically meaningful models result that fit the data about equally well, the investigator should suspend judgment and look for additional empirical and theoretical evidence to decide between the rival models. Make sure, however, that the rival models are really different, that is, the effects that occur in one model and not in the other are not too small and negligible from a substantive point of view. Attention to statistical significance tends to divert attention from the size of the effects.

Whenever possible, divide the sample into two random parts, carry out the search procedure on one part and apply the results to the other to see whether the model still fits and the included effects remain significant. This is an especially useful strategy to avoid overfitting. Because during the search process the same data are continually used to formulate as well as test the hypotheses, there is a serious risk that certain (higher order) effects are peculiar to this particular sample and explain only sampling variance instead of systematic variance. Bonett and Bentler (1983) have provided further information on this.

To follow both the principle of theoretical meaningfulness and the principle of parsimony, a good choice for the baseline model with which to start the selection procedure is the most parsimonious model that is theoretically meaningful. If the baseline model is rejected, effects are added, one at a time, on the basis of theoretical considerations and inspection of the residual frequencies. (Variants of modification indices that are available in the LISREL program and that tell us what improvement of fit to expect when a particular restriction is released are not routinely available in log-linear programs, but are in principle applicable in log-linear modeling as well — see note 8.)

If the baseline model (or another model) need not be rejected, carry the search for better models a little further to minimize the danger of overlooking important effects.

After accepting a model, start deleting nonsignificant effects to arrive at a more parsimonious (but well-fitting) model, again on the basis of theoretical considerations, but also making use of the standardized effects.

The final model may still not be the true population model, but careful application of the above principles should optimize our chances of finding it.

Usually the data explorations sketched above rely heavily on the outcomes of various statistical tests, that is, on the significance levels obtained by the adjusted residuals, the standardized effects, and, especially, the conditional and unconditional L^2-statistics. However, in a strict sense it is not possible to justify such use in this exploratory context: A large number of tests are being used which are not independent of each other and most important they are applied ex post facto, after inspecting the data. In this situation there is no way of telling what risks we run when accepting or rejecting a particular model in terms of the probabilities of error types I and II.[9]

Therefore, it has been proposed to base the decision to reject or accept particular models not on the significance levels obtained by the test statistics, but on the values of L^2 or Pearson-χ^2 themselves. These test statistics are then used as purely *descriptive measures* of the discrepancy between a model and the empirical reality, between estimated expected and observed frequencies. Bollen (1989) and Mulaik et al. (1989) have provided a critical review of several descriptive goodness-of-fit measures, only some of which will be discussed below.

There are two problems with treating L^2 and Pearson-χ^2 themselves as descriptive measures. One is the dependency of their sizes on the sample size N. The other, and more important problem is the lack of a clear demarcation line, comparable to $p < .05$, that separates acceptable models from unacceptable ones. Adding parameters to a model results in smaller values of the test statistics, and the "best" model would always be the saturated model with $L^2 = 0$.

The first difficulty can be remedied by computing L^2 and Pearson-χ^2 on the relative frequencies, that is, on p and $\hat{\pi}$, or, what amounts to the same, by dividing the test statistics computed in the usual way by N. This statistic is called the *effect size* and is denoted as e. In terms of L^2:

$$e = \frac{L^2}{N}$$

$$w = \sqrt{e}$$

(2.32)

The effect size e plays an important role in the determination of the power of conditional and unconditional χ^2-tests (Bonett & Bentler, 1983; Cohen, 1977), but it can also be used in a purely descriptive context. Cohen recommended working with $w = \sqrt{e}$, which in a 2×2 table is equal to the association coefficient ϕ if w is based on the Pearson-χ^2 test for independence (Hays, 1981: 556). A possible demarcation line then might be: If w is larger than 0.10, there is a serious discrepancy between the model and the data, and the model has to be rejected; in the case of conditional testing, in favor of the more unrestricted unsaturated model, or, in the case of unconditional testing, in favor of the saturated model. Of course, blindly applying such a demarcation line without considering the theoretical meaning of the model is as bad as blindly applying a .05 significance level.

Comparisons between models in terms of L^2 or Pearson-χ^2 that take the number of parameters into account are possible by using the F-statistic. In terms of L^2:

$$F = \frac{L^2}{df}$$

(2.33)

F can be treated as a genuine test statistic, as it follows approximately the F-distribution with degrees of freedom $df_1 = df$ used in equation 2.33 and $df_2 = \infty$, if the model is true (Goodman, 1971a, 1975; Haberman, 1978: 17; Wheaton, 1987). Its expected value is then one, as the expected value for L^2 is the number of degrees of freedom. Models with values close to one are therefore to be preferred above models with larger values for F. Wheaton (1987) expresses serious doubts as to the usefulness of F, but as a descriptive measure it does take into account that a reasonable fit has to do both with small discrepancies between expected and observed data and with the number of parameters that are used to acquire small discrepancies, that is, with the complexity of the model.

Goodman (1972a, 1972b) developed a set of measures to choose between hierarchically nested models (and compared them somewhat unfortunately with partial and multiple correlations coefficients). The general form of these measures is in terms of the L^2- or e-statistic:

$$R' = \frac{L_r^2 - L_u^2}{L_r^2} = \frac{e_r - e_u}{e_r} \qquad (2.34)$$

High values for R' (e.g., 0.80 — Zahn & Fein, 1979) indicate that the extra effects which were added to the unrestricted model relative to the restricted model account for a considerable part (80%) of the discrepancies that existed between the expected frequencies of the restricted model and the observed frequencies. High values of R' (close to one) suggest accepting the unrestricted model; low values (close to zero) give preference to the restricted model.

A disadvantage of R' is that it does not take into account the complexity of the model, that is, the number of extra parameters. Therefore Bonett and Bentler (1983) have recommended two variants of R', based on the F-statistic of which the most useful seems to be:[10]

$$\hat{\delta} = \left(\frac{L_r^2}{df_r} - \frac{L_u^2}{df_u} \right) \Big/ \frac{L_r^2}{df_r} = \frac{F_r - F_u}{F_r} \qquad (2.35)$$

Low values of $\hat{\delta}$ point to choosing the restricted, high values to selecting the unrestricted model. Whereas R' obtains higher and higher values when more and more parameters are added to the unrestricted model, this is not the case for $\hat{\delta}$. The coefficient $\hat{\delta}$ may even become negative when the reduction in L^2 is offset by the loss in degrees of freedom. Usually $\hat{\delta}$ will be much smaller than R', at any rate $\hat{\delta} \leq R'$. For choosing among models, $\hat{\delta}$ appears to be of greater practical value than R'.

Using descriptive or inferential measures of goodness-of-fit, the object of the selection procedures outlined above is to find the true population model, the one that generated the observed data. Akaike (1973) has developed a somewhat different approach to model selection in which he abandons the notion of a true model and tries to integrate model selection and evaluation, employing information theoretical notions. His estimation procedures are identical to the usual maximum likelihood procedures, but his model selection criterion is different.

From among a set of competing models, not necessarily hierarchically nested, the best one (not the "true" one) should be chosen, that is, the one that provides the greatest amount of information about the real world and will tell us most about future observations. Two aspects of a model are relevant in this respect. On the one hand, the estimated expected frequencies should be close to the observed frequencies, on the other hand, the model should be as parsimonious as possible, as the least complex model is the model with the highest information gain. A model should take the essential characteristics of the sample into account but ignore its idiosyncratic features.

The operationalization and quantification of these concepts resulted in the model selection criterion, Akaike's Information Criterion (AIC), which indicates to what extent a particular model deviates from reality, "penalizing" the model according to its degree of complexity. In principle the model with the lowest AIC must be chosen. Interesting interpretations of AIC within a Bayesian context exist. An overview of the potentialities of this approach, together with several modifications of the AIC.index, was presented in a special issue of *Psychometrika* (Akaike, 1987; Bozdogan, 1987; Sclove, 1987). Although a thorough treatment is outside the scope of this book, one of the AIC variants, BIC, will be briefly discussed.

Raftery (1986a, 1986b) was one of the first to point to the possibilities of the AIC approach within the context of log-linear modeling. He proposed the BIC index − Bayesian Information Criterion − which is similar to an AIC variant developed by Schwarz (1978) and can easily be computed from the standard output for log-linear models:

$$BIC \equiv L^2 - (\ln N)(\text{df}) \qquad (2.36)$$

where L^2 is the usual test statistic, N the sample size, and df the number of degrees of freedom belonging to L^2. The original AIC formula has ln2 instead of $\ln N$, but is otherwise identical.

For each model that is taken into consideration, BIC must be computed, using the ordinary unconditional L^2 statistic. The model with the lowest BIC value should be chosen. A model that does not adequately fit the observed frequencies produces a high L^2 value and is not likely to be selected. On the other hand, a model that produces too good a fit at the expense of having a large number of parameters and, accordingly, a small number of degrees of freedom is penalized, as then the second part of the right hand side of equation 2.36 will be small.

As Raftery (1986a) showed, BIC may be interpreted within a Bayesian context. Selecting a model means answering the question:

given the data, which of two models M_0 or M_1 is more likely to be the true model? The . . . question can be answered by calculating the posterior odds for M_0 against M_1, defined by

$$B = \frac{\text{Prob } [M_0 \text{ is the true model given the data}]}{\text{Prob } [M_1 \text{ is the true model given the data}]}$$

(p. 145)

In large samples BIC approximately equals $-2\ln B$; thus BIC is monotonically related to the posterior odds of preferring model M_0 above M_1.

> If BIC is negative, (and consequently ln B is positive and B itself greater than 1 - J.H.) . . . we should accept M0 in the sense of preferring it to the saturated model. If we are comparing several models, we should prefer the one with the lowest BIC value. (pp.145-146)

BIC appears to work very well in practice. Together with (conditional) tests and descriptive measures such as $\hat{\delta}$, it gives the investigator a powerful tool to choose the best model from among the set of theoretically meaningful models.

2.6 The Collapsibility Theorem

In the first sections it was stipulated that, in principle, log-linear effects take on different values when other variables are introduced and that the effects in the marginal tables are usually different from the corresponding effect in the full table. The collapsibility theorem, which will be needed in many places in this book, tells us exactly under what circumstances the effects of hierarchical models change when additional variables are being introduced and under what circumstances neglecting certain variables and analyzing particular marginal tables lead to different conclusions. Bishop et al. (1975) formulated the collapsibility theorem as follows (with some slight modifications because of differences in notation):

> Suppose the variables in a multidimensional table are divided into three mutually exclusive groups. One group is collapsible with respect to the τ-terms involving a second group, but not with respect to the τ-terms involving the third group, if and only if the first two groups are independent of each other (that is, the τ-terms linking them are 1). (p. 47)

To illustrate this theorem, let us take a six-dimensional table *ABCDEF*. We set up a particular hierarchical model for this table and would like to know whether the parameters pertaining to the variables C, D, E, and F take on different values if we collapse the table over A and B.

Let us assume that the model is such that neither A nor B is directly related to C or D; that E and F are directly related to A, B, or both; and

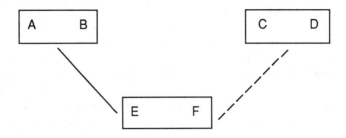

Figure 2.1.

that no requirements are made for the relations between E, F on the one hand and C, D on the other. Models that satisfy these conditions are, among others, $\{ABEF, CDEF\}$ and $\{AB, AE, BF, EF, CD\}$. A schematic representation might be as in Figure 2.1.

The six variables are thus divided into three mutually exclusive groups:

1. the variables to be collapsed, here A and B;
2. the variables which according to the postulated hierarchical model are independent of the variables from the first group, here, C and D;
3. the remaining variables, here E and F.

The collapsibility theorem implies that the values of all parameters which have superscripts containing one or more variables of the second group (C, D) will not change when the table is collapsed over one or more categories of the variables of the first group (A, B). It does not matter whether the parameters τ_{kl}^{CD}, τ_{km}^{CE}, τ_{kn}^{CF}, τ_{lm}^{DE}, τ_{ln}^{DF}, and all higher order effects in which the superscripts C or D occur such as τ_{kmn}^{CEF} are determined by means of table $ABCDEF$ or the marginal table $CDEF$.

However, the collapsibility theorem also implies that all parameters which have superscripts containing variables only from the third group (E, F) such as τ_{mn}^{EF} will have different values in the collapsed table $CDEF$ and the original table $ABCDEF$.

It should be kept in mind that the collapsibility theorem pertains only to the effect parameters themselves, but not to the standard errors of the estimates and therefore also not to the standardized effects. The latter will in principle always assume different values after collapsing.

2.7 The Effect Model

In the previous sections dealing with the frequency model, it was assumed that all variables were at the same causal level. Yet, often the distinction between dependent and independent variables will come into the interpretation of the results. When the data in Table 2.3 on the relations among age, religious membership, and voting were used as an example, the connotation that age and religion are determinants of voting behavior could not be avoided even when the terms cause and effect as such were not used.

This implicit preference to give the variables a causal order and to interpret their relationships in a causal sense can be expressed explicitly in *effect models*. In these models the quantities to be explained are not the sizes of the cell frequencies, but the sizes of the (conditional) odds of belonging to a certain category of a particular dependent variable, for example, the odds of being a voter rather than a nonvoter.[11] As the logarithms of the odds are indicated by the terms log odds or logit, these effect models are often called *logit models*. The effect parameters of a logit model have a causal connotation contrary to the "effect" parameters of the frequency model.

Effect models may be used both in a "modified multiple regression approach" and in a "modified path analysis approach" (Goodman, 1972a, 1973b). In the first approach, analogous to the standard multiple regression model, there is one dependent variable and several independent ones and we are interested in the direct effects each of the independent variables have on the dependent variable. This model is discussed in Section 2.7.1 for the case of a dichotomous dependent variable and in Section 2.7.2 for the general case of a polytomous dependent variable.

The modified path analysis approach is analogous to the standard path analytic models in which a causal model is set up for the relations among all variables and in which particular variables may serve both as dependent and as independent variables.

The treatment of the effect model will be considerably shorter than that of the frequency model. The effect model can be reformulated in terms of the frequency model so that all that was said above regarding model fitting, etc. can also be applied to effect models.

2.7.1 The Modified Multiple Regression Approach: Effect Models With A Dichotomous Dependent Variable

Let us go back to Table 2.3 but now explicitly from the perspective of voting (C) as being dependent upon age (A) and religious membership (B). The question is: Do the odds of being a voter rather than a nonvoter (represented in the last column of Table 2.3) vary according to age and religion?

Conditional odds and logits are denoted as follows (bearing in mind that $\pi_{i1}^{AC} / \pi_{i2}^{AC} = F_{i1}^{AC} / F_{i2}^{AC}$, etc.):

$$\Omega_{ij}^{AB\overline{C}} = \frac{F_{ij1}^{ABC}}{F_{ij2}^{ABC}} \qquad \hat{\Omega}_{ij}^{AB\overline{C}} = \frac{\hat{F}_{ij1}^{ABC}}{\hat{F}_{ij2}^{ABC}} \qquad \omega_{ij}^{AB\overline{C}} = \frac{f_{ij1}^{ABC}}{f_{ij2}^{ABC}}$$

$$\Phi_{ij}^{AB\overline{C}} = \ln\Omega_{ij}^{AB\overline{C}} \qquad \hat{\Phi}_{ij}^{AB\overline{C}} = \ln\hat{\Omega}_{ij}^{AB\overline{C}} \qquad \phi_{ij}^{AB\overline{C}} = \ln\omega_{ij}^{AB\overline{C}} \qquad (2.37)$$

where a bar is placed above the dependent variable.

Sometimes the dependent variable is provided with a subscript k/k', where $k \neq k'$ ($\Omega_{ijk/k'}^{AB\overline{C}}$) to indicate that the odds refer to the probabilities of belonging to category k rather than k'. If $k = 1$ and $k' = 2$, this subscript is almost always omitted. Thus, in Table 2.3, $\omega_{11}^{AB\overline{C}} = 1.979$.

As with the frequency model, the effect model can be saturated or unsaturated. Also, identifying restrictions must be imposed on the parameters of the effect model. Unless otherwise stated, the sum of the logit parameters β over any subscript will be set to zero and the product of the parameters of the multiplicative model γ over any subscript will be set to one. Other parametrizations are also possible here.

The *saturated effect model* for a three dimensional table ABC, where C is the dependent variable, is in multiplicative and logarithmic form:

$$\Omega_{ij}^{AB\overline{C}} = \gamma^{\overline{C}} \gamma_i^{A\overline{C}} \gamma_j^{B\overline{C}} \gamma_{ij}^{AB\overline{C}}$$

$$\Phi_{ij}^{AB\overline{C}} = \beta^{\overline{C}} + \beta_i^{A\overline{C}} + \beta_j^{B\overline{C}} + \beta_{ij}^{AB\overline{C}} \qquad (2.38)$$

Analogous to the frequency model, the conditional odds in Table 2.3 can be described in terms of an overall effect, two one-variable effects, and a two-variable effect. The overall effect $\gamma^{\overline{C}}$ or $\beta^{\overline{C}}$ indicates how large the (log-)odds of being a voter instead of a nonvoter are on average in the categories of the independent variables, in this case the variables Age and Religious Membership, where the average is calculated as the geometric mean of the conditional odds or as the arithmetic mean of the logits. The one-variable effects of age, γ_1^{AC} and γ_2^{AC}, indicate to what extent the partial odds voter/nonvoter for the younger and the older age groups respectively deviate from the overall effect, where the partial odds are the average conditional odds voter/nonvoter among members and nonmembers of a religious denomination. The one-variable effects of religious membership have an analogous interpretation when age is held constant. The two-variable effects $\gamma_{ij}^{AB\overline{C}}$ show whether the effects of age are different among members and nonmembers or whether the effects of religion are different among the old and the young.

To obtain the maximum likelihood estimates of the parameters in model 2.38, the maximum likelihood estimates $\hat{\Omega}$ or $\hat{\Phi}$ are needed. For the saturated effect model these are the observed (log-)odds ω and ϕ. Analogous to the frequency model, it is possible to write down closed form expressions for the parameter estimates of the effect model in terms of the observed odds. This will not be done, as the parameters of the effect model may also be obtained from the parameters of the frequency model. This can easily be seen from the following equation:

$$
\begin{aligned}
\Omega_{ij\,1/2}^{AB\overline{C}} = \frac{F_{ij1}^{ABC}}{F_{ij2}^{ABC}} &= \frac{\eta\tau_i^A\tau_j^B\tau_{ij}^{AB}\tau_1^C\tau_{i1}^{AC}\tau_{j1}^{BC}\tau_{ij1}^{ABC}}{\eta\tau_i^A\tau_j^B\tau_{ij}^{AB}\tau_2^C\tau_{i2}^{AC}\tau_{j2}^{BC}\tau_{ij2}^{ABC}} \\[2mm]
&= \frac{\tau_1^C}{\tau_2^C}\cdot\frac{\tau_{i1}^{AC}}{\tau_{i2}^{AC}}\cdot\frac{\tau_{j1}^{BC}}{\tau_{j2}^{BC}}\cdot\frac{\tau_{ij1}^{ABC}}{\tau_{ij2}^{ABC}} \\[2mm]
&= \gamma_{1/2}^{\overline{C}}\,\gamma_{i\,1/2}^{A\overline{C}}\,\gamma_{j\,1/2}^{B\overline{C}}\,\gamma_{ij\,1/2}^{AB\overline{C}}
\end{aligned}
\tag{2.39}
$$

If C is dichotomous and if effects are expressed in terms of deviations from the mean effect, then from equation 2.39 it follows that $\gamma = \tau^2$, for example, $\tau_{i1}^{AC}/\tau_{i2}^{AC} = \tau_{i1}^{AC}/(1/\tau_{i1}^{AC}) = (\tau_{i1}^{AC})^2 = (\gamma_i^{AC})$. The same line of reasoning leads to: $\beta = 2\lambda$.

Because of this, the results in Table 2.4 can be used very simply to calculate the parameter estimates of the saturated effect model for Table 2.3: One just squares the appropriate τ parameter values or doubles the corresponding λ's. The standardized effects remain the same because the logit parameters β are twice the values of the log-linear λ parameters, but the standard errors of the β estimates are also twice the standard errors of λ.

From Table 2.4, using $\hat{\tau}_1^C$, it may be concluded that the average odds of being a voter in Table 2.3 are $1.443^2 = 2.082$; the odds of being a nonvoter equal $1/2.082 = 0.480$. There are on average twice as many voters as nonvoters.

From $\hat{\tau}_{11}^{AC}$ it can be inferred that among younger people, the partial odds voter/nonvoter are $0.837^2 = 0.701$ times lower than the overall partial odds, whereas among the older people the partial odds are $1/.701 = 1.427$ higher. As concluded before: Regardless of religious membership, the younger are much less inclined to vote than the older. Similar reasoning, based on $\hat{\tau}_{11}^{BC}$, leads to the conclusion that holding age constant, the tendency to vote is much stronger among religious than among non-religious people: $\hat{\gamma}_1^{B\overline{C}} = 1.331^2 = 1.772$ whereas $\hat{\gamma}_2^{B\overline{C}} = 0.564$.

The three-variable interaction term $\hat{\gamma}_{11}^{AB\overline{C}} = (\hat{\tau}_{111}^{ABC})^2 = 0.875^2 = 0.766$ shows that the relation between age and voting is different among members and nonmembers. The conditional effects of age on voting among the religious people are $\hat{\gamma}_1^{AC/B} = \hat{\gamma}_1^{AC} \times \hat{\gamma}_{11}^{AB\overline{C}} = 0.701 \times 0.766 = 0.536$ and $\hat{\gamma}_2^{A\overline{C}/B} = 1/0.536 = 1.863$. The corresponding conditional effects among the nonreligious people are $\hat{\gamma}_{1\ 2}^{A\overline{C}/B} = \hat{\gamma}_1^{A\overline{C}}/\hat{\gamma}_{11}^{AB\overline{C}} = 0.701 / 0.766 = 0.915$ and $\hat{\gamma}_{1\ 2}^{A\overline{C}/B} = 1.093$. Although the nature of the relationship does not vary, the effects of age are much stronger among the members of a religious denomination than among the nonmembers. Similar computations teach us that the effects of religious membership on voting are very strong among the old ($\hat{\gamma}_{1\ 2}^{B\overline{C}/A} = 2.313$) and in the same direction, but definitely weaker, among the young ($\hat{\gamma}_{1\ 1}^{B\overline{C}/A} = 1.356$).

Unsaturated effect models can also be defined. The maximum likelihood estimates for these effect models can be obtained directly by solving (iteratively) the appropriate maximum likelihood equations or by using, in the manner described above, the parameters of the corresponding frequency model. Let us take the following unsaturated effect model:

$$\Omega_{ij}^{AB\overline{C}} = \gamma^{\overline{C}} \gamma_j^{B\overline{C}} \qquad (2.40)$$

In terms of our example, equation 2.40 implies that religious membership has a direct influence on voting, but that age, holding religion constant, has no direct effect on it.

Model 2.40 does not provide explicit information about the relation between age and religion. In effect models, no restrictions are imposed on the relations among the independent variables, or rather these relations are taken for granted, as they are observed. The maximum likelihood estimates of the effect model are such that the observed multivariate frequency distribution of the independent variables is reproduced exactly.

This is of great importance for choosing the appropriate frequency model to obtain the estimates for the parameters of the effect model. One might think that in the case of equation 2.40 where only parameters with the subscripts C and BC appear, frequency model $\{BC\}$ is applicable and that squaring the τ_{11}^{BC} parameter of this model provides the correct maximum likelihood estimate of $\gamma_1^{B\overline{C}}$ in equation 2.40. However, this is not true. The maximum likelihood estimates for the frequency model $\{BC\}$ do not reproduce the observed frequency distribution AB as the estimates for the effect model would, and the estimate $(\tau_{11}^{BC})^2$ in frequency model $\{BC\}$ is not identical to the estimate for $\gamma_1^{B\overline{C}}$ in effect model 2.40.

Actually, frequency model $\{BC\}$ imposes rather strange restrictions upon the data. It implies that on average the variable A is uniformly distributed within the categories of BC, and that A and B are independent of each other when C is held constant. These kinds of restrictions were not intended when formulating equation 2.40.

The estimates of the parameters in equation 2.40 have to be obtained by means of the frequency model $\{AB, BC\}$ in which a direct relation exists between B and C, in which A and C are not directly related to each other, and in which the observed frequency distribution of AB is reproduced exactly.

In general, the effects γ can be obtained by means of a frequency model in which, besides the corresponding parameters of the effect model, all those parameters have been included that are necessary to reproduce the observed multivariate frequency distribution of the independent variables. (If some of the cells of this multivariate frequency distribution have zero entries, the corresponding logits and the effect parameters are undefined. In that case, one might consider deleting some higher order effects among the independent variables leading to nonzero estimated expected frequencies for the observed empty cells. Fienberg, 1980, Section 6.2, discussed this possibility; see also Section 2.9.)

The test statistics L^2 and Pearson-χ^2 for the unsaturated effect model are the same as for the corresponding frequency model, as are the degrees of freedom. In general, testing and fitting procedures for effect models are the same as those for frequency models.

2.7.2 The Modified Multiple Regression Approach: Effect Models with a Polytomous Dependent Variable

From the previous section it may seem that effect models can be used only in the case of a dichotomous dependent variable, because the effect model can help determine to what extent the odds of belonging to one category of the dependent variable rather than to another depend upon the values of the independent variables. However, the effect model is perfectly applicable in the case of a polytomous dependent variable although the results are somewhat harder to describe than for a dichotomous dependent variable.

Suppose that in the age-religion-voting example, voting indicated party preference, with seven political parties to choose among. We could also set up a model like the one in equation 2.40 for this situation in which age does not have a direct effect on voting:

$$\Omega^{AB\overline{C}}_{ijk/k'} = \gamma^{\overline{C}}_{k/k'} \cdot \gamma^{B\overline{C}}_{jk/k'} \qquad k \neq k' \tag{2.41}$$

We assume that the effect model 2.41 applies to all possible odds that may be formulated for the dependent variable. For seven categories there are 21 odds of which at most $7 - 1 = 6$ are independent of each other. We could for instance investigate the odds $C = 1$ rather than $C = 2$; $C = 1$ rather than $C = 3$; etc., up to and including $C = 1$ rather than $C = 7$. Or we could analyze the odds $C = 1$ rather than $C = 2$; $C = 2$ rather than $C = 3$; and so on up to and including $C = 6$ rather than $C = 7$. All other odds may be derived from each of these basic sets of six independent odds.[12]

To obtain the estimated effect parameters, six separate analyses could be carried out. In each analysis only the cell frequencies pertaining to the relevant odds are considered. Because the odds in the basic set are asymptotically independent of each other, the model may be evaluated in its totality by summing the L^2-values and degrees of freedom obtained for each separate analysis (Fienberg, 1980, Section 6.6).

Given that equation 2.41 applies for all odds, one can also use the frequency model $\{AB, BC\}$ for the complete table to obtain the test statistics and the maximum likelihood estimates for all parameters of the entire effect model. For example: $\gamma^{BC}_{j1/2} = \tau^{BC}_{j1}/\tau^{BC}_{j2}$, $\gamma^{BC}_{j1/3} = \tau^{BC}_{j1}/\tau^{BC}_{j3}$, and, in general, $\gamma^{BC}_{jk/k'} = \tau^{BC}_{jk}/\tau^{BC}_{jk'}$.

When the same model equation does not apply to all odds, either separate analyses have to be carried out or a nonhierarchical frequency model for the complete table has to be used. Let us assume that of the seven political parties, three are religious and four nonreligious. One might be of the opinion that religion determines the odds of voting for a religious rather than a nonreligious party, but that religion becomes irrelevant when choosing between any two nonreligious parties or between any two religious parties.

In such cases one might employ six separate effect analyses, as the hierarchical frequency model $\{AB, BC\}$ for the complete table does not impose the right kind of restrictions and does not provide the correct estimates. But the restrictions referred to can also be translated into a nonhierarchical frequency model for the complete table imposing the appropriate equality constraints on certain τ_{jk}^{BC} parameters. The parameters of such a nonhierarchical frequency model can be obtained routinely by means of computer programs using the Newton-Raphson algorithm.

A simple example of an effect model with a polytomous dependent variable can be given through the analysis of the Total column in Table 2.1. This column shows the observed frequencies of table ABC, where, as before, A refers to Age, B to Religious Denomination, but C now stands for Party Preference with categories 1 = preference for a Left Wing Party, 2 = preference for a Right Wing Party, and 3 = preference for a Christian Democratic Party.

It is assumed that effect model 2.41 is valid for the population, the model in which age does not have a direct effect on the odds of choosing one party rather than another, but religion does. The corresponding frequency model is $\{AB, BC\}$. It need not be rejected for this data: $L^2 = 3.44$, df = 4, $p = 0.487$. The relevant estimated parameters of the frequency model are the one-variable effects of party preference: $\hat{\tau}_1^C = 0.860$, $\hat{\tau}_2^C = 0.871$, and $\hat{\tau}_3^C = 1.336$; and the effects of religion on party preference: $\hat{\tau}_{11}^{BC} = 0.744$, $\hat{\tau}_{12}^{BC} = 0.940$, and $\hat{\tau}_{13}^{BC} = 1.430$. The one-variable effects are, as usual, not particularly interesting, and no attention will be paid to them.

The effect of religion on the odds of choosing a left wing party rather than a right wing party, $\hat{\gamma}_{11/2}^{BC}$, equals $\hat{\tau}_{11}^{BC}/\hat{\tau}_{12}^{BC} = 0.744/0.940 = 0.791$. The odds that religious people express a preference for a left wing party instead of a right wing party are about 0.8 times the average; for nonreligious people, the odds are $1/0.791 = 1.264$ times the average odds.

The effects of religion on the odds of preferring a left wing party above a christian democratic party, $\hat{\gamma}_{11/3}^{BC}$, are $\hat{\tau}_{11}^{BC}/\hat{\tau}_{13}^{BC} = 0.744/1.430 = 0.520$. The odds that religious people express preference for a left wing party rather than a christian-democratic party are about half the average odds; for nonreligious people the odds are about twice (1.922) the average odds.

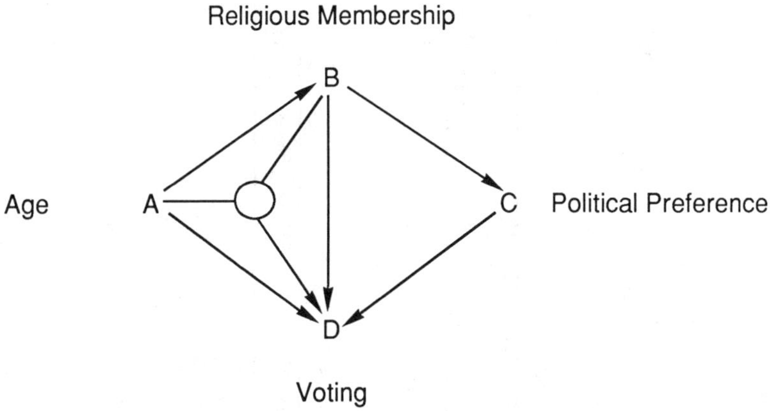

Religious Membership

Figure 2.2. Path Diagram for the Data in Table 2.1

The influence of religion on the odds right wing/christian democratic can be derived from the results given so far for the other two odds, or also from the relevant $\hat{\tau}$ parameters. Either way it follows that: $\hat{\gamma}^{BC}_{12/3} = 0.657$ and $\hat{\gamma}^{B\overline{C}}_{22/3}, = 1/0.657 = 1.521$.

Compared to nonreligious people, religious people have a very strong preference for the christian democratic parties and tend to avoid the right wing parties, but avoid the left wing ones even more.

2.7.3 The Modified Path Analysis Approach

In much the same way as the standard multiple regression models can be extended to path analytic models in which a system of multiple regression equations is considered simultaneously, it is possible to extend the modified regression model discussed above (Goodman, 1973a, 1973b).

Let us finally look at all the data in Table 2.1. It is assumed that the true model that has generated these observed data is the causal model represented in Figure 2.2 (where a knot represents an interaction effect between the connected variables).

One might regard variable D, Voting Behavior, as the ultimate dependent variable. According to the model in Figure 2.2, all other variables are supposed to have direct effects on voting; moreover, the effects of age and religious denomination on D interact with each other. All these

TABLE 2.5 Modified Path Analysis

Table	Observed Frequencies	Frequency Model	Effect Model	L^2	df	p
AB	f_{ij++}^{ABCD}	$\{AB\}$	H_1: $\Omega_i^{A\bar{B}} = \gamma^B \gamma_i^{A\bar{B}}$	0	0	0
ABC	f_{ijk+}^{ABCD}	$\{AB, BC\}$	H_2: $\Omega_{ij/k/k'}^{AB\bar{C}} = \gamma_{k/k'}^{\bar{C}} \gamma_{jk/k'}^{B\bar{C}}$	3.44	4	0.487
ABCD	f_{ijkl}^{ABCD}	$\{ABC, ABD, CD\}$	H_3: $\Omega_{ijk}^{ABC\bar{D}} = \gamma^{\bar{D}} \gamma_i^{A\bar{D}} \gamma_j^{B\bar{D}} \gamma_k^{C\bar{D}} \gamma_{ij}^{AB\bar{D}}$	4.41	6	0.621
			H^*: $H_1 \cap H_2 \cap H_3$	7.85	10	0.643

SOURCE: Table 2.1 (model: Fig. 2.2)

effects on D may be represented by the effect model that is denoted as H_3 in Table 2.5 or by the corresponding frequency model $\{ABC, ABD, CD\}$. However, in this effect model H_3, no account is taken of the fact that the causal model in Figure 2.2 also implies certain restriction on the relations between the independent variables A, B, and C.

According to the path diagram, there is no direct relation between A and C. This cannot be taken into account by postulating for table $ABCD$ frequency model $\{BC, ABD, CD\}$ instead of $\{ABC, ABD, CD\}$. The expected frequencies of model $\{BC, ABD, CD\}$ do not satisfy the restrictions implied by Figure 2.2. In much the same vein as has been argued before, the causal model in Figure 2.2 does not imply that there is no direct relation between A and C in table $ABCD$, holding B and D constant, as is implied by model $\{BC, ABD, CD\}$. If the path diagram reflects the true state of affairs, it is nonsensical to hold D constant: D is causally dependent upon A, B, and C and consequently cannot influence the relations among A, B, and C. What is meant by this path diagram is that there is no direct relationship between A and C in the table ABC, holding B constant.

The right way to proceed is to set up the marginal table ABC and postulate for this table the effect model we have used above (equation 2.41), which is denoted as H_2 in Table 2.5. Equivalently, frequency model $\{AB, BC\}$ can be used.

Holding D constant when looking into the relationship between A and C (τ_{ik}^{AC}) in table $ABCD$ is not just superfluous, but potentially harmful, as it may lead to the wrong conclusions. It follows from the collapsibility theorem that, if the path model is true, collapsing over D will change the value of the τ_{ik}^{AC} parameter. And so when τ_{ik}^{AC} equals one in table ABC, it will not do so in the full table $ABCD$.

Through similar reasoning it will be clear that the strength of the relationship between A and B and its statistical significance have to be investigated in table AB. In this case it is "hypothesized" that the saturated model applies for table AB, denoted as H_1 in Table 2.5.

Adhering to the adage that what is causally posterior cannot influence that which is causally prior, these kinds of path analytic models have to be approached stepwise. The path model in Figure 2.2 then actually implies that all three hypotheses H_1, H_2, and H_3 are true at the same time and that if this composite hypothesis, denoted as H^*, need not be rejected, the strength of the various effects have to be determined via several different (marginal) tables.

According to Goodman (1973b), H^*, the model in its totality, can be evaluated by using the test-statistic L^*, being the sum of the separate L^2-values for each of the hypotheses, and df^*, the sum of the separate df. If H^* is true, that is, if all separate hypotheses are true, L^* approximates the χ^2-distribution with df^* degrees of freedom.

The test results for each of the separate hypotheses, as well as for H^*, are presented in Table 2.5. It is clear that H^* need not be rejected; the causal model in Figure 2.2 may be accepted for the data in Table 2.1.

The (stepwise) computation of the parameter estimates has in fact for the larger part already been carried out in the previous section. From table AB it was learned that compared to the younger ones, older people are more religious (Section 2.3). It was concluded from table ABC that members of a religious denomination, in contrast to nonmembers, have a definite preference for christian democratic parties, and stay away from the right wing, but especially the left wing parties (Section 2.7.2).

There remain the direct effects on voting as estimated for table $ABCD$ according to hypothesis H_3.[13] Applying the frequency model $\{ABC, ABD, CD\}$ and converting the τ into γ parameters yields the following results. The direct effects of age are such that, on average, younger people are less inclined to vote than older people ($\hat{\gamma}_1^{A\overline{D}} = 0.658$). The direct effects of religious membership are somewhat stronger: $\hat{\gamma}_1^{B\overline{D}} = 1.774$ (and $\hat{\gamma}_2^{B\overline{D}} = 0.563$); members are more inclined to vote than nonmembers.

The influences of these two independent variables on voting are not independent of each other: $\hat{\gamma}_{11}^{AB\overline{D}} = 0.794$. The effects of age are in the same direction among members and nonmembers, but stronger among the former than among the latter. Looked at from the other side, the effects of religion go in the same direction among the young and the old, but they are stronger among the latter than among the former.

Finally, there are the direct effects of party preference on voting: $\hat{\gamma}_1^{C\overline{D}} = 0.340$; $\hat{\gamma}_2^{C\overline{D}} = 3.445$; $\hat{\gamma}_3^{C\overline{D}} = 0.856$. Adherents of the left seldom vote, whereas the proponents of the right have a strong tendency to vote; the christian democratics occupy a middle position in this respect.

The way the "modified path analytic models" are handled is very similar to what is done in the standard application of path analysis. (For an illuminating discussion of this, see Kiiveri & Speed, 1982.) Whereas the standard path analytic model consists of a system of multiple regression equations, the effect model in the path analytic variant is a system of ordinary logit models. The logit models in their turn are equivalent to frequency models. Consequently, what has been noted about the latter also applies here. For example, various hierarchically nested hypotheses H^* can be formulated and tested (conditionally) against each other. In this way a number of restrictions for several (sub)models can be tested simultaneously.

However, the analogy with standard path analysis may not be pressed too far (Brier, 1978; Fienberg, 1980). An important difference is that with log-linear models it is not possible to partition the total (marginal) association between two variables in direct, indirect, and spurious effects, where an indirect effect is the product of the several direct effects that form a particular causal path. This is possible in standard path analysis because the same (intervening) variables appear as dependent and independent variables in the equations. With logit models, however, the dependent "variables" are the odds which do not return as independent variables in other equations.

Some have argued that although systems of logit equations may have less potential in certain respects than the standard path analysis models, they are more flexible in other respects. Goodman (1973a) emphasized that, unlike standard path analysis, his modified path analysis approach can be used for causal models which are not recursive, but in which loops, that is, causal feedbacks, occur. He based this possibility on the fact that the γ parameters are symmetric in the sense of $\gamma_i^{AB} = \gamma_i^{BA}$. For example, in a 2×2 table AB, the effect of $A = 1$ on the odds $B = 1/B = 2$ is the same as the effect of $B = 1$ on the odds $A = 1/A = 2$, that is, both are equal to $(\tau_{11}^{AB})^2$ as may be derived from equations like equation 2.39. However, getting the same result when estimating the effect of A on B as when estimating the effect of B on A is different from estimating both the effect of A on B and the effect of B on A simultaneously in one nonrecursive model.

In conclusion, it should be pointed out that the estimated expected frequencies $\hat{F}^*$ for H^* can be computed from the estimated expected frequencies of the separate (sub)models. These frequencies $\hat{F}^*$ can also be used to calculate L^*. In the case of the causal model in Figure 2.1 and the hypotheses mentioned in Table 2.5, $\hat{F}^*$ is defined as follows:

$$\hat{F}^*_{ijkl} = \hat{F}_{ij} \frac{\hat{F}_{ijk}}{f_{ij}} \frac{\hat{F}_{ijkl}}{f_{ijk}} \qquad (2.42)$$

where $\hat{F}_{ij}$ are the estimated expected frequencies under H_1, $\hat{F}_{ijk}$ under H_2, and $\hat{F}_{ijkl}$ under H_3.

The estimated effect parameters for the whole model can be calculated using the appropriate (marginal) frequencies $\hat{F}^*$ or, as illustrated above, using the estimated expected frequencies for each separate model, with the same results.

Formula 2.42 can be rendered in terms of probabilities and proportions. To understand this reformulation, it has to be remembered that, for the estimated expected frequencies obtained under the hypothesis H_2 in Table 2.5, $\hat{F}_{ij+}^{ABC} = f_{ij+}^{ABC}$ and, under H_3, $\hat{F}_{ijk+}^{ABCD} = f_{ijk+}^{ABCD}$. Furthermore, a bar over a superscript of a probability refers to a conditional probability, so that, for example, $\pi_{ijk}^{A B \overline{C}}$ is the conditional probability that someone belongs to the category k of variable C, given that this person belongs to $A = i$ and $B = j$.

$$\hat{\Pi}_{ijkl}^* = \hat{\Pi}_{ij} \; \frac{\hat{\Pi}_{ijk}}{P_{ij}} \; \frac{\hat{\Pi}_{ijkl}}{P_{ijk}} = \hat{\Pi}_{ij} \; \frac{\hat{\Pi}_{ijk}}{\hat{\Pi}_{ij+}} \; \frac{\hat{\Pi}_{ijkl}}{\hat{\Pi}_{ijk+}} = \hat{\Pi}_{ij}^{AB} \hat{\Pi}_{ijk}^{AB\overline{C}} \hat{\Pi}_{ijkl}^{ABC\overline{D}} \qquad (2.43)$$

That the stepwise approach is an essential characteristic of modified path analytic models is embodied in this formula. The estimated probability that a randomly selected individual obtains the scoring pattern (i, j, k, l) on the joint variable $ABCD$ if H^* is valid for the population is equal to the estimated probability that, H_1 being true, this individual belongs to $A = i$ and $B = j$; multiplied by the estimated conditional probability that, H_2 being true, this person belongs to $C = k$, given $AB = (i, j)$; multiplied by the estimated conditional probability that, H_3 being true, this person belongs to $D = l$, given $ABC = (i, j, k)$.

2.8 Ordinal Categorical Variables

In the log-linear models dealt with so far, no assumptions have been made about the level of measurement of the variables: The variables have been treated as nominal level variables. However, the categories of some variables may have an intrinsic order which one wants to take into account in the log-linear analyses.

In the data in Table 2.6, which are from the same fictitious data set as Table 2.1, we find the relation between Educational Level (A) and Preference for a Left Wing Political Party (B). Ordinary log-linear analysis would not take the ordinal character of educational level into account. But it might well be that the researcher expects support for the politically left wing to decrease monotonically with education. There are

TABLE 2.6 Educational Level and Preference for a Left Wing Political Party

	A. Educational Level					
B. Left Wing Party Preference	Low 1	2	3	4	High 5	Total
1. Yes	69	67	44	26	5	211
2. No	131	133	106	124	45	539
3. Total	200	200	150	150	50	750
Odds Yes/No	0.527	0.504	0.415	0.210	0.111	
$\hat{\gamma}_i^{A\bar{B}}$	1.736	1.661	1.369	0.691	0.367	0.306
$\hat{\beta}_i^{A\bar{B}}$	0.552	0.507	0.314	-0.369	-1.004	-1.193

SOURCE: As for Table 2.1

several ways in which such hypotheses and expectations may be taken
into account (Agresti, 1989; see also note 12). A widely adopted practice
is to perform an ordinary log-linear analysis and to check afterwards if
the outcomes corroborate the hypotheses. For example, for Table 2.6 it
is expected that the odds yes/no decline (weakly) monotonically as the
level of education increases. Translated into the terminology of the
saturated effect model with left wing party preference as the dependent
variable, declining odds imply:

$$\gamma_i^{A\bar{B}} \geq \gamma_{i'}^{A\bar{B}} \qquad \text{for all } i < i' \qquad (2.44)$$

As can be seen from the last two rows of Table 2.6 where the estimates
of the relevant parameters effect parameters are presented, the hypothesis embodied in property 2.44 is confirmed.

Employing this kind of post hoc comparison is, however, not very
satisfactory. What should be done when the effect parameters are not in
agreement with restriction 2.44? Should the monotonicity hypothesis be
rejected? It might well be that the violations of the hypothesis are only
the result of sampling fluctuations. What is needed is a method to impose
a priori *inequality restrictions* like property 2.44 on the parameters and
to obtain estimates under the condition that the order restrictions are
valid for the population.

Agresti, Chuang, and Kezouh (1987) developed a method for imposing (weak) order restrictions on the parameters using isotonic regression
procedures (see also Dykstra & Lemke, 1988). Although this procedure
has not been completely worked out for higher than two dimensional

tables and there are still problems with regard to testing these models statistically, it does appear promising.

Agresti (1984), Clogg (1982a, 1982b), Goodman (1984, 1985, 1987), Haberman (1978, 1979), Hout (1983), and many others have suggested the use of methods in which ordinal level data are treated as if measured at interval level. Although they call it an ordinal level approach, assumptions are made which make sense only at interval level of measurement. The models they developed are usually denoted by the name Goodman gave them, namely, association models, in two variants: log-linear and log-bilinear association models.

Log-linear association models are more readily understood when drawing parallels with the standard regression model. Let us apply a simple standard regression analysis to Table 2.6, with Party Preference as the dependent variable Y (1 = yes, 0 = no) and Educational Level as the independent variable X (1 = low, . . . , 5 = high). Using the ordinary least squares method, the following result is obtained:

$$\hat{Y} = a + bX = 0.432 - 0.059X$$

The predicted $\hat{Y}$ scores (actually the predicted proportions "Yes") for the successive intervals of X are 0.372, 0.313, 0.254, 0.194, and 0.135. These predicted scores are perfectly linearly related to X. For each increase in educational level with one category, the predicted proportion of people who prefer a left wing party drops by $b = 0.059$; the difference between the predicted proportions of supporters of the left is .059 for each pair of adjacent categories of educational level.

In terms of the multiplicative effect model, similar restrictions on the data in Table 2.6 imply that the odds yes/no become smaller by a constant factor for each increase in educational level. Let us assume that the effects of education are such that the successive odds yes/no with each increase in educational level diminish by a constant factor 0.9. If the odds for the first category of education were 0.5, then the odds for the second category would be $(0.5)(0.9) = 0.45$, for the third category $(0.5)(0.9)^2 = 0.405$, and so on. These odds are perfectly related to X according to an exponential curve, $c(0.9)^i$, where 0.9 is the effect parameter (comparable to b in the regression equation), where i is the score on education (X in the regression equation), and c is a constant, for example, the overall partial odds yes/no (comparable to the intercept a).

Taking the logarithms of these quantities leads to a linear relationship between education and party preference: $\ln c + i \ln 0.9$. The differences between the logits of each pair of adjacent categories of education are constant.

Estimating the parameters of such a model for Table 2.6 involves the following steps. We start as it were with the saturated effect model:

$$\Omega_i^{A\overline{B}} = \gamma^{\overline{B}} \gamma_i^{A\overline{B}}$$

but then immediately apply the restriction that the effects of education on the odds yes/no, $\gamma_i^{A\overline{B}}$, are such that the estimated expected odds lie on an exponential curve:

$$\gamma_i^{A\overline{B}} = (\gamma^*)^i \tag{2.45}$$

In terms of the logit equation this becomes:

$$\Phi_i^{A\overline{B}} = \beta^{\overline{B}} + \beta_i^{A\overline{B}} \tag{2.46}$$

where

$$\beta_i^{A\overline{B}} = (\beta^*)i$$

Because of the relations between the parameters of the effect and the frequency model, equations 2.45 and 2.46 imply identical linear restrictions on the τ parameters of the frequency model. Since these restrictions can be easily translated into a design matrix, effect models with these kinds of restrictions can be handled by the standard programs for log-linear analysis that use a design matrix (for example, Haberman's program FREQ — see Haberman, 1978).

Crucial in these log-linear association models are of course the scores that are assigned to the categories of education, that is, the exponent i in equation 2.45. Assigning the scores $-2, -1, 0, 1, 2$ (as will be done here) first of all indicates that the successive levels of education are equidistant from each other; should one think that, for example, the highest level is "much higher," this category should be given a relatively higher score, for example, 3. Changing the distances between the categories will affect the estimate for γ^* in equation 2.45. But comparing distances between educational levels makes sense only if it is assumed that education has been measured at interval level, and not just at ordinal level.

Using the scores -2 through 2 ensures that the parameters $\gamma_i^{A\overline{B}}$ sum to zero as usual. This implies that, as usual, $\gamma^{\overline{B}}$ can be interpreted as the overall partial odds. If a set of scores is assigned that does not sum to zero, such as 1, 2, 3, 4, 5, $\gamma^{\overline{B}}$ can no longer be interpreted as the overall partial odds, although the value of γ^* will remain the same. A general discussion of when log-linear parameters can be uniquely determined and when and how they vary under different coding schemes is given by Long (1984).

Although the saturated effect model has zero degrees of freedom, because of restriction 2.45, there are now degrees of freedom to test the model and evaluate whether or not the relation between education and party preference follows a simple exponential curve (is linear in the log odds). Because there is just one parameter γ^* left to be estimated instead of $(2-1)(5-1) = 4$ parameters γ_i^{AB}, there are three degrees of freedom for this particular association model when applied to Table 2.6.

The test results are $L^2 = 4.779$, df $= 3$, $p = 0.19$ (Pearson-$\chi^2 = 4.720$). The hypothesis that the odds yes/no decrease by a constant factor when educational level increases need not be rejected. The overall partial odds are estimated as $\hat{\gamma}^B = 0.328$, which is rather close to the value obtained for the saturated model (0.306 — Table 2.6). The most interesting effect parameter is estimated as $\hat{\gamma}^* = 0.734$. The odds decrease by a constant factor 0.734; starting with the odds for the first and lowest educational level, each successive odds is 0.734 times smaller than the immediately preceding one. Remembering that the coding scheme for i was $-2, 1, 0, 2, 1$, the $\hat{\gamma}_i^{AB}$ parameters are $\hat{\gamma}_1^{AB} = 1.857$, $\hat{\gamma}_2^{AB} = 1.363$, $\hat{\gamma}_3^{AB} = 1$, $\hat{\gamma}_4^{AB} = 0.734$, $\hat{\gamma}_5^{AB} = 0.538$ (and the corresponding β_i^{AB} parameters: 0.310, 0.155, 0, 0.155, 0.310).

These kinds of association models are called log-linear association models because they can be formulated as ordinary log-linear models with linear restrictions on the log-linear parameters. This is not the case for *log-bilinear association models*.

Let us assume that the association model dealt with above had to be rejected. One might then decide to give preference to the saturated model, concluding that the assumption of linearity with regard to the log-linear effect parameters was false, or one might maintain that the relation is linear but that the coding scheme used for educational level, that is, the exponent i in equation 2.45, was wrong. In the latter case, one might try out several different coding schemes or try to estimate the appropriate coding from the data, assuming a linear relationship. The latter is what is essentially done in log-bilinear association models.

If we look at the ß parameters of the saturated model in the last row of Table 2.6, the relation between education and party preference does not appear to be exactly linear. Linear relationship would imply that the difference between all adjacent ß's is constant. As can be seen, the difference between belonging to educational level 1 instead of 2 does not make much of a difference for the odds yes/no, which is rather in sharp contrast to levels 4 and 5. The relation between education and party preference can be made exactly linear by recoding the educational levels in such a way that levels 1 and 2 are very close to each other, and levels 4 and 5 are rather wide apart.

In log-bilinear models, an optimal solution is obtained for the spacing of the categories in the sense that maximum likelihood estimates for the exponent i in equation 2.45 are obtained, assuming a linear effect of education on the logits yes/no. In log-bilinear models, it is not only the usual effect parameters that are estimated, but also the distances between the categories of the variable(s).

Log-bilinear models are especially useful in two-dimensional tables, because the scores of the variable(s) are derived from the bivariate frequency distribution. In multidimensional tables the scores of the variable(s) might vary depending on which variables are compared. Some kind of overall optimal solution has to be found. This raises questions about the precise meaning of the estimated scores.

Log-bilinear models will not be used in this book. However, those interested in a thorough investigation of one particular table, for example, a social mobility table, are well advised to consult the literature cited above.

2.9 Some Problems: Small Sample Size and Polytomous Variables

The most persistent problem inherent in any form of table analysis is the problem of *small sample size* and sparse data. As has been said in Section 2.4, if many expected cell frequencies of the multivariate table have low values (< 5 or < 1), it cannot be expected that the test statistics L^2 and Pearson-χ^2 approximate the theoretical χ^2-distribution (Haberman, 1988a). Moreover, the standardized effects and the adjusted residuals will not approximate the standard normal distribution. A further problem is that the power of these tests has to be expected to be low. Finally, small samples are more apt to cause zero cells in the marginals to be reproduced, which in turn leads to expected cell frequencies equal to zero and to not being able to estimate some of the log-linear effects (they assume the boundary values $+\infty$ or $-\infty$).

There is little that can be done about these problems. Alternative χ^2-test statistics have been proposed (Zelterman, 1987). However, their advantage over the traditional statistics is not clear in many circumstances and they are not routinely available from standard computer programs. Also the problem of low power is not solved.

Searching for more data is obviously the best recommendation. If these are unavailable, several other strategies might be tried. Collapsing tables over categories or variables is the route most investigators take. It follows from the collapsibility theorem that this will generally influence the parameter estimates (see also Reynolds, 1977). However, as long as one realizes that the relations found pertain to the table actually used, this strategy may well be the wisest, and many times the only viable one.

Because conditional tests can be expected to have more power and to give a better approximation to the χ^2-distribution, if possible, conditional rather than unconditional tests should be given preference. Moreover, both L^2 and Pearson-χ^2 should be computed because if they have different values this is a sign not to rely too heavily on either one of them.

It has become almost a routine to solve the problem of estimated expected cell frequencies equal to zero by adding 0.5 to all observed cell frequencies. This practice seems to be based on the recommendation of Gart and Zweifel (1967) who proved that for the saturated model more stable estimates $\hat{F}$ can be obtained by using $\hat{F} = f + 0.5$. However, this is not a practice to be recommended for all models under all circumstances. There have been suggestions that it is sometimes better to correct in other ways, for example, to add 0.25 or to subtract 0.25 (Cox, 1977; Gart, Pettigrew, & Thomas, 1985; Hauck, Anderson, & Leaky, 1982). Others, using a Bayesian approach, have suggested letting the constant(s) to be added to the observed cell frequencies depend on the number of parameters and the nature of the unsaturated model the investigator has in mind (Bishop et al., 1975, Chapter 12; Clogg & Eliason, 1987; Clogg, Rubin, Schenker, Schultz, & Weidman, 1986; Smith, 1976). Probably the best thing the investigator can do is use a few reasonable approaches for eliminating the empty cells and compare the results of the several corrections. One should be very careful with the interpretations of those effects that vary wildly among the several approaches.

Another major problem in log-linear analysis has to do with the fact that in the case of *polytomous variables* many parameters are needed to describe the relations among the variables. For example, in a 5 × 5 table AB there are 25 (and 16 independent) parameters to describe the association between A and B. Although this is more or less inherent to working with nominal level variables, one measure which expresses the association between two variables is sometimes very convenient.

Some researchers recommend fitting a particular log-linear model and then applying some arbitrary association measure like Goodman and

Kruskal's (1954) τ or γ, or even the product-moment correlation coefficient r to the estimated expected frequencies (Smith, 1976). The arguments against this approach are obvious: These measures of association make different assumptions about the data than do log-linear models. For example, if according to the log-linear model the relation between age (A) and voting (C) is the same among religious and nonreligious people (B), that is, $\tau_{ijk}^{ABC} = 1$ and $\tau_{ik1}^{AC/B} = \tau_{ik2}^{AC/B}$, this does not necessarily imply that Goodman and Kruskal's τ for the relation between age and voting will be the same for religious and nonreligious people when computed on the estimated expected frequencies $\hat{F}$.

Magidson (1981), Haberman (1982), and Kim (1984b) recommended supplementing the log-linear analysis by using an overall measure of association derived from an information-theoretical approach. Although there are many similarities between the informationtheoretical approach and log-linear modeling, the two are not identical, and the argument used above can be leveled against their proposals as well.

Kawasaki and Zimmermann tried to overcome this problem by computing any association measure not on the estimated expected frequencies, but on the effect parameters τ_{ij}^{AB}, τ_{ik}^{AC}, etc. themselves (Kawasaki, 1982; Kawasaki & Zimmermann, 1981). Although lack of interaction in log-linear terms will then also lead to lack of interaction according to the association measures chosen, this still does not solve the possible incompatibility of the underlying models. For instance, using percentage differences d or product-moment correlations r implies the use of an additive model. Such an additive model, consistently applied from the beginning, would lead in general to different estimated expected frequencies, and thus to other values for d or r.

Clogg (1979a, Appendix E) proposed normed τ parameters, Q, bounded by +1 and −1. The main disadvantage of this proposal is that these Q parameters cannot be interpreted straightforwardly in terms of odds and odds ratios as the τ parameters can.

A very simple solution, if one overall measure of association is needed, is to use an overall measure which reflects either the mean level of the relevant effects or the maximum difference within the relevant set:

$$\tau_{max}^* = \frac{\tau_{max}}{\tau_{min}} \qquad \lambda_{max}^* = \lambda_{max} - \lambda_{min} \qquad (2.47)$$

$$\tau_{mean} = \left(\prod_{i=1}^{I} \tau_i' \right)^{1/I} \qquad \lambda_{mean} = \frac{1}{I} \sum_{i=1}^{I} |\lambda_i| \qquad (2.48)$$

where $\tau_i' = \tau_i$ if $\tau_i \geq 1$ and $\tau_i' = 1/\tau_i$ if $\tau_i < 1$.

The main disadvantage of the maximum effect 2.47 is its sensitivity to outliers; the main disadvantage of the mean effect 2.48, especially in large tables, is that many effects may be equal to each other and will therefore lie in the neighborhood of "no effect" causing τ_{mean} to have a rather low value, maybe despite the existence of a few large systematic effects.

Clogg and Eliason (1987) in an excellent review article discussed many other common problems, along with possible solutions. But here as elsewhere in this chapter, only some of the main points have been mentioned. Other problems and possibilities of log-linear modeling, such as working with indirectly observed latent variables and handling tables with a priori or structural empty cells (not to be confused with the sampling zeros discussed above), will be dealt with in later chapters.

Notes

1. General expositions on log-linear modeling are available at all levels of difficulty. Haberman wrote a very exhaustive but mathematically difficult book (Haberman, 1974). Comprehensive but much more accessible are his later books (Haberman, 1978, 1979). At the same level is the book by Bishop et al. (1975). On a more introductory level are the books by Fienberg (1980), Knoke and Burke (1980), and Reynolds (1977). Those readers who find this introductory chapter too demanding are well advised to study the Knoke and Burke book first.

2. More about the maximum likelihood principle can be found in Hays (1981, Section 5.5) and Mood et al. (1974: 276-286). Bishop et al. (1975: 58) discuss the advantages of using maximum likelihood estimators in the context of log-linear modeling.

3. Maximum likelihood estimates of the log-linear parameters can also be obtained when the sampling distribution is the Poisson distribution (Bishop et al., 1975: 62). This distribution arises when observations have been made during a fixed period without knowing in advance the total number of observations. Survey data seldom follow a Poisson distribution. Sometimes they do not follow the (product-)multinomial distribution either. In survey analysis, many other sampling methods such as cluster sampling and multistage sampling are used besides (stratified) simple random sampling. Strictly speaking, the estimation and testing procedures proposed in this book have to be adapted in such cases. This does not usually happen. In regression analysis this neglect of more complicated sampling schemes generally leads to underestimation of standard errors and overestimation of the size of the test statistics. These more complex schemes may also lead to analogous results in log-linear analysis and so to enlarging the error of falsely rejecting the null hypothesis. When it is known how to weight the observed frequencies resulting from these complex sampling schemes to mirror the probabilities in the population, Clogg and Eliason (1987: 21-28) gave a general and practical procedure to cope with these more complex sampling schemes. Fundamental work in this area that promises to be very useful was presented by Rao and Thomas (1988); see also Lee, Forthofer, and Lorimor (1989, especially Section 7).

4. The phrase "according to the log-linear model" has been added for a particular reason. From the fact that all parameters τ_{ijk}^{ABC} are equal to one, one rightly concludes that the conditional two-variable effects τ (or λ) are the same. But it does not imply that there will be no (higher order) interaction in terms of any conditional measure of association. For example, contrary to what some researchers seem to believe, the absence of interaction effects in the (multiplicative) log-linear model does not guarantee the absence of interaction effects in the (additive) regression model. Nor does the presence of interaction effects in the log-linear model make it automatically necessary to incorporate interaction terms in the multiple regression equation.

5. These limitations of IPF refer to the standard procedure outlined here. It is possible to supplement IPF with a procedure to calculate the standard errors using the δ method or Fisher's information matrix (Bishop et al., 1975, Section 14.6; McCullagh & Nelder, 1983) and there is a generalized variant of IPF suited for nonhierarchical models (Darroch & Ratcliff, 1972). Besides the fact that the generalized variant usually converges rather slowly, there is the problem that these supplements and variants are not regularly available in standard computer programs. On the other hand, imposing restrictions to ensure linear relationships between variables can be handled rather easily by ECTA and particularly by Eliason's CDAS (see Chapter 8).

6. Several computer programs, for example, CDAS, MULTIQUAL, and GLIM have built-in facilities for setting up design matrices.

7. For hierarchical models the conditional test statistic can be written in a form which is more directly related to the non-conditional test statistic (Bishop et al., 1975: 127):

$$L_{r/u}^2 = 2 \sum_i \sum_j [\hat{F}_{ij-u} \ln(\hat{F}_{ij-u} / \hat{F}_{ij-r})]$$

8. Although social scientists are inclined to ignore the issue of power when applying χ^2-tests, several good, accessible, and rather exhaustive reviews of it are available. (See Cohen, 1977, Erdfelder, 1984, Milligan, 1980, and, especially, Bonett & Bentler, 1983.) Also very relevant is the work Saris and Satorra, and Matsueda and Bielby have done within the context of LISREL modeling, that can easily be applied as well to log-linear modeling (Matsueda & Bielby, 1986; Saris & Satorra, 1988; Saris et al., 1987; Saris & Stronkhorst, 1984, Chapter 11; Satorra & Saris, 1985). They showed how to base the decision to reject or accept a particular model not only on the significance level α (probability of type I error), but also on the power of the test to detect certain effects or effects of at least a particular size (in the log-linear context, for example $|\lambda| > 0.2$).

9. Aitkin (1980) among others presented "a formal simultaneous test that may be applied to the entire model selection procedure, including the testing of models chosen after inspection of the data" (p. 173). However, this kind of test procedure usually involves too rigid and automatic an application of statistical decision rules. Moreover, the probability of falsely accepting a particular null hypothesis is generally very great.

10. The other measure is $\delta = (F_r - F_u)/(F_r - 1)$. In contrast to δ, δ can be larger than one. δ attains the value one when $L_u^2 = df_u$. The other two indices, δ and R' attain the value one when the unrestricted model fits the data perfectly.

11. Although it is natural within the context of the multiplicative log-linear model to take the (conditional) odds as the quantities to be explained, sometimes it is more desirable to take other quantities, for example, the (conditional) proportions of people that belong to a particular category of the dependent variable. For example, one might explain the variation in percentage of voters (%Yes in Table 2.3) as a function of age,

religion, and political preference instead of explaining the odds voter/nonvoter. Clogg and Eliason (1987) discussed the differences between taking the odds and taking the proportions as dependent quantities and showed how standard computer programs for carrying out log-linear analyses can be used to take the (conditional) proportions as the quantities to be explained.

12. Fienberg (1980, Section 6.6) stated the possibility of making use of "continuation odds," especially useful when the dependent variable is measured at ordinal level. The successive odds are defined as follows if the dependent variable has seven categories. First, one compares the chances of belonging to category 1 rather than to the categories 2 through 7, the latter being obtained by summing the probabilities of belonging to each of the categories 2 through 7. Next, one compares the chances of belonging to category 2 rather than to the categories 3 through 7. And so on, up to and including comparing the odds of belonging to category 6 rather than to 7. These odds are asymptotically independent of each other and the results of the separate analyses can be combined. However, because of the summing of frequencies (probabilities), it is not possible to translate such a model into an ordinary log-linear frequency model. Other ways of dealing with ordinal variables will be discussed in Section 2.8.

13. Application of the collapsibility theorem tells us when the corresponding estimated parameters of the various (sub)models differ. For instance, the (unstandardized) parameter τ_{ij}^{AB} will be the same in model $\{AB\}$ for table AB, as in model $\{AB, BC\}$ for table ABC. However, one has to be careful. It may appear to some as a logical consequence of the path diagram and the collapsibility theorem that the effect of A on D may also be computed from table ABD, as from $ABCD$, with the same results. So seemingly the results for table ABD, denoted in Section 2.3 as ABC, can be used again to determine the influence of Age (A) on Voting Behavior (D). But this is not true.

Effect model H_3 for table $ABCD$ is equivalent to frequency model $\{ABC, ABD, CD\}$. According to the collapsibility model, the effect parameters with superscript AD will change when the table is collapsed over C. After all, in model $\{ABC, ABD, CD\}$, C is related both to A, and D.

3. Latent Class Analysis and Log-Linear Models with Latent Variables

3.1 Introduction

In the social sciences there is often a large gap between the languages of theory and research (Blalock, 1968). The concepts that are used at the theoretical level cannot be observed directly in reality. For example, "We do not 'see' a person's 'loyalty' or a group's 'solidarity,' but we infer these by watching the individual or group respond to various stimuli" (Blalock, 1982: 26). The gap has to be bridged through careful theoretical reasoning, resulting in *measurement models* in which the relations between the theoretical concepts and the indicators are made explicit. Such measurement models usually contain both directly observed manifest variables — the indicators — and not directly observed latent variables — the theoretical variables.

Scaling techniques are based on general measurement models that can be used for the measurement of a wide variety of phenomena. One of the oldest and most frequently used scaling models is the factor analysis model, applicable when all variables have been measured at interval or ratio level. In recent years, the factor analysis approach has been incorporated into the more general structural modeling approach with latent variables, of which LISREL has become the best known (Cuttance & Ecob, 1987; Jöreskog & Sörbom, 1979; Long, 1983a, 1983b). LISREL offers a very flexible framework for studying the sizes, sources and consequences of unreliability and invalidity, while at the same time correcting the relevant parameter estimates for the unreliability and invalidity found (Zeller & Carmines, 1980).

When the variables are categorical and measured at nominal level only, Lazarsfeld's latent class analysis offers an excellent framework for

93

attacking the problems of unreliability and invalidity, very much comparable to factor analysis (Lazarsfeld, 1950a, 1950b; Lazarsfeld & Henry, 1968; Wiggins, 1955, 1973).[1]

Until rather recently, very few applications of latent class analysis were to be found, in spite of its great potentialities. This certainly had to do with the fact that the estimation and testing procedures Lazarsfeld and others developed were not very satisfactory: They were arbitrary in important ways, did not make use of all available data, or were not flexible enough to handle a wide variety of latent class models.

A breakthrough was accomplished by Haberman and Goodman (Clogg, 1981a; Goodman, 1974a, 1974b; Haberman, 1976, 1977, 1979; Langeheine & Rost, 1988; McCutcheon, 1987). They formulated the latent class model in terms of a log-linear model with latent variables and showed how the maximum likelihood estimates of the parameters of this log-linear model could be obtained. Moreover, through their approach it is very easy to modify the standard latent class model with one latent variable into a model with several polytomous latent variables.

The basic latent class model with one latent variable will be introduced in Section 3.2.1. It will be shown how this basic model corresponds with a log-linear model with one latent variable. The EM-algorithm for obtaining the maximum likelihood estimates of the parameters of the basic model will be discussed in Section 3.2.2.

Many kinds of a priori restrictions may be imposed upon the parameters. Through these restrictions various assumptions about the nature of our data can be tested, such as the supposition that particular manifest, directly observed, variables are equivalent indicators of the underlying latent variable, or that some are error-free measures of it. How to impose such restrictions will be the topic of Section 3.2.3.

At the end of Section 3.2, attention will be paid to the identifiability of the parameters of latent class models and to problems of model fitting and testing (Section 3.2.4).

Once a well-fitting model that lends itself to a sound theoretical interpretation has been obtained, the question is how to investigate the relationship of the latent variable to other, "external" variables, for example, to age and gender. Analogous to computing factor scores, one may compute the score on the underlying latent class variable for each individual and relate these individual scores to the "external" variables. Or, following a procedure similar to what is done in the LISREL model, one may expand the original latent class model by including the "external" variables in the model and investigate the relationship between the latent variable and the "external" variables without computing individual latent scores. Although the first approach is usually recommended and has been chosen by most investigators, arguments for preferring the latter method will be presented in Section 3.3.

Extensions to latent class models having more than one latent variable will be discussed in Section 3.4. The relations between the several latent variables may be investigated by means of saturated and unsaturated log-linear models. For a number of practical reasons, investigators have very seldom put restrictions on the relations between the latent variables and have usually postulated the saturated model. However, it is possible to define unsaturated models as well, also in the form of modified path models.

In this way, log-linear modeling with latent variables leads to what could be called "a modified LISREL approach" (Hagenaars, 1985, 1988a; Heinen, & Hamers, 1980), having much in common with the original LISREL model. Direct effects between the indicators can be handled taking care of test-retest effects, correlated error terms, etc. (Section 3.5); simultaneous analyses in several groups are possible (Section 3.6); and, in general, causal models with categorical latent variables can be evaluated (Section 3.7).

How to deal with latent and manifest variables whose categories are ordered will be discussed in the last section of this chapter.

All examples and applications in this chapter involve "simple" cross-sectional data; in the next chapters the usefulness of log-linear models with latent variables for the more complex analyses of longitudinal data will be shown.

3.2 Latent Class Analysis: Log-linear Models with One Latent Variable

3.2.1 The Basic Model

The main elements of the standard latent class model are a set of manifest variables whose scores are directly observed and a latent variable whose scores are not directly observed. The manifest variables are regarded as indicators of an underlying theoretical variable represented by the latent variable. In principle, all variables, both latent and manifest, are treated as categorical nominal-level variables. The categories of the latent variable are usually called the "latent classes." The several manifest and latent variables may have different numbers of categories.

To take a concrete example, Hagenaars and Halman (1989) carried out a latent class analysis on a set of nine items expressing the respondents' views on important economic and political issues. Seven items were dichotomous (e.g., Desirability of Societal Change: desirable, not desirable; Trust In Parliament: trust, no trust); one was trichotomous (Political Left/right Selfplacement: left, middle, right); and one item had four categories (Who Should Be Running Business: owners alone, owners and employees together, employees alone, don't know who's to decide).

A latent class analysis was carried out to see whether these nine items could be regarded as indicators of one underlying political typology. The results of the analysis pointed to the existence of a latent variable with four categories. The four latent classes could be interpreted meaningfully as four basic political types: the Progressives, the Conservatives, the Individualists, and the Politically Noninvolved.

According to the latent class model, each person in the population belongs to one and only one of the latent classes. In terms of the example, each person belongs to one and only one of the four basic political types.

The manifest variables are the indicators of the latent variable, but they are very seldom perfect indicators, completely reliable and valid. For example, members of the conservative latent class did not choose with absolute certainty the conservative alternative on each item, but only with a particular probability, albeit a probability that is higher than for the progressives. The position a person has on the underlying latent variable does systematically influence this person's score on the manifest variables, but this relation is not perfect and deterministic. Other outside factors will also influence the observed scores and will make the relation between latent and manifest variables only probabilistic. In this way, measurement error is reckoned with.

That the latent variable is the underlying fundamental variable is expressed by assuming that each person's score on each of the indicators is systematically determined only by the latent variable and not by the scores on the other indicators. Members of the progressive latent class tend to choose the progressive alternative on each item only and just because they are progressive; conservatives tend to choose the conservative alternatives only and just because they belong to the conservative latent class, etc.

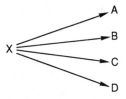

Figure 3.1. The Basic Latent Class Model with One Latent Variable X and Four Manifest Variables, A, B, C, and D

The nature of the relationships among the manifest and latent variables assumed in the latent class model may be depicted as in Figure 3.1. In this figure there are four manifest variables A through D, and one latent variable X.

There are no direct relations among the manifest variables. They are associated with each other, but only because each one of them is directly related to X. The latent variable X accounts completely for the associations found in tables AB, AC, etc. In terms of Lazarsfeld's (1955) elaboration terminology, the associations between the manifest variables are spurious, and holding X constant will make the manifest relations disappear.

This is the basic assumption of *local independence*. Local independence implies that within each latent class, for $X = t$, the conditional response probability of obtaining the score i on the manifest variable A is independent of the conditional probability of obtaining the score $B = j$, $C = k$, and $D = l$.

The latent class model may be represented in terms of log-linear modeling or in terms of *Lazarsfeld's original parametrization*. Starting with the latter and following Goodman's notation (Goodman, 1974a, 1974b), the basic equations of the latent class model in the case of four manifest variables A through D and one latent variable X with T categories may be stated as follows:

$$\pi_{ijkl}^{ABCD} = \sum_{t=1}^{T} \pi_{ijklt}^{ABCDX} \tag{3.1}$$

where

$$\pi_{ijklt}^{ABCDX} = \pi_{ijklt}^{\overline{ABCDX}} = \pi_t^{X} \pi_{it}^{\overline{AX}} \pi_{jt}^{\overline{BX}} \pi_{kt}^{\overline{CX}} \pi_{lt}^{\overline{DX}} \tag{3.2}$$

The notation is the same as in the previous chapter, be it that X is a latent variable with not directly observable scores. π_{ijkl}^{ABCD} denotes the probability that in the population a randomly selected individual scores (i, j, k, l) on the joint variable $ABCD$. The probability of obtaining the score (i, j, k, l, t) on the joint variable $ABCDX$ is indicated by π_{ijklt}^{ABCDX}, and is the probability of belonging to $X = t$. The parameter $\pi_{it}^{\overline{AX}}$ is a conditional response probability, namely, the probability that an individual obtains the score $A = i$, given this person belongs to latent class t of X; the remaining conditional response probabilities in equation 3.2 have similar meanings. All parameters in equations 3.1 and 3.2 are probabilities and subject to the standard restrictions: They cannot be smaller than zero or larger than one and their sum is one after summation over the appropriate subscripts, for example, $\Sigma_i \pi_t^X = \Sigma_i \pi_{it}^{AX} = 1$.

Equation 3.1 implies that the population may be divided into T exhaustive and mutually exclusive latent classes; each person belongs to one and only one latent class. In this sense, equation 3.1 expresses the existence of the latent variable X.

The assumption of local independence is essential to equation 3.2. From elementary rules of probability theory, it follows that the probability π_{ijklt}^{ABCDX} of obtaining the score (i, j, k, l, t) on the joint variable $ABCDX$ equals the probability π_t^X of belonging to $X = t$, times the conditional response probability $\pi_{ijklt}^{\overline{ABCDX}}$ of obtaining the score (i, j, k, l) on the joint variable $ABCD$, given $X = t$. According to the local independence assumption, the conditional response probabilities $\pi_{it}^{\overline{AX}}$, $\pi_{jt}^{\overline{BX}}$, $\pi_{kt}^{\overline{CX}}$, and $\pi_{lt}^{\overline{DX}}$ are independent of each other. So $\pi_{ijklt}^{\overline{ABCDX}}$ can be obtained by simply multiplying these conditional probabilities.

The (independence) restrictions that are imposed by the latent class model on the observed data (Kiiveri & Speed, 1982) may be rendered in a completely equivalent form in terms of the *log-linear model*. For the case of four manifest variables A through D and one latent variable X, the equation for the basic latent class model presented in Figure 3.1 is in multiplicative and log-linear form:

$$F_{ijklt}^{ABCDX} = \eta \tau_i^A \tau_j^B \tau_k^C \tau_l^D \tau_t^X \tau_{it}^{AX} \tau_{jt}^{BX} \tau_{kt}^{CX} \tau_{lt}^{DX}$$

$$G_{ijklt}^{ABCDX} = \theta + \lambda_i^A + \lambda_j^B + \lambda_k^C + \lambda_l^D + \lambda_t^X + \lambda_{it}^{AX} + \lambda_{jt}^{BX} + \lambda_{kt}^{CX} + \lambda_{lt}^{DX}$$

(3.3)

where the parameters are subjected to the usual identifying restrictions: Unless stated otherwise, the product of the τ parameters, multiplied over

TABLE 3.1   Essential Responsibility of Government Toward Issues:
The Netherlands

A. Equal Rights Men/Women	B. Good Education	C. Good Medical Care	D. Equal Rights Guest workers		Total
			1. Yes	2. No	
1. Yes	1. Yes	1. Yes	59	56	115
		2. No	14	36	50
	2. No	1. Yes	7	15	22
		2. No	4	23	27
2. No	1. Yes	1. Yes	75	161	236
		2. No	22	115	137
	2. No	1. Yes	8	68	76
		2. No	22	123	145
		Total	211	597	808

NOTE: As mentioned by Barnes and Kaase (1979), the respondents were asked to consider a list of ten issues facing society, four of which have been used here: A. "Guaranteeing equal rights for men and women;" B. "Providing a good education;" C. "Providing good medical care;" D. "Providing equal rights for guest [foreign] workers." Then the question was posed: "How much responsibility does government in general have toward this problem? An essential responsibility; an important responsibility; some responsibility; no responsibility at all." In this table, the category "1. Yes" includes the answer "an essential responsibility" and the category "2. No" all other alternatives.
SOURCE: Political Action 1974—The Netherlands; see Barnes and Kaase (1979)

any subscript, equals one and the λ parameters analogously sum to zero (see equation 2.11).

Again, there are direct relations between the latent variable and each of the manifest variables, but the parameters concerning the relations among the manifest variables are all absent. The expected frequency $F_{ijkl\,t}^{ABCDX}$ for this hierarchical log-linear model $\{AX, BX, CX, DX\}$ is identical to $\pi_{ijkl\,t}^{ABCDX}$ in equation 3.2 times the sample size N; the τ parameters in equation 3.3 can be rewritten in terms of the π parameters at the right hand side of equation 3.2.

An empirical example may further clarify the precise meaning of the parameters. The example used above concerning the four basic political types is not practical, as it involves a huge nine-dimensional observed frequency table with 1536 cells, too large to present here. A much simpler but relevant example is possible using the data presented in Table 3.1.

These data come from the Political Action Study, an international comparative survey conducted in eight countries (Barnes & Kaase,

1979).[2] Table 3.1 contains data from the Netherlands; data from other countries will be analyzed in later sections of this chapter.

Among other things, the Political Action Study focuses on the political agenda: What issues do the respondents think politicians should put on their agenda and how do the respondents feel about the way politicians handle these issues? The respondents were asked which of a list of ten issues have to be considered as a responsibility of government. Four issues have been selected here: Guaranteeing Equal Rights for Men and Women (A), Providing a Good Education (B), Providing Good Medical Care (C), and Providing Equal Rights for Guest (Foreign) Workers (D). The issues B (education) and C (medical care) are more or less generally accepted tasks of government in the modern welfare state. As may be inferred from Table 3.1, 67% think that education and 56% that medical care is an *essential* responsibility of government. The issues A (equal rights men/women) and D (equal rights guest workers) are more controversial in this respect. Only 26% endorse the opinion that either of them ought to be an essential part of government's efforts. In terms of Inglehart's generational modernization theory (Inglehart, 1977, 1979), B and C might represent "materialist" values that are part of a modern society and are accepted by all, whereas A and D might represent "postmaterialist" values belonging to postmodern society, endorsed especially by the younger generations.

However, by emphasizing the difference between items A and D on the one hand and items B and C on the other, what the items have in common may be overlooked. It is just possible that they all can be conceived of as indicators of one underlying latent dichotomous variable X, one (latent) class of which consists of proponents of a considerably large task for government in general and the other of people who generally see a more limited task for government.

This possibility can be investigated using a latent class model such as the one depicted in Figure 3.1, with a dichotomous latent variable X explaining the observed relations among the manifest variables A through D as found in the observed frequency Table 3.1.

In order to test this latent class model and obtain the maximum likelihood estimates of the model parameters (equation 3.2 or 3.3), it is necessary to find the maximum likelihood estimates $\hat{F}_{ijklt}^{ABCDX}$ or, what amounts to the same, $\hat{\pi}_{ijklt}^{ABCDX}$. This is somewhat more complicated than in regular log-linear models without latent variables, because of the presence of the unobserved scores on X. Methods for finding the maxi-

TABLE 3.2   Two Latent Class Model For Data On Government's Responsibilities: The Netherlands

Latent Class		A. Equal Rights Men/Women $\hat{\pi}_{i\,t}^{\bar{A}X}$		B. Good Education $\hat{\pi}_{j\,t}^{\bar{B}X}$		C. Good Medical Care $\hat{\pi}_{k\,t}^{\bar{C}X}$		D. Equal Rights Guest Workers $\hat{\pi}_{i\,t}^{\bar{D}X}$	
X_t	$\hat{\pi}_t^X$	1. Yes	2. No	1. Yes	2. No	1. Yes	2. No	1. Yes	2. No
1	.410	.404	.596	.951	.049	.851	.149	.465	.535
2	.590	.168	.832	.468	.532	.351	.649	.120	.880

$L^2 = 13.99$, df = 6, $p = .03$,
Pearson-$\chi^2 = 13.97$
SOURCE: Table 3.1

mum likelihood estimates will be discussed in the next section; only the outcomes of the latent class analysis for Table 3.1 will be presented here.

The test outcomes are presented at the bottom of Table 3.2. Employing a significance level of .01, one would accept the model; with a level of .05, the model would be rejected. For the time being the model will be accepted, but there are at least some doubts as to the validity of the proposed latent class model for these data. In the next sections a better fitting model will be obtained.

According to the estimates $\hat{\pi}_t^X$ in Table 3.2, a proportion of 0.41 of the (Dutch) population belongs to latent class 1 and 0.59 to class 2, their sum of course being one. As in factor analysis, the meaning of the latent variable is derived from the relations between the latent and the manifest variables. Latent class 1 may be interpreted as consisting of the true proponents of an ample task for the government and latent class 2 of the true supporters of a more limited task. This can be inferred from the estimates of the conditional probabilities in Table 3.2.

According to these estimates, the chance that the members of latent class 1 say that it is an essential responsibility of government to guarantee equal rights for men and women, $\hat{\pi}_{11}^{AX}$, equals .404; the probability $\hat{\pi}_{21}^{AX}$ that they do not think so is 1 – .404 = .596. Although the estimated proportion of members of class 1 who score Yes on this item is not high, even less than .50, the corresponding proportion among the members of latent class 2 is even smaller: .168.

More or less the same pattern arises with regard to the item on providing equal rights for guest workers. For latent class 1 the probability of scoring Yes is just under .50 (.465), but for latent class 2 it is only

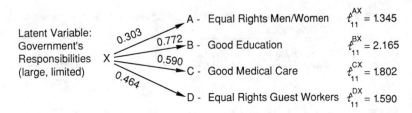

Figure 3.2. Log-Linear Effects ($\hat{\lambda}$) for the Basic Latent Class Model Applied to the Data on Government's Responsibilities in Table 3.1

.120. In both latent classes only a minority thinks that guaranteeing equal rights for men and women or for guest workers is an essential responsibility of the government, reflecting the fact, noted above, that these items are not "popular." But at the same time it is clear that the members of latent class 1 are more inclined to see guaranteeing equal rights as a task of government than the members of latent class 2.

The more "popular" items B (providing good education) and C (providing good medical care) are seen by an overwhelming majority of latent class 1 as an essential task of government. In latent class 1, the probabilities of scoring Yes are .951 and .851 respectively. In latent class 2, these probabilities drop to .468 and .351 respectively (although they still are not as low as for the items on equal rights). So we see that for all items there is a greater (conditional) probability that the people in latent class 1 will say that the issue concerned is an essential responsibility of government.

These relationships between the latent and the manifest variables can be expressed in log-linear terms. The parameter estimates are presented in Figure 3.2, the $\hat{\tau}$ parameters at the right, and the corresponding $\hat{\lambda}$ parameters next to the arrows.

Should one prefer the $\hat{\gamma}$ and $\hat{\beta}$ parameters of the effect model, those can easily be obtained by squaring the $\hat{\tau}$ parameters or multiplying the $\hat{\lambda}$ parameters by two.

The parameter estimates in Figure 3.2 can be calculated in the normal way by using the estimated expected frequencies $\hat{F}_{ijklt}^{ABCDX}$ (see Section 3.2.2), but also by using the parameter estimates of the conditional response probabilities in Table 3.2. Let us take $\hat{\tau}_{11}^{AX}$ as an example. According to the collapsibility theorem (Section 2.6), the parameter estimate $\hat{\tau}_{11}^{AX}$ in the log-linear model {AX, BX, CX, DX} can be obtained by using the frequencies $\hat{F}_{ijklt}^{ABCDX}$ in the full table $ABCDX$, but also, with the same results, by using the frequencies $\hat{F}_{i+++t}^{ABCDX}$ in the collapsed table AX. This collapsed table AX can be found in Table 3.2 under the heading "Equal Rights Men/ Women," be it "percentaged" horizontally. But as has been explained in the previous chapter (Sections 2.2, 2.3), this kind of transformation of the absolute frequencies does not influence the two-variable parameter estimates, as it involves multiplying all cells of a particular row with a constant. In this way, the log-linear parameters can be expressed in terms of the conditional response probabilities in equation 3.2. For example, using the formulas for estimating the parameters of log-linear models (equation 2.14) it follows that:

$$(\hat{\tau}_{11}^{AX})^4 = (\hat{\pi}_{11}^{\overline{AX}}\hat{\pi}_{22}^{\overline{AX}})/(\hat{\pi}_{12}^{\overline{AX}}\hat{\pi}_{21}^{\overline{AX}})$$

As expected, all parameters in Figure 3.2 are positive: people in latent class 1 $(X = 1)$ are more inclined to respond Yes on each item than the people in latent class 2 $(X = 2)$. Item B (Good Education) has the strongest relation with X, item A (men/women) the weakest; B is a more reliable indicator of X than A. Or, stated otherwise, the distinction between the two latent classes has more to do with the distinction whether or not government has to provide good education than with the distinction whether or not government has to guarantee equal rights for men and women.

3.2.2 Obtaining the Maximum Likelihood Estimates: The EM-Algorithm

The maximum likelihood estimates $\hat{F}_{ijklt}^{ABCDX}$ or $\hat{\pi}_{ijklt}^{ABCDX}$ of the basic latent class model $\{AX, BX, CX, DX\}$ are obtained according to the same principles as the estimates for log-linear models without latent variables: The estimated expected frequencies $\hat{F}$ (and $\hat{\pi}$) are computed in such a way that they are exactly in conformity with the restrictions implied by the postulated model and are as close as possible to the observed frequencies in terms of the L^2-statistic.

If X was not a latent variable, the procedures outlined in Section 2.4 for ordinary log-linear models would have been directly applicable and the maximum likelihood estimates for model $\{AX, BX, CX, DX\}$ would be computed by means of the observed frequencies f_{ijklt}^{ABCDX} (or the observed proportions p_{ijklt}^{ABCDX}) or more specifically by means of the sufficient statistics $f_{it}^{AX}, f_{jt}^{BX}, f_{kt}^{CX}$, and f_{lt}^{DX}. However, X is latent and its scores are not directly observed. Consequently, f^{AX}_{it}, f^{BX}_{jt}, etc. are not directly known and observed frequencies.

Haberman (1979), however, indicates that "the same maximum likelihood equations apply as in the ordinary case in which all frequency counts are directly observed, except that the unobserved counts are replaced by their estimated conditional expected values given the observed marginal totals" (p. 543). If these "estimated conditional expected values" are denoted as $\hat{f}$, it follows that the estimated sufficient statistics for model $\{AX, BX, CX, DX\}$ are $\hat{f}^{AX}_{it}$, $\hat{f}^{BX}_{jt}$, $\hat{f}^{CX}_{kt}$, and $\hat{f}^{DX}_{lt}$, which are exactly reproduced by the estimated expected frequencies $\hat{F}^{ABCDX}_{ijklt}$.

The unobserved frequencies f^{ABCDX}_{ijklt} are estimated as follows:

$$\hat{f}^{ABCDX}_{ijklt} = (\hat{F}^{ABCDX}_{ijklt}/\hat{F}^{ABCD}_{ijkl})f^{ABCD}_{ijkl} = \hat{\pi}^{ABCD\bar{X}}_{ijklt} f^{ABCD}_{ijkl} \tag{3.4}$$

or in terms of proportions:

$$\hat{p}^{ABCDX}_{ijklt} = (\hat{\pi}^{ABCDX}_{ijklt}/\hat{\pi}^{ABCD}_{ijkl})p^{ABCD}_{ijkl} = \hat{\pi}^{ABCD\bar{X}}_{ijklt} p^{ABCD}_{ijkl} \tag{3.5}$$

where $\hat{\pi}^{ABCD\bar{X}}_{ijklt}$ indicates the estimated conditional probability that an individual in the population belongs to $X = t$, given that her or his score on $ABCD$ is (i, j, k, l). Note that the estimates $\hat{f}^{ABCDX}_{ijklt}$ are such that: $\hat{f}^{ABCDX}_{ijkl+} = f^{ABCD}_{ijkl}$.

In order to obtain the parameter estimates of a particular log-linear model with latent variables, one has to find the maximum likelihood estimates $\hat{F}$, using particular marginals estimates $\hat{f}$, which in their turn have to be obtained by using the estimates $\hat{F}$. Both $\hat{f}$ and $\hat{F}$ can be simultaneously obtained through variants of the Newton-Raphson procedure and the Iterative Proportional Fitting (IPF) algorithm discussed in the previous chapter. Haberman presented a variant of the Newton-Raphson procedure, the scoring algorithm, along with a computer program called LAT (Haberman, 1979, Section 10.3, Appendix A2), a variant of the IPF-algorithm, the EM-algorithm (Haberman, 1979, Section 10.2), and, more recently, a modified Newton-Raphson algorithm implemented in the program NEWTON (Haberman, 1988b).

The relative advantages and disadvantages of the *Newton-Raphson (and Scoring) algorithm* and the *EM-algorithm* are generally comparable to the relative (dis)advantages of Newton-Raphson and IPF in the ordinary case of log-linear models without latent variables. The Newton-Raphson algorithm converges with fewer iterations than the EM-algorithm, but each iteration generally requires much more computing time. The Newton-Raphson algorithm automatically provides the estimated standard errors as a by-product of the computations and the EM-algorithm does not (although, after having obtained the maximum likelihood estimates by means of the EM-algorithm, standard errors can be estimated by finding the information matrix — Grover, 1987; Meilijson, 1989). When using the Newton-Raphson algorithm, an often complicated design matrix has to be provided which, however, at the same time, offers the possibility to impose all kinds of restrictions on the parameters and thus to evaluate a large variety of (nonhierarchical) log-linear models.

The major disadvantage of the scoring algorithm (LAT) compared to the EM-algorithm is that LAT requires very good initial estimates to start the iterative process. Otherwise the iterations will not converge. When using LAT, it often appeared that convergence was possible only when the initial estimates were in the neighborhood of the final solution and, in fact, would be considered identical to the final estimates for all practical purposes. For the EM-algorithm very rough initial estimates suffice, in many cases even random initial estimates.

Haberman's recent modified Newton-Raphson algorithm (NEWTON) is a large improvement on LAT. Although not as robust as programs using the EM-algorithm, NEWTON converges to the maximum likelihood estimates using initial parameter estimates that are rather crude approximations of the final estimates. In our (very limited) experience, only in a very few cases had several sets of initial estimates to be used before convergence was reached. NEWTON, however, has one serious disadvantage, it does consume a lot of computing time, generally much more than programs using the EM-algorithm. Because estimating the parameters of log-linear models with latent variables already requires much more computing time, using whatever algorithm, than estimating the parameters of ordinary log-linear models, this is a serious drawback in practical applications.

Because of all this, the EM-algorithm will be preferred for the estimation of the parameters of log-linear models with latent variables.

The EM-algorithm is an iterative procedure in which each iteration consists of two steps, the E-step and the M-step.[3] In general terms, one may describe the E-step as the step where the "Expected sufficient statistics" are computed using the current parameter estimates obtained in the previous iteration, whereas in the M-step the likelihood function is Maximized by using the newly obtained estimates of the sufficient statistics for the computation of new parameter estimates.

To obtain the parameter estimates of model $\{AX, BX, CX, DX\}$, initial values are given to all log-linear parameters in equation 3.3 in such a way that the identifying restrictions are satisfied, meaning mostly that the product of the τ parameters over any subscript equal one. As has been said, very rough initial estimates suffice. For example, set all one-variable τ parameters equal to 1 and all two-variable effects equal to 2 or -2. Using these initial parameter estimates, one can compute initial estimates $\hat{F}_{ijklt}^{ABCDX}(0)$ by means of equation 3.3.

The estimates $\hat{F}(0)$ are used in the E-step to find by means of equation 3.4 (initial) estimates $\hat{f}_{ijklt}^{ABCDX}(0)$ and thereby (initial) estimates $\hat{f}(0)$ of the sufficient statistics $\hat{f}_{it}^{AX}$, $\hat{f}_{jt}^{BX}$, $\hat{f}_{kt}^{CX}$, and $\hat{f}_{lt}^{DX}$. In the M-step the initial estimated expected frequencies $\hat{F}(0)$ are improved by successively adjusting them to estimated sufficient statistics, that is, to the marginal frequencies $\hat{f}(0)$ to be reproduced by the model. The computations carried out in the M-step are identical to the computations carried out in the IPF procedure in the case of a log-linear model without latent variables.

The results of the M-step, the improved estimates $\hat{F}(1)$, are used in the next iteration to improve the estimates $\hat{f}(0)$ of the previous E-step; the improved estimates $\hat{f}(1)$ are used to improve the estimates $\hat{F}(1)$; and so on. This continues until the outcomes converge. Usually a large number of iterations is required, far more than in the case of log-linear models without latent variables.

The final estimates $\hat{F}$ can be used to calculate the maximum likelihood estimates of the effect parameters by means of equations similar to equation 2.13. Summing the final estimates $\hat{F}_{ijklt}^{ABCDX}$ over the latent variable results in the estimated expected observed frequencies $\hat{F}_{ijkl}^{ABCD}$, which can be used to test the model by comparing them to the observed frequencies f_{ijkl}^{ABCD} by means of the regular test statistics L^2 or Pearson-χ^2.

Goodman presented a variant of the EM-algorithm in terms of the latent class parametrization in equation 3.2 (Goodman, 1974a, 1974b). *Goodman's algorithm* has been implemented in the programs MLLSA (Clogg, 1981a) and LCAG (Hagenaars, 1988a; Hagenaars & Luijkx, 1987). The advantage of this parametrization is that it is more easily understood by most people than the log-linear formulation. Moreover, Goodman showed how to obtain a large number of restricted models (see Section 3.2.3) equivalent to nonhierarchical log-linear models. For these reasons, log-linear models with latent variables will be dealt with mainly by means of Goodman's algorithm.

Goodman's algorithm starts with assigning initial values to the parameters at the right hand side of equation 3.2, taking into account that these parameters are probabilities; after summation over the appropriate subscript they have to sum to one and they cannot exceed the boundaries zero and one. In general, the extreme values zero and one should be avoided, because once these boundary values are reached, they will remain the same during all further iterations. Again, these initial values may be very rough approximations of the final estimates, for example, a high probability equals .80 and a low probability .20. Very often even "random" initial estimates suffice without seriously affecting the number of iterations.

By means of these initial parameter values and equation 3.2, initial estimates $\hat{\pi}_{ijklt}^{ABCD\overline{X}}(0)$ are computed. As before in the E-step, the estimated sufficient statistics are computed, now denoted as $\hat{p}(0)$ and calculated by means of equation 3.5.

Next, in the M-step, improved parameter estimates are obtained using the current estimates $\hat{p}$. First, a new estimate of $\hat{\pi}_t^X$ is obtained. Second, new estimates of the conditional response probabilities are computed by means of $\hat{p}$ and the newly obtained estimates $\hat{\pi}_t^X$. The formulas used are:

$$\hat{\pi}_t^X = \sum_{i=1}^{I} \sum_{j=1}^{J} \sum_{k=1}^{K} \sum_{l=1}^{L} \hat{p}_{ijklt}^{ABCDX}$$

$$\hat{\pi}_{it}^{\overline{A}X} = \left(\sum_{j=1}^{J} \sum_{k=1}^{K} \sum_{l=1}^{L} \hat{p}_{ijklt}^{ABCDX} \right) / \hat{\pi}_t^X \qquad (3.6)$$

$$\hat{\pi}_{jt}^{\overline{B}X} = \left(\sum_{i=1}^{I} \sum_{k=1}^{K} \sum_{l=1}^{L} \hat{p}_{ijklt}^{ABCDX} \right) / \hat{\pi}_t^X$$

and similar formulas for the remaining conditional response probabilities.

By means of the newly estimated parameters, improved estimates $\hat{\pi}_{ijklt}^{ABCDX}(1)$ can be calculated, again using equation 3.2. As some simple calculations may show, successive application of equations 3.6 and 3.2 implies that the new estimates $\hat{\pi}(1)$ reproduce exactly the current estimates of the sufficient statistics, the marginal proportions $\hat{p}(0)$: $\hat{p}_{it}^{AX}$, $\hat{p}_{jt}^{BX}$, $\hat{p}_{kt}^{CX}$, and $\hat{p}_{jt}^{DX}$ (Haberman, 1979, Exercise 10; Hagenaars, 1985: 76; Kiiveri & Speed, 1982: 266).

The new estimates $\hat{\pi}(1)$ are used in the next E-step to find improved estimates $\hat{p}(1)$, which can be used in the following M-step to improve the estimates $\hat{\pi}(1)$, and so on until the results converge.

The final parameter estimates thus obtained are the maximum likelihood estimates or "terminal maxima." Goodman's algorithm avoids "impossible" outcomes in the sense of estimates of probabilities that are larger than one or smaller than zero. However, during the iterations one or more values may approach the limits zero or one. The final solution is then the maximum likelihood solution, given that the parameters concerned are truly zero or one in the population.

The EM-algorithm provides a rather safe road to the obtainment of the maximum likelihood estimates. Convergence to the maximum likelihood is generally achieved (Haberman, 1976). However, there are some pitfalls. Sometimes the iterations converge to a less than optimal solution, to a local maximum instead of the global maximum of the likelihood function. Such a local maximum can be discovered by repeating the procedure with other starting values. If this results in different parameter values and a smaller value for the test statistic L^2, the result with the smallest L^2-value is the maximum likelihood estimate sought.

It is also possible that different parameter estimates are obtained, but that the L^2-value remains the same. In this case there is an identification problem: There are no unique estimates for all parameters. This possibility will be discussed further in Section 3.2.4.

3.2.3 Restricted Models

The standard latent class model with one latent variable discussed above can be modified and extended in many ways (Goodman, 1974a, 1974b, Rindskopf, 1983). Most importantly, it can be reformulated to encompass models with two or more latent variables. This will be shown in Section 3.4. But the one latent variable model itself may also appear in many forms, distinguished from each other in terms of different restrictions on the model's parameters. The kinds of restrictions meant here will be discussed in this section, mainly in terms of the latent class parametrization and Goodman's algorithm. Goodman (1974a, 1974b) showed how, within his variant of the EM-algorithm, equality restrictions can be imposed on the latent class parameters, where the equality restrictions imply setting certain parameter estimates equal to each other or to a particular constant.

First of all the restrictions may involve the *probability distribution of the latent variable X*, that is, the parameters π_t^X. Sometimes one has definite expectations about the size of a particular latent class. For example, in an election study the true proportion of nonvoters, or of voters for particular parties, may be known from recent national elections. The investigator may want to fix the proportions belonging to the relevant categories of the latent variable "party support" in agreement with these expected proportions. At other times, an investigator may expect that the sizes of particular latent classes are equal to each other.

Such expectations can be incorporated in Goodman's algorithm. For each iteration, the improved estimates of the parameters are computed in the M-step in the usual way as indicated in equation 3.6. But before these new parameter estimates are used to obtain new estimates for π_{ijklt}^{ABCDX}, the restrictions are imposed upon the parameters. If one or more parameters π_t^X are supposed to have a fixed value, these fixed values are assigned to $\hat{\pi}_t^X$, if necessary along with a proportional rescaling of the other $\hat{\pi}_t^X$ parameters to ensure that the sum of all $\hat{\pi}_t^X$ parameters equals one. If a set of two or more π_t^X parameters are supposed to have equal values, each $\hat{\pi}_t^X$ parameter of this set gets a value equal to the arithmetic mean of all current estimates of the π_t^X parameters in this set.

Let us look at a simple example. It was concluded from the estimates $\hat{\pi}_t^X$ in Table 3.2 that there are somewhat more opponents of an ample task for government ($\hat{\pi}_1^X = .590$) than proponents ($\hat{\pi}_2^X = .410$). One might wonder whether this difference in size is significant. A standard latent class analysis is carried out on the data in Table 3.1, but with the restriction that $\pi_1^X = \pi_2^X = .50$. The test results are: $L^2 = 16.04$, df = 7, $p = .03$ (Pearson-$\chi^2 = 15.59$). As the original model, this one may be accepted, employing a .01 significance level. Furthermore, conditional testing shows that the restricted model need not be rejected in favor of the original unrestricted model ($L_{r/u}^2 = 16.04 - 13.99 = 2.05$, df$_{r/u} = 7 - 6 = 1$, $p = .15$). So, there is no reason to reject the hypothesis that there are equal numbers of opponents and proponents of an large task for government.

The restrictions on the *conditional response probabilities* $\bar{\pi}_{it}^{AX}$, etc. may also assume two forms: A fixed value may be given to one or more conditional probabilities, or conditional probabilities may be set equal to each other.

In Goodman's algorithm, fixed values are assigned to the relevant conditional probabilities during each iteration at the end of the M-step, that is, at the same point where the fixed values for the $\hat{\pi}_t^X$ parameters are assigned, along with rescaling some other conditional parameters where necessary to guarantee summation to one. The fixed values used most often are zero and one. The assignment of the fixed values zero or one to particular conditional probabilities need not necessarily occur at the end of each M-step, but can also be realized by setting the initial estimates of these parameters equal to zero or one, as these extreme values will not change during the iterations.

The values zero and one can be used to indicate that particular manifest variables or categories are perfect indicators of the latent variable or particular latent classes. For instance, if it is assumed that the proponents of an ample task for government ($X = 1$) will absolutely be of the opinion that providing a good education is an essential task of government ($B = 1$), the restriction $\pi_{11}^{BX} = 1$ (and $\pi_{21}^{BX} = 0$) is applied.

Setting particular conditional probabilities equal to each other occurs at the end of the M-step by assigning to each of them the weighted arithmetic mean of the estimates obtained in equation 3.6, where the weighting occurs by means of $\hat{\pi}_t^X$, the estimated sizes of the latent classes to which the conditional probabilities belong.

Equalities between conditional probabilities can be used to show that particular manifest variables are equivalent, equally reliable, "parallel" indicators of the latent variable. For instance, given the nature of the indicators in Table 3.1, it might be assumed that A (men/women) and D (guest workers) are parallel items, as are B (education) and C (medical care). The probability that someone from latent class 1 considers providing equal rights for men and women an essential responsibility of the government is the same as the probability that this person considers providing equal rights for guest workers an essential responsibility, etc. This would imply the restrictions:

$$\pi_{11}^{\overline{A}X} = \pi_{11}^{\overline{D}X} \qquad \pi_{12}^{\overline{A}X} = \pi_{12}^{\overline{D}X} \qquad \pi_{11}^{\overline{B}X} = \pi_{11}^{\overline{C}X} \qquad \pi_{12}^{\overline{B}X} = \pi_{12}^{\overline{C}X}$$

Applying these extra restrictions to the latent class model presented in Table 3.2 results in a bad fit: $L^2 = 45.91$, df $= 10$, $p = .00$ (Pearson-$\chi^2 = 43.85$). A and D, and B and C are not strictly equivalent indicators.

Another frequently used application of equality constraints on the conditional response probabilities is to indicate that the error rates are the same in several latent classes. In terms of our example, it might be assumed that the probability for the proponents of an ample task for the government ($X = 1$) to give the "wrong" answer, that is, to say that providing good medical care is *not* an essential task for the government ($C = 2$), is the same as the probability that the opponents of an ample government task ($X = 2$) say that medical care has to be an essential part of government's agenda ($C = 1$): $\pi_{21}^{CX} = \pi_{12}^{CX}$. (For the data in Table 3.1 this is not an acceptable assumption: $L^2 = 22.44$, df $= 7$, $p = .00$.)

The restricted models above are all formulated in terms of constraints on the latent class parameters in equation 3.2. But all these restrictions can also be rendered in the form of *restrictions on the log-linear parameters* in equation 3.3, leading to the same estimated expected frequencies. The restrictions on the probability distribution of the latent variable have to be formulated within the context of path analytic models with latent variables and will be dealt with later. Setting particular conditional response probabilities equal to one or zero leads to structural zero expected cell frequencies in table $ABCDX$, because particular combinations of scores on the manifest and latent variables do not occur. The log-linear parameters in equation 3.3 have to be estimated under the condition that the relevant estimated expected cell frequencies $\hat{F}_{ijklt}^{ABCDX}$ are zero (Haberman, 1979, p. 554).

Making the manifest variables A and D and B and C completely equivalent, parallel indicators, as done above, implies that in equation 3.3 the following restrictions should be applied: $\tau_i^A = \tau_i^D$, $\tau_j^B = \tau_j^C$, $\tau_{it}^{AX} = \tau_{it}^{DX}, \tau_{jt}^{BX} = \tau_{jt}^{CX}$.[4] Setting the reliabilities of C equal for both latent classes results in a model that has close connections with the simple non-hierarchical model discussed in Section 2.4 (equation 2.22), and can be obtained by setting $\tau_k^C = 1$ in equation 3.3.

3.2.4 Identifiability: Testing and Fitting

The parameters of log-linear models with latent variables cannot always be uniquely determined. Sometimes the system of equations 2.1 and 2.2 (or 2.3) has more than one solution for some or all parameters. This means that different sets of solutions can be obtained, each of which yields the same estimated expected frequencies $\hat{F}_{ijklt}^{ABCDX}$ and, consequently, the same values for the test statistics L^2 and Pearson-χ^2.

The problem of *identifiability* was extensively dealt with by Goodman (1974b). A *necessary condition* for identifiability is that the number of parameters to be independently estimated is not greater than the number of independent knowns, the observed cell frequencies. However, for a great many models in which this condition is satisfied, it nevertheless holds that the parameters cannot be uniquely determined.

Goodman (1974b) indicated a *sufficient condition* for local identifiability. One has to compute a matrix of partial derivatives of the nonredundant observed probabilities with respect to the nonredundant model parameters. A model is locally identifiable if this matrix has full column rank. If this condition is not fulfilled, then there exist in the neighborhood of the solution found, other solutions which would lead to exactly the same value of the test statistics L^2 and Pearson-χ^2.

An unidentifiable model can be made identifiable by imposing restrictions. The reverse, however, is also true: An identifiable model can be rendered unidentifiable by the introduction of extra restrictions (Goodman, 1974b, and an example in Section 3.4.2).

Some programs based on the EM-algorithm, like Clogg's MLLSA (Clogg, 1955), provide information about the fulfillment of the sufficient condition. In programs making use of the Newton-Raphson or scoring algorithm, like Haberman's NEWTON and LAT (Haberman, 1988b and 1979, respectively), nonidentifiability will be noticed because the covariance matrix of the estimates is not of full rank, and one or more of the standard errors of the parameter estimates will be rendered as exactly 0. Otherwise, the identifiability of the model parameters can be investigated by changing the obtained parameter estimates somewhat, and using the changed parameters as initial estimates for a new run of the program. For example, a few "low" values are changed into "high" ones and vice versa. Generally, when a model is not identifiable, the second run will give different parameter estimates but the same estimated expected frequencies $\hat{F}_{ijklt}^{ABCDX}$.

If a model is identifiable, it may be tested by means of the usual χ^2-*statistics*, L^2 and Pearson-χ^2, comparing the estimated expected frequencies $\hat{F}_{ijklt}^{ABCDX}$ ($= N\hat{\pi}_{ijklt}^{ABCDX}$) with the observed frequencies f_{ijkl}^{ABCD} ($= Np_{ijkl}^{ABCD}$) as in equation 2.21.

The number of *degrees of freedom* df of an identifiable model is equal to the number of independent knowns, the cell frequencies, minus the number of independent parameters to be estimated. The number of degrees of freedom for the basic latent class model in Table 3.2, equals 6. Table 3.1 contains 16 observed cell frequencies or proportions. Because the proportions sum to one, there are 15 "independent knowns." Given that several of the estimated probabilities in Table 3.2 also have to sum to one, there are 9 independent "unknowns," and df = 6.

The number of degrees of freedom may also be determined using the log-linear parametrization. Given the usual identifying restrictions and counting the overall effect as a parameter to be estimated, there are 10 independent parameters to be estimated. Sixteen cell frequencies lead to df = 16 − 10 = 6.

The *testing and fitting procedures* for ordinary log-linear models can also be used in the case of log-linear models with latent variables. All that has been said in Section 2.5 is similarly applicable here.

3.3 Association between Latent and External Variables: Latent Class Scores and Quasi-latent Variables

A logical follow up to the latent class analysis carried out for Table 3.1 is to investigate what kind of people are opposed and what kind of people advocate an ample task for the government. Are the proponents of an ample task more to be found among the younger or among the older generations? Are leftist people more of the opinion that the government has broad responsibilities than other people? To answer these questions, one has to explore the relations of the latent variable X with manifest variables such as age and political preference. These manifest variables will be denoted as "external" variables, because not being indicators of X, they are not part of the original latent class model.

As with factor analysis, this exploration can be carried out in two ways. In the first place, analogous to the computation and use of factor scores, it is possible to compute for each individual a predicted latent score $\tilde{X}$ on the basis of the scores on the manifest variables A through D, and relate $\tilde{X}$ to the scores on the external variables. An alternative is to incorporate the external variables into the latent class model, make them "internal," and investigate the relations between the "external" variables and the latent variable X within the model. This is actually very similar to what is done when carrying out (linear) causal analysis with, for example, Jöreskog's LISREL (Jöreskog & Sörbom, 1979). Both approaches have their (dis)advantages, as will be discussed below.

The possibility of assigning each individual a particular *predicted score* $\tilde{X}$, in other words, assigning each individual to one particular class by means of the scores on the manifest variables, has from the very beginning played an important role in latent class analysis. Lazarsfeld and Henry (1968) called it "one of the most valuable aspects of latent structure analysis" (p. 36). More recently, Clogg (1981a), Goodman

(1974a), Madden and Dillon (1982), and McCutcheon (1987) echoed this sentiment.

The point of departure for the computation of $\tilde{X}$ is the conditional probability that an individual from the population belongs to $X = t$, given the scoring pattern (i, j, k, l) on the manifest variables.

$$\pi_{ijkl\,t}^{ABCD\overline{X}} = \pi_{ijkl\,t}^{ABCDX} / \pi_{ijkl}^{ABCD} \tag{3.7}$$

After this conditional probability for each latent class t is computed, the individual may be assigned to the latent class for which this conditional probability is largest. Denoting this (conditional) modal latent class as t^*, it follows for each individual:

$$\tilde{X} = t^* \qquad \text{given } (i, j, k, l) \text{ for } ABCD \tag{3.8}$$

In contrast to X, $\tilde{X}$ is an "observed" variable and can be related in the usual way to external variables like age and political preference.

However, in general $\tilde{X}$ will not be a perfect substitute for X. Except for the extreme cases where the estimated conditional probability $\pi_{ijkl\,t^*}^{ABCDX}$ equals 1, there is a chance that the individual does not belong to the modal class, that he or she is assigned to the wrong latent class. The probability of misclassification $\varepsilon_{ijkl}^{ABCD}$ for a particular individual with observed scoring pattern (i, j, k, l) can be computed as follows, where t^* refers to the conditional modal latent class:

$$\varepsilon_{ijkl}^{ABCD} = 1 - \pi_{ijkl\,t^*}^{ABCD\overline{X}} \tag{3.9}$$

The overall probability of a misclassification for all individuals, E, is then defined as:

$$E = \sum_{i=1}^{I} \sum_{j=1}^{J} \sum_{k=1}^{K} \sum_{l=1}^{L} \pi_{ijkl}^{ABCD} \varepsilon_{ijkl}^{ABCD} \tag{3.10}$$

E expresses what proportion of the population is expected to be misclassified following the (conditional) modal assignment rule and as such E is a measure of how adequate $\tilde{X}$ is as a substitute of X. The stronger the manifest variables are related to X, the better X can be predicted on the basis of the observed scores and the closer $\tilde{X}$ will be to X. If E is large, one has to reckon with the possibility that the distributions of X and $\tilde{X}$ differ and that the relations between the external variables and $\tilde{X}$ are not the same as between the external variables and X.

Clogg (1981a) suggested a modification of E resulting in Goodman and Kruskal's measure of association λ which in this case indicates how much better X can be predicted with the aid of the joint variable $ABCD$ than without it (Goodman & Kruskal, 1954). The formula for λ is

$$\lambda_{X.ABCD} = \frac{(1 - \pi_{t*}^X) - E}{(1 - \pi_{t*}^X)} \qquad (3.11)$$

where π_{t*}^X refers to the unconditional modal class t^* of X.

By replacing the parameters on the right hand side of the equations 3.7 through 3.11 with their maximum likelihood estimators, the maximum likelihood estimators for the various parameters on the left hand side of the equations are found: $\hat{\varepsilon}$, $\hat{E}$, and $\hat{\lambda}$; the substitute for X will be denoted as $\tilde{X}$ when it is based on the population parameters, and as $\tilde{X}'$ when it is computed by means of the maximum likelihood estimates.

For the example used so far, Table 3.2, $\hat{E} = 0.17$ and $\hat{\lambda} = .67$. If we were to assign people to the latent classes 1 and 2 using the modal assignment rule, we may expect to classify $(100 - 17)\% = 83\%$ of the total sample correctly, which, according to λ, is 67% better than if we would have assigned all respondents to the (marginal) modal class 2.

It is good to keep in mind that the measures $\hat{E}$ and $\hat{\lambda}$ are not measures of goodness-of-fit like L^2 and Pearson-χ^2. It is quite possible that a model which has to be rejected according to the χ^2-test yields a lower value of $\hat{E}$ or a higher value of $\hat{\lambda}$ than a model that does fit the data.

If $\tilde{X}$ is a poor substitute of X according to λ and E, it is uncertain, as was said above, whether the external variables are related in the same way to $\tilde{X}$ as to X. Underlying this is a fundamental problem which has rarely if ever been discussed in the latent class literature. This problem has to do with the *identifiability of the individual scores on X* (not $\tilde{X}$ or $\tilde{X}'$) and is very similar to the problem of indeterminacy of factor scores (not: component scores) discussed by Steiger (1979a, 1979b) and Steiger and Schönemann (1978), among others.

The problem is that knowing the observed scores for each individual and knowing the values of the latent class parameters, that is, the values of π_{ijklt}^{ABCDX}, is generally not enough to determine exactly for each individual what is the score on X. Unless $E = 0$, several different sets of X scores may be assigned to the individuals, all in agreement with the values of the model parameters and the observed scores, but not identical with each other and, in some not even extreme cases, negatively correlated with each other (Hagenaars, 1985).

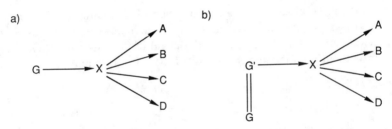

Figure 3.3. Latent Class Model with an "External Variable" G (Generation)

When different sets of X scores exist that are far from perfectly related to each other, these sets may be differently related to the external variables, even with opposite signs. Without additional assumptions there is no way of telling which set of latent scores is the best — they are all in agreement with the model — and what the true association is between the latent and the external variables.

Moreover, when the individual latent scores X are not identifiable, then the meaning of the predicted individual scores $\tilde{X}$ is dubious. If $E \neq 0$, the set of $\tilde{X}$ scores is not exactly equivalent to one of the admissable set of X scores, and, even if it were equivalent, it would still be just one of the several possible admissable sets of X scores.

How severe this problem is depends mainly on the strength of the relations between the latent and the manifest variables. If one or more manifest variables are perfectly related to X, with response probabilities equal to one and zero and $E = 0$, there actually is no problem. Only one admissable set of X values exists and $\tilde{X}$ is identical to X. But then X is not really latent and could be replaced by one or more of the manifest variables. If the manifest and latent variables are only weakly related to each other, the latitude for the set of X scores is large and the $\tilde{X}$ scores will differ from any admissable set of X scores.

The way to avoid these problems is to extend the original latent class model and set up an identifiable model in which the relations of the external variables with the latent and manifest variables are specified.

If we would like to know whether or not the older generations are more in favor of an ample task for the government than the younger generations, we could set up a model like the one in Figure 3.3a. This model is an *extension of the basic model* used before (Figures 3.1, 3.2), in that the "external" variable Generation (G) has been added. G has a direct influence on X, but the relations between the external variable G and the manifest variables A through D are completely mediated by X. This extended latent class model may be rendered in log-linear terms as follows:

$$F^{ABCDXG}_{ijkltm} = \eta\tau^A_i\tau^B_j\tau^C_k\tau^X_l\tau^G_m\tau^{AX}_{il}\tau^{BX}_{jl}\tau^{CX}_{kl}\tau^{DX}_{ll}\tau^{XG}_{lm} \qquad (3.12)$$

The parameters of this log-linear model $\{AX, BX, CX, DX, GX\}$ can be estimated by means of the Newton-Raphson algorithm or the EM-algorithm using the observed frequency table $ABCDG$ and the estimated sufficient statistics $\hat{f}^{AX}_{il}, \hat{f}^{BX}_{jl}, \hat{f}^{CX}_{kl}, \hat{f}^{DX}_{ll}$, and $\hat{f}^{XG}_{lm}$. The parameter estimate $\hat{\tau}^{XG}_{lm}$ is of special interest as it shows the relation between G and the latent variable X we were after. In terms of latent class analysis and Goodman's algorithm, the parameter estimates can be obtained by considering the model in Figure 3.3a as an ordinary basic latent class model with one latent variable X and five manifest variables A through D, and G. The probabilities π^{ABCDXG}_{ijkltm} of this latent class model multiplied by the sample size N are identical to F^{ABCDXG}_{ijkltm} in equation 3.12.

Additional external variables may be introduced into the latent class model, for example, H and I, by combining them into one joint variable GHI and considering this joint variable as an additional indicator of X. A disadvantage of introducing external variables in this manner is that it is impossible to exclude certain higher order interactions among G, H, I, and X. It is possible, however, to define more parsimonious latent class models by introducing *quasi-latent variables*. In Sections 3.5 through 3.7 many examples of this will be presented. Here, the principle of introducing quasi-latent variables for defining parsimonious latent class models will be briefly discussed.

First, the extended latent class model in Figure 3.3a is transformed into the model in Figure 3.3b. In the latter model, there are two latent variables X and G', and five manifest variables, namely, the indicators A through D, and the external variable G. If X has T categories and G' has M categories, the model in Figure 3.3b may be conceived of as a basic latent class model with $T \times M$ latent classes which form the categories of the one (joint) latent variable XG'. The relation between G' and X can be determined from the probabilities $\pi^{XG'}_{tm'}$. What is special about this model is that G and G' are perfectly related to each other (and that the scores on A through D depend only on X and not on G', as in model 3.3a).

G' is a quasi-latent variable: If somebody has the score m on G', then the score on G will also be m, with absolute certainty. This implies the following restrictions on the conditional response probabilities $\pi_{m t\, m'}^{GXG'}$:

$$\pi_{m t\, m'}^{\overline{G}XG'} = \begin{cases} 1 & \text{if } m = m' \\ 0 & \text{if } m \neq m' \end{cases} \qquad (3.13)$$

Because of these restrictions, G and G' are in fact identical variables and the relations between G' and X will be the same as those between G and X.

To ensure that the scores on the manifest variables are dependent only on X and that G' (and consequently G) and the manifest variables are conditionally independent of each other holding X constant, the following additional restrictions are imposed on the conditional response probabilities:

$$
\begin{aligned}
\pi_{i\,t\,1}^{\overline{A}XG'} &= \pi_{i\,t\,2}^{\overline{A}XG'} = \ldots = \pi_{i\,t\,M'}^{\overline{A}XG'} \\[4pt]
\pi_{j\,t\,1}^{\overline{B}XG'} &= \pi_{j\,t\,2}^{\overline{B}XG'} = \ldots = \pi_{j\,t\,M'}^{\overline{B}XG'} \\[4pt]
\pi_{k\,t\,1}^{\overline{C}XG'} &= \pi_{k\,t\,2}^{\overline{C}XG'} = \ldots = \pi_{k\,t\,M'}^{\overline{C}XG'} \\[4pt]
\pi_{l\,t\,1}^{\overline{D}XG'} &= \pi_{l\,t\,2}^{\overline{D}XG'} = \ldots = \pi_{l\,t\,M'}^{\overline{D}XG'}
\end{aligned}
\qquad (3.14)
$$

All these restrictions can easily be applied using standard programs for latent class modeling. The probabilities $\pi_{ijkl\,m t\,m'}^{ABCDGXG'}$ when summed over m' are identical to the probabilities $\pi_{ijkl\,m t}^{ABCDGX}$ obtained for the model in Figure 3.3a. By imposing restrictions on $\pi_{t\,m'}^{XG'}$ the relation between G' (or G) and X can be made to conform to particular constraints, for example, independence of X and G' (G), that is, $\tau_{t\,m'}^{XG'} = 1$.

If the model in Figure 3.3 is valid for the population, the parameters $\hat{\pi}_{t\,m'}^{XG'}$ will give the maximum likelihood estimate of the nature and strength of the relationship between the latent variable X and the external variable G' without the problems that are connected with using the "observed" distribution of $\tilde{X}'$ and G. However, it is necessary to work with the complete observed cross-classification of all indicators and external variables (here: the observed table $ABCDG$). In the other procedure, the simpler cross-classifications of all indicators and of $\tilde{X}$ and the external variables are used separately (here: tables $ABCD$ and $\tilde{X}'G$). With a small sample the latter may have to be preferred. Therefore, further research into the extent of the indeterminacy of the individual latent scores X and the association between X and $\tilde{X}$ remains desirable.

3.4 Models with Two or More Latent Variables

3.4.1 Saturated Models for Latent Variables

In the previous Section, a model was discussed with two latent variables, one of which was quasi-latent. Using some of the same principles, latent class models can be defined in which two or more variables occur that are all really latent.

Testing the model for the data on government's responsibilities (Tables 3.1 and 3.2) provided a borderline result and it was not clear whether to accept the latent class model or not. One way to improve the model would be to increase the number of categories of the one latent variable X (along with introducing restrictions on the parameters to make the model identifiable).[5] Another way would be to increase the number of latent variables.

Instead of letting the four manifest variables A through D be the indicators of one dichotomous latent variable X (proponents versus opponents of an ample task for government in general), they might be regarded as the indicators of two latent variables Y and Z. The dichotomous latent variable Y represents the way people feel about government having an essential responsibility with regard to "ideal" matters, indicated by the postmaterialists issues "equal rights for men and women" (A) and "equal rights for guest workers" (D), whereas the dichotomous variable Z represents people's feeling about government taking essential responsibility with regard to "material" matters, indicated by "providing good education" (B) and "providing medical care" (C).

The two latent variable model is depicted in Figure 3.4. In log-linear terms, this latent class model corresponds with the log-linear model {YZ, YA, YD, ZB, ZC}:

$$F^{ABCDYZ}_{ijklrs} = \eta \tau^A_i \tau^B_j \tau^C_k \tau^D_l \tau^Y_r \tau^Z_s \tau^{AY}_{ir} \tau^{DY}_{lr} \tau^{BZ}_{js} \tau^{CZ}_{ks} \tau^{YZ}_{rs} \qquad (3.15)$$

This model can be handled by the EM-algorithm described in Section 3.2.2. One starts with initial estimates for the parameters in equation 3.15 to obtain the initial estimates $\hat{F}^{ABCDYZ}_{ijklrs}(0)$. These estimates $\hat{F}(0)$ are used in the E-step to obtain estimates for the sufficient statistics $\hat{f}^{AY}_{ir}, \hat{f}^{DY}_{lr},$

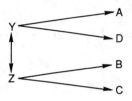

Figure 3.4. Latent Class Model with Two Latent Variables Y and Z and Four Manifest Variables A, B, C, and D

$\hat{f}^{BZ}_{js}$, $\hat{f}^{CZ}_{ks}$, and $\hat{f}^{YZ}_{rs}$. These current estimates $\hat{f}$ are used in the M-step to improve the estimates $\hat{F}(0)$, etc.

To use Goodman's algorithm, the two latent variables model is conceived of as a variant of the basic latent class model (Figure 3.1). The one latent variable X now has four categories: the 2×2 categories of the joint latent variable YZ and $\pi^X_t \equiv \pi^{YZ}_{rs}$, $\pi^{AX}_{it} \equiv \pi^{AYZ}_{irs}$, etc.[6] Because the scores on A and D are directly dependent only on Y, and the scores on B and C only on Z, the following restrictions should be imposed on the conditional response probabilities:

$$\pi^{\overline{A}YZ}_{i\,r1} = \pi^{\overline{A}YZ}_{i\,r2} \qquad \pi^{\overline{D}YZ}_{l\,r1} = \pi^{\overline{D}YZ}_{l\,r2}$$

$$\pi^{\overline{B}YZ}_{j\,1s} = \pi^{\overline{B}YZ}_{j\,2s} \qquad \pi^{\overline{C}YZ}_{k\,1s} = \pi^{\overline{C}YZ}_{k\,2s} \tag{3.16}$$

The results of application of this model to the data in Table 3.1 are presented in Table 3.3 and Figure 3.5. The log-linear effects in Figure 3.5 may be calculated on the basis of $\hat{F}^{ABCDYZ}_{ijklrs}$, or by means of the estimates $\hat{\pi}^{YZ}_{rs}$, π^{AYZ}_{irs}, etc., in Table 3.3, as follows from the collapsibility theorem.

Taking all restrictions into account, there are 11 latent class parameters to be independently estimated and $15 - 11 = 4$ degrees of freedom left to test the model. The test results are mentioned under Table 3.3, showing that the two latent variable model is a very acceptable one. Although a conditional test is not allowed (see note 5), given the outcomes of the unconditional tests, the two latent variable model is to be preferred. From the estimates in Table 3.3 it can be learned that 26.8% $(.259 + .009)$ of the respondents think that government has an essential task with regard to "ideal" matters $(Y = 1)$, and 43.5% $(.259 + .176)$ take this stand with regard to "material" matters $(Z = 1)$.

TABLE 3.3 Two Latent Variable Model For Data On Government's Responsibilities: The Netherlands

Latent Class			$\hat{\pi}_{rs}^{YZ}$	A. Equal Rights Men/Women $\hat{\pi}_{lrs}^{\bar{A}YZ}$		B. Good Education $\hat{\pi}_{jrs}^{\bar{B}YZ}$		C. Good Medical Care $\hat{\pi}_{krs}^{\bar{C}YZ}$		D. Equal Rights Guest Workers $\hat{\pi}_{lrs}^{\bar{D}YZ}$	
X_t	Y_r	Z_s		1. Yes	2. No	1. Yes	2. No	1. Yes	2. No	1. Yes	2. No
1	1	1	.259	.507	.493	.948	.052	.852	.148	.654	.346
2	1	2	.009	.507	.493	.449	.551	.327	.673	.654	.346
3	2	1	.176	.176	.824	.948	.052	.852	.148	.118	.882
4	2	2	.556	.176	.824	.449	.551	.327	.673	.118	.882

$L^2 = 5.76$, df = 4, $p = .22$,
Pearson-$\chi^2 = 5.75$
SOURCE: Table 3.1

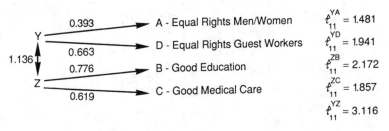

Figure 3.5. Log-Linear Effects ($\hat{\lambda}$) for the Two Latent Variable Model Applied to the Data on Government's Responsibilities in Table 3.1

As can be seen from the effects in Figure 3.5, B and C are strongly related to Z. D is also a good indicator of Y, but the relation between Y and A is much weaker.

The most interesting relation is between Y and Z represented by $\hat{\pi}_{rs}^{YZ}$ in Table 3.3 and estimated as $\hat{\tau}_{11}^{YZ} = 3.116$ (Figure 3.5). This is a very strong relationship, much stronger than the relations between the indicators of Y and the indicators of Z. If from Table 3.1 the marginal tables AB, AC, BD, CD are set up and the two-variable effect for each marginal table is calculated, the relation B-D appears to be the strongest: $\hat{\tau}_{11}^{BD} = 1.267$. The relation between the opinions of people with regard to government's responsibilities in "material" matters and those with regard to government's responsibilities in "ideal" matters is about 2.5 times stronger on the latent level than on the manifest level. The association between the manifest variables is corrected for measurement error, analogous to the standard "correction for attenuation" of the classical test theory.

The very strong association between the latent variables is mainly caused by the very small proportion of people who are in latent class 2 ($YZ = 12$): There are almost no individuals who are of the opinion that the government has a great responsibility with regard to "ideal" matters but has to adopt a more reserved attitude with respect to "material" matters. Formulated in terms of a cumulative (Guttman) scale: "Ideal" matters are more difficult than "material" matters.

It is possible to set up a model in which the small latent class 2 is actually empty. A and D are then the indicators of the latent variable Y, and B and C of the latent variable Z, where the latent variables Y and Z form a perfect Guttman scale. The analysis carried out before (Table 3.3) has to be repeated, but leaving out the second latent class $YZ = 12$: $\pi_{12}^{YZ} = 0$, and leaving out of course all conditional probabilities and restrictions referring to this class.[7] In terms of the log-linear parametrization in equation 3.15, this restriction means carrying out the usual analysis but setting all estimates $F_{i\,j\,kl\,12}^{ABCDYZ}$ to zero.

Not surprisingly, the test results are favorable: $L^2 = 5.75$, df $= 5$, $p = .33$ (Pearson-$\chi^2 = 5.80$). The parameter estimates are practically the same as those presented in Table 3.3, except of course the estimates referring to latent class 2, which have been deleted. So the final conclusion of the analysis is that there are three groups of people. Those who think that government bears an essential responsibility with regard to "ideal" and "material" matters (latent class 1 in Table 3.3); those who are of the opinion that the government has responsibilities in "material" but not in "ideal" matters (latent class 3 in Table 3.3); and those who are of the opinion that government has nothing to do with either "ideal" or "material" matters (latent class in Table 3.3).

In a very special way, the two latent variable analysis has been reduced to a one latent variable analysis, showing the flexibility and versatility of latent class analysis (without implying, of course, that all two or more latent variable models can be reduced to one latent variable models).

Extensions to more than two latent variables follow easily from the above principles. The one latent variable X of the basic model is considered as the joint variable of the three or more latent variables and restrictions are made on the conditional response probabilities to ensure that the responses on certain manifest variables are influenced only by those latent variables of which they are the indicators. This procedure has only one disadvantage: no restrictions are imposed on the relations between the latent variables. In fact, the saturated model is always assumed to describe the relations between the latent variables. How to set up unsaturated models is the topic of the next section.

3.4.2 Unsaturated and Modified Path Analysis Models for Latent Variables: A Modified LISREL Approach

How to impose equality restrictions on the π_t^X parameters in Goodman's variant of the EM-algorithm has been dealt with in Section 3.2.2. However, when the probability distribution of the one latent variable X actually represents a multivariate distribution of two or more latent variables, for example, W, Y, and Z, these equality constraints are of little use to express that the multivariate distribution satisfies a particular unsaturated log-linear model, for example, the no three-variable interaction model $\{WY, WZ, YZ\}$.

In general terms, the EM-algorithm can easily be adapted to handle unsaturated models for the relations between latent variables. In the three latent variable case, it essentially amounts to replacing the (estimated) sufficient statistic $\hat{f}_{qrs}^{WYZ}$, by, for example, the (estimated) sufficient statistics $\hat{f}_{qr}^{WY}, \hat{f}_{qs}^{WZ}, \hat{f}_{rs}^{YZ}$.

In Goodman's variant of the EM-algorithm, this may be achieved as follows. At the end of each M-step, at the same place where, if necessary, the equality restrictions are imposed on the parameters, the current estimates $\hat{\pi}_t^X \equiv \hat{\pi}_{qrs}^{WYZ}$ obtained by means of equation 3.6, denoted as $\hat{\pi}_{qrs}^{WYZ}(0)$, are used to improve by means of iterative proportional fitting the best estimates of $\hat{\pi}_{qrs}^{WYZ}$ obtained so far that correspond with model $\{WY, WZ, YZ\}$. That is, our best estimates of $\hat{\pi}_{qrs}^{WYZ}$ obtained so far that correspond with model $\{WY, WZ, YZ\}$ are iteratively adjusted to the relevant marginals of the current estimates $\hat{\pi}_{qrs}^{WYZ}(0)$, namely, $\hat{\pi}_{qr+}^{WYZ}(0)$, $\hat{\pi}_{q+s}^{WYZ}(0)$, and $\hat{\pi}_{+rs}^{WYZ}(0)$. The resulting estimates $\hat{\pi}_{qrs}^{WYZ}(1)$ are used in the next E-step; and so on. What is done is very much like imposing equality restrictions in the manner Goodman described, the main difference being that imposing unsaturated hierarchical models on the probability distribution of the latent variables is in itself an iterative (proportional fitting) procedure.

In the LCAG program, Goodman's algorithm has been implemented with the additional possibility of defining unsaturated hierarchical log-linear models for the relations among latent variables, where the postulated unsaturated model may take the form of the modified path model explained in Section 2.7.3 (Hagenaars, 1988a; Hagenaars et al., 1980; Hagenaars and Luijkx, 1987).

If the causal structure among the latent variables W, Y, and Z is such that Z is causally dependent on W and Y (without the three-variable interaction WYZ) and W and Y themselves are statistically independent of each other, a modified path analysis approach is needed. The model required for the latent marginal table WY is $\{W, Y\}$, and model $\{WY, WZ, YZ\}$ applies for the complete latent table WYZ. The procedure to impose unsaturated models described above has to be applied to the probabilities of each of the separate (marginal) tables at the end of the M-step, after which the newly estimated probabilities corresponding to each of the separate submodels and tables have to be combined into a new estimate of $\pi *^{WYZ}_{qrs}$, satisfying the complete modified path model (Section 2.7.3, equations 2.42, 2.43). This improved estimate of $\pi *^{WYZ}_{qrs}$ is used in the next E-step; and so on.

The result of this extension of Goodman's algorithm is that, using a program such as LCAG, the relations among categorical variables can be investigated by means of modified path models with latent variables in much the same vein as LISREL deals with the relations among continuous variables. In the remainder of this chapter and in the next chapters, the extreme usefulness and versatility of this feature will be shown. Only a very simple example will be presented in this Section.

According to the outcomes presented in Figure 3.5, the latent variables Y (responsibility of the government toward "ideal" matters) and Z (toward "material" matters) are very strongly related to each other: $\hat{\lambda}^{YZ}_{11} = 1.136$. Let us, nevertheless, by way of example test the hypothesis that these latent variables are independent of each other and are, in terms of factor analysis, orthogonal of each other. Independence of Y and Z implies that in equation 3.15 $\tau^{YZ}_{11} = 1$. Because the estimates for the independence model can be rendered in "closed form expressions," independence between Y and Z can be stated as : $\pi^{YZ}_{rs} = \pi^{YZ}_{r+}\pi^{YZ}_{+s}$.

Carrying out the two latent variable model (equation 3.15) again for the data in Table 3.1, but now with the additional restriction that the relation between Y and Z conforms to model $\{Y, Z\}$, results in the following values of the test statistics: $L^2 = 63.09$ and Pearson-$\chi^2 = 59.61$. The number of degrees of freedom at first appears to be five: Because of the restriction $\tau^{YZ}_{rs} = 1$, one degree of freedom has been gained compared with the model with correlated dichotomous latent variables.

However, the model is not identifiable. Different initial estimates yield different parameter estimates, all resulting in the same value of L^2 and Pearson-χ^2. This is a striking example of what has been said before: Introducing extra restrictions can render an identifiable model unidentifiable.

Additional restrictions, for example, the assumption that the "error rates" are the same for both latent classes (Section 3.2.3) can make it identifiable again. The conditional probabilities of a corresponding, "correct" answer are assumed to be the same for all latent classes:

$$\pi^{\overline{A}YZ}_{11s} = \pi^{\overline{A}YZ}_{22s} \qquad \pi^{\overline{D}YZ}_{11s} = \pi^{\overline{D}YZ}_{22s}$$

$$\pi^{\overline{B}YZ}_{1r1} = \pi^{\overline{B}YZ}_{2r2} \qquad \pi^{\overline{C}YZ}_{1r1} = \pi^{\overline{C}YZ}_{2r2} \qquad (3.17)$$

Imposing these additional restrictions, along with independence between Y and Z, gives the same values of the test statistics ($L^2 = 63.09$), but now the model is identifiable with df = 9. Not surprisingly, this model has to be rejected and the conclusion is that $\hat{\lambda}_{11}^{YZ}$ in Figure 3.5 is statistically significant.

3.5 Local Dependence Models: Direct Effects between Indicators

Basic to the latent class model is the assumption that the scores on the manifest variables, the indicators, are systematically dependent only on the latent positions the respondents have and that consequently the associations among the manifest variables are completely explained by the latent variable(s). But there may be circumstances in which this assumption of local independence is not realistic. Omitted variables and correlated response error due to interviewer effects, test-retest effects, response consistency effects, social desirability, "yea/nay-saying" tendencies, etc. (Bradburn, 1983; Turner & Martin, 1984, Vol. 1, Part III) may result in associations among the manifest variables which cannot be attributed to the influence of the underlying latent variable(s). The latent variable(s) can explain only part of the total associations among the indicators, but even holding the latent variable(s) constant, some association among the indicators will be left.

In much the same vein as correlations among the error terms can be introduced in LISREL models, direct relations among the manifest variables can be taken into account in latent class models, thus relaxing, or rather circumventing, the local independence assumptions (Hagenaars, 1988b). Because within the framework of this book correlated response errors in the form of reinterview or test-retest effects stand to the fore, local dependence models will be extensively dealt with in the second chapter on Panel Analysis, Section 5.5.2. Here, only the basic approach to modeling direct effects among indicators will be given, using Figure 3.6.

Figure 3.6a represents a basic latent class model with one latent variable X and four indicators A through D in which there exists a direct effect between A and B. As it stands, this model cannot be dealt with by the standard latent class approach. But when reformulated as in Figure 3.6b, it is a "standard" latent class model and can be handled by Goodman's algorithm using all the features discussed in previous sections.

a)   b)

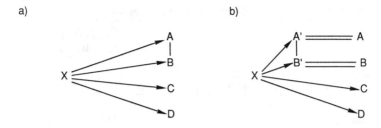

Figure 3.6. A Basic Latent Class Model with Direct Effects Between Indicators

There are three latent variables X, A', and B' and four indicators A through D in Figure 3.6b. The conditional response probabilities have to be constrained in such a manner that A depends only on A', B only on B', and C and D only on X (see Section 3.4.1, equation 3.16). Moreover, the conditional response probabilities concerning A and B are either zero or one, because A' and B' are quasi-latent variables (see Section 3.3, equation 3.13). Finally, the estimated probabilities $\hat{\pi}_{t\,i'\,j'}^{XA'B'}$ have to be brought into agreement with model $\{XA', XB', A'B'\}$. With all these restrictions the model in Figure 3.6b is identical to the model in Figure 3.6a and can be used to estimate the strength of the direct effect $A - B$, as well as all other effects in Figure 3.6a.

3.6 Simultaneous Analyses in Several Groups

The data on government's responsibilities in Table 3.1 were taken from the Political Action Study in which, beside the Netherlands, seven other countries participated. In four countries, the Netherlands, Austria, Germany, and Switzerland, exactly the same question was used to determine government's responsibilities in "ideal" and "material" matters. In the other countries the indicator D referred to equal rights for racial minorities.

The two latent variable analysis that was carried out for the Netherlands and reported in Section 3.4.1 was also conducted for Austria, Germany, and Switzerland. Comparative analyses among the four countries revealed that the same Guttman pattern for the latent variables was found in all four countries, that the proportions of supporters and opponents of an ample task for government regarding "ideal" and "material" topics did not differ significantly between the four countries, but the

TABLE 3.4   Responsibility of Government Toward Issues

| A. Equal Rights Men/Women | B. Good Education | C. Good Medical Care | D. Equal Rights Guest Workers | | |
			1. Yes	2. No	Total
Germany					
1. Yes	1. Yes	1. Yes	416	123	539
		2. No	92	26	118
	2. No	1. Yes	133	69	202
		2. No	159	85	244
2. No	1. Yes	1. Yes	52	46	98
		2. No	18	24	42
	2. No	1. Yes	27	32	59
		2. No	54	69	123
		Total	951	474	1425
Switzerland					
1. Yes	1. Yes	1. Yes	119	28	147
		2. No	56	25	81
	2. No	1. Yes	68	26	94
		2. No	84	54	138
2. No	1. Yes	1. Yes	18	15	33
		2. No	18	9	27
	2. No	1. Yes	17	6	23
		2. No	32	48	80
		Total	412	211	623

NOTE: For items A and D, the category "1. Yes" includes the answers "an essential responsibility" and "an important responsibility;" the category "2. No" all other alternatives. For the items B and C the category "1. Yes" includes the answer "an essential responsibility," and the category "2. No" all other alternatives. See also the Note under Table 3.1.
SOURCE: Political Action 1974 - Germany and Switzerland; see Barnes and Kaase (1979)

relations between the latent and the manifest variables varied among them (Hagenaars, 1985, Section 3.5).

How such comparative analyses have to be carried out and what kinds of difficulties have to be overcome will be illustrated using the data in Table 3.4. Comparative analyses are in themselves already rather complex. Therefore, the following exposition will be simplified by restricting the analysis to two countries, Germany and Switzerland. Moreover, the structural empty cells caused by the Guttman property were avoided

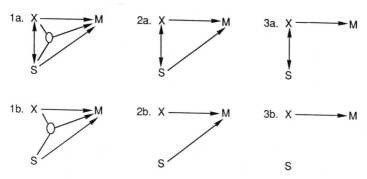

Figure 3.7. Relations among Latent Variables (*X*), Manifest Variables (*M*), and Stratifying Factor (*S*)

by dichotomizing the indicators differently, as mentioned under Table 3.4, making them "equally difficult."

Clogg and Goodman (1984, 1985, 1986) discussed the principles of carrying out simultaneous latent class analyses for several groups. Hagenaars (1984, 1988a) advocated a very similar but somewhat more general approach which will be explained below. In this exposition use will be made of Figure 3.7, in which *X* represents the set of latent variables (here: *Y* and *Z*), *M* the set of manifest variables, that is, the indicators of *X* (here: *A* through *D*), and *S* the set of stratifying factors or group variables (here: the dichotomous variable "Country").

An obvious start for the comparative analysis between Germany and Switzerland is to carry out the two latent variable analysis for each country separately. From the test outcomes presented under Table 3.5 it can be concluded that the two latent variable model explains the data well in Germany and in Switzerland. A test that the two-variable model holds simultaneously for both countries can simply be obtained by summing the L^2 values and the df for the separate tests. The simultaneous test confirms the expectation that in both countries the two latent variable model is valid.

This kind of comparative analysis is depicted in Figure 3.7(1a). In this figure, there is a direct relation between the latent variable *X* and the manifest variable *M*. Moreover, the stratifier *S* is directly related to *X* and *M*, and, because of the three-variable interaction *SXM*, the relationship between *X* and *M* varies among the categories of *S*. In terms of our example, it is only assumed that in both countries the specified two latent

TABLE 3.5 Two Latent Variable Model For Data On Government's Responsibilities: Germany and Switzerland

Latent Class			$\hat{\pi}^{YZ}_{rs}$	A. Equal Rights Men/Women $\hat{\pi}^{AYZ}_{ijrs}$		B. Good Education $\hat{\pi}^{BYZ}_{jrs}$		C. Good Medical Care $\hat{\pi}^{CYZ}_{krs}$		D. Equal Rights Guest Workers $\hat{\pi}^{DYZ}_{lrs}$	
X_t	Y_r	Z_s		1. Yes	2. No	1. Yes	2. No	1. Yes	2. No	1. Yes	2. No
Germany											
1	1	1	.469	.943	.057	.839	.161	.879	.121	.819	.181
2	1	2	.167	.943	.057	.180	.820	.292	.708	.819	.181
3	2	1	.107	.480	.520	.839	.161	.879	.121	.402	.598
4	2	2	.257	.480	.520	.180	.820	.292	.708	.402	.598
			$\hat{\lambda}^{YZ}_{11} = 0.477$	$\hat{\lambda}^{AY}_{11} = 0.720$		$\hat{\lambda}^{BZ}_{11} = 0.791$		$\hat{\lambda}^{CZ}_{11} = 0.718$		$\hat{\lambda}^{DY}_{11} = 0.487$	
Switzerland											
1	1	1	.628	.849	.151	.607	.393	.674	.326	.797	.203
2	1	2	.081	.849	.151	.115	.885	.002	.998	.797	.203
3	2	1	.079	.468	.532	.607	.393	.674	.326	.332	.668
4	2	2	.212	.468	.532	.115	.885	.002	.998	.332	.668
			$\hat{\lambda}^{YZ}_{11} = 0.758$	$\hat{\lambda}^{AY}_{11} = 0.464$		$\hat{\lambda}^{BZ}_{11} = 0.618$		$\hat{\lambda}^{CZ}_{11} = 1.791$		$\hat{\lambda}^{DY}_{11} = 0.517$	

Germany: $L^2 = 6.46$, df = 4, $p = .17$, Pearson-$\chi^2 = 6.44$
Switzerland: $L^2 = 7.21$, df = 4, $p = .13$, Pearson-$\chi^2 = 7.18$
Both countries simultaneously: $L^2 = 13.67$, df = 8, $p = .09$
SOURCE: Table 3.4

variable model applies without implying that some or all of the parameters are equal for the two countries. It is postulated that, in each country, model $\{YZ, AY, DY, BZ, CZ\}$ is valid without additional restrictions. If the dichotomous variable Country is indicated by S, carrying out separate analyses for each country is similar to applying model $\{SYZ, SAY, SDY, SBZ, SCZ\}$ to the whole data set in Table 3.4.

Because the two latent variable model is valid for both countries, it might be interesting to see whether or not there are more proponents of an ample task for government in Germany than in Switzerland. From Table 3.5 it can be learned that in Switzerland about 70% is of the opinion that government bears responsibility for "ideal" matters and about the same percentage feels the same about "material" matters ($\hat{\pi}_{1+}^{YZ} = .709$, $\hat{\pi}_{+1}^{YZ} = .707$). In Germany the corresponding percentages of supporters of an ample task for government are somewhat lower and both about 60% ($\hat{\pi}_{1+}^{YZ} = .636$, $\hat{\pi}_{+1}^{YZ} = .576$). The question arises whether these differences are significant.

Put more generally: What is the relationship between the latent variables Y and Z, and the country variable S. As was mentioned before, in the separate analyses the saturated model $\{SYZ\}$ is postulated for the relations among S, Y and Z. Several more parsimonious models might be defined, for example, the no three-variable interaction model $\{YZ, SY, SZ\}$ or even $\{YZ, S\}$. This latter model implies that the relative distribution of the joint latent variable YZ is exactly the same in Germany and Switzerland. How to fit such more parsimonious models by means of the (extended) Goodman algorithm will be thoroughly explained because, once this is understood, testing and fitting other kinds of models relevant for cross-cultural (and cross-temporal) research follow easily.

The starting point is the five dimensional Table $SABCD$, Table 3.4, where $S = 1$ refers to Germany and $S = 2$ to Switzerland, and A through D have obvious meanings. The latent variable X has eight latent classes formed by the $2 \times 2 \times 2$ categories of the joint latent variable YZS'. S' is a quasi-latent variable, and perfectly related to S:

$$\pi_{mrsm'}^{\bar{S}\,YZS'} = \begin{cases} 1 & \text{if } m = m' \\ 0 & \text{if } m \neq m' \end{cases} \tag{3.18}$$

Furthermore, we need restrictions to ensure that in each country A and D depend only on Y, and B and C only on Z:

$$\pi_{i\,r\,11}^{\overline{AYZS'}} = \pi_{i\,r\,21}^{\overline{AYZS'}} \qquad \pi_{i\,r\,12}^{\overline{AYZS'}} = \pi_{i\,r\,22}^{\overline{AYZS'}}$$

$$\pi_{j\,1s\,1}^{\overline{BYZS'}} = \pi_{j\,2s\,1}^{\overline{BYZS'}} \qquad \pi_{j\,1s\,2}^{\overline{BYZS'}} = \pi_{j\,2s\,2}^{\overline{BYZS'}} \qquad\qquad (3.19)$$

$$\pi_{k\,1s\,1}^{\overline{CYZS'}} = \pi_{k\,2s\,1}^{\overline{CYZS'}} \qquad \pi_{k\,1s\,2}^{\overline{CYZS'}} = \pi_{k\,2s\,2}^{\overline{CYZS'}}$$

$$\pi_{l\,r\,11}^{\overline{DYZS'}} = \pi_{l\,r\,21}^{\overline{DYZS'}} \qquad \pi_{l\,r\,12}^{\overline{DYZS'}} = \pi_{l\,r\,22}^{\overline{DYZS'}}$$

Without additional constraints we now have model $\{SYZ, SYA, SYD, SZB,$ $SZC\}$ resulting in the outcomes of Table 3.5. Using the LCAG program, it is possible to put constraints on the parameters $\pi_{rsm'}^{YZS'}$, for example, the constraints implied by the unsaturated model $\{YZ, S'Y, S'Z\}$. Together with the restrictions in equations 3.18, 3.19, this extra restriction leads to a modified path model in which model $\{YZ, SY, SZ\}$ is postulated for the marginal table SYZ and for the complete table $SYZABCD$, model $\{SYZ, SYA, SYD, SZB, SZC\}$.

The most extreme model in this respect is model $\{S, YZ\}$ in which S is made completely independent of YZ in the marginal table SYZ. In Figure 3.7, this model is represented as 1b. In terms of the LCAG program it means restricting the parameters $\pi_{rsm'}^{YZS'}$ to correspond with model $\{S', YZ\}$. Somewhat surprisingly, this model fits the data in Table 3.4: $L^2 = 17.23$, df $= 11$, $p = .10$ (Pearson-$\chi^2 = 16.78$). Compared with the model in Table 3.5, the fit is not significantly worse: $L^2_{r/u} = 17.23 - 13.67 = 3.56$, $df_{r/u} = 11 - 8 = 3$, $p = .31$. The differences between Germany and Switzerland with respect to the (bivariate) distribution of the latent variables "responsibilities of the government" can be attributed to sampling fluctuations and need no profound explanation in terms of differences in political culture, that is, if the latent variables have the same meaning in Germany and Switzerland.

In latent structure models the meaning of a latent variable has to a large extent to be derived from the relationships the latent variable has with the observed variables. Although the general pattern of the outcomes in Table 3.5 is the same for Germany and Switzerland, there are some conspicuous differences, for example, with regard to item C. Conclusions to the effect that the (joint) distribution of the latent variables is the same in both countries can be more easily justified if they follow from a model in which the relations between the latent and the manifest variables is the same in both countries.

Ideally, we like to compare models such as 3a and 3b in Figure 3.7 in which S is related to M only through X. In terms of our example, the model in Figure 3.7(3a) represents model {*SYZ*, *YA*, *YD*, *ZB*, *ZC*}. Its parameters can be estimated by means of Goodman's algorithm. As before, eight latent classes are postulated, consisting of the categories of the (quasi-)latent variables *YZS'* and the restrictions in equations 3.18, 3.19 are applied. Moreover, because *S* is not directly related to the indicators *A* through *D*, the following extra constraints hold:

$$\pi_{irs1}^{\overline{A}YZS'} = \pi_{irs2}^{\overline{A}YZS'} \qquad \pi_{jrs1}^{\overline{B}YZS'} = \pi_{jrs2}^{\overline{B}YZS'}$$

$$\pi_{krs1}^{\overline{C}YZS'} = \pi_{krs2}^{\overline{C}YZS'} \qquad \pi_{lrs1}^{\overline{D}YZS'} = \pi_{lrs2}^{\overline{D}YZS'} \tag{3.20}$$

Model {*SYZ*, *YA*, *YD*, *ZB*, *ZC*} does not fit the data in Table 3.4: $L^2 = 35.30$, df = 16, $p = .004$, Pearson-$\chi^2 = 33.25$. The relations between *S* and the manifest variables are not completely mediated by the latent variables. However, that does not necessarily mean that the relations between the manifest and the latent variables in Germany are different from Switzerland, making it difficult to compare the latent distributions for both countries. It may be the case that for some historical and political reasons the government in one country traditionally occupies itself with educational or medical affairs. This could lead to a relatively large proportion of respondents in this country who are of the (manifest) opinion that education and medical care are essential or important tasks of government, without making the relations between the manifest and the latent variables different for both countries.

A possible model along these lines is depicted in Figure 3.7(2a, 2b). In terms of our example, model 2a is the log-linear model {*SYZ*, *YA*, *YD*, *ZB*, *ZC*, *SA*, *SB*, *SC*, *SD*}. LCAG can be used to obtain the maximum likelihood estimates of this model although with some difficulties because the restrictions implied cannot be rendered in the form of equalities between (conditional) probabilities. There will be seven latent variables: *Y* and *Z* plus five quasi-latent variables *S'*, *A'*, *B'*, *C'*, and *D'*. The relations between the quasi-latent variables and their manifest counterparts are made perfect by setting the relevant conditional probabilities to zero or one, in the usual manner described above. The probabilities $\pi_{rsm'i'j'k'l'}^{YZS'A'B'C'D'}$ are restricted to conform to model {*S'YZ*, *YA'*, *YD'*, *ZB'*, *ZC'*, *S'A'*, *S'B'*, *S'C'*, *S'D'*}. This model fits the data rather well: $L^2 = 18.56$, df = 12, $p = .10$, and not significantly worse than model {*SYZ*, *SYA*, *SYD*, *SZB*, *SZC*} in Table 3.5: $L_{r/u}^2 = 18.56 - 13.67 = 4.89$, df$_{r/u} = 12 - 8 = 4$, $p = .30$.

It may be concluded that the relations between the indicators *A* through *D* and the latent variables *Y* and *Z* are the same in Germany and

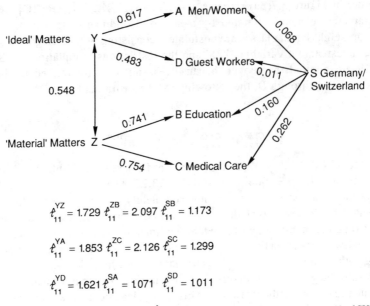

$t_{11}^{YZ} = 1.729 \ t_{11}^{ZB} = 2.097 \ t_{11}^{SB} = 1.173$

$t_{11}^{YA} = 1.853 \ t_{11}^{ZC} = 2.126 \ t_{11}^{SC} = 1.299$

$t_{11}^{YD} = 1.621 \ t_{11}^{SA} = 1.071 \ t_{11}^{SD} = 1.011$

Figure 3.8. Log-Linear Effects ($\hat{\lambda}$) for the Modified Path Model $\{YZ, S\}$, $\{SYZ,$ $YA, YD, ZB, ZC, SA, SB, SC, SD\}$; Data on Government's Responsibilities from Table 3.4

Switzerland. But the model does not impose restrictions on the distribution of the latent variables Y and Z for both countries. We could formulate more parsimonious (modified path) models in which the complete table $S'YZA'B'C'D'$ (or equivalently $SYZABCD$) is in agreement with model $\{S'YZ, YA', YD', ZB', ZC', S'A', S'B', S'C', S'D'\}$ (or $\{SYZ, YA, YD, ZB, ZC, SA, SB, SC, SD\}$), but the marginal table $S'YZ$ (or SYZ) satisfies more restricted models than the saturated one. The least restrictive is the no three-factor interaction model $\{YZ, S'Y, S'Z\}$ (or $\{YZ, SY, SZ\}$); then models $\{YZ, S'Z\}$ and $\{YZ, S'Y\}$ and, finally, model $\{YZ, S'\}$ in which the joint distribution of Y and Z is the same in both countries (see Figure 3.7[2b]).

All four of these more restricted models fit the data, yielding significance levels of about $p = .10$. However, conditional L^2 tests, BIC, and Bonett and Bentler's $\hat{\delta}$ (see Section 2.5) all point to preferring the most restricted model for the marginal table $S'YZ$, namely, $\{YZ, S'\}$. Moreover the parameter estimates for the most restricted model and the second best model $\{YZ, S'Z\}$ appeared to be almost the same and the extra parameter $\hat{\lambda}_{11}^{SZ}$ negligibly small (−.11).

As shown in Figure 3.8, according to this most restricted model, the relations between the latent and the manifest variables, and among the latent variables themselves, follow the pattern already presented in Table 3.5, albeit that it is now concluded that all these relations and the joint distribution of the latent variables are the same in Germany and Switzerland. The effects of S on the "ideal" indicators A and D are almost zero, and the terms SA and SD can probably be deleted from the model. The effect of S on B is somewhat stronger; the influence on C is the strongest, though still not very large. People in Germany are somewhat more inclined to say that education and medical care are responsibilities of the government than people in Switzerland.

Cross-cultural research is complicated and full of pitfalls. Achieving comparability of indices is a vexing problem. The models proposed in Figure 3.7 do not solve all difficulties, but they can help the investigator to evaluate better the comparability of indices and to detect sources of incomparability.

3.7 Causal Models With Latent Variables

The latent class model is a measurement model and, as such, seldom an objective in itself. The results of a latent class analysis are to be used in connection with the other concepts and variables of a particular investigation. The principles of how to carry out these extended analyses have been presented in previous sections. In this section, these principles will be applied to investigate an elaborate modified path model with a latent variable, using the data in Table 3.6.

The Column Total of this table shows for Germany the multivariate frequency distribution of the by now familiar indicators of government responsibilities A through D. In contrast to Table 3.4, Germany, the indicators in Table 3.6 are dichotomized according to whether or not the item is considered to be an essential responsibility of government. This is the same cutting point that was used in Table 3.1 for the Netherlands and, consequently, the same latent Guttman scale is to be expected.

Applying to the Column Total the same two latent variable model as for the Netherlands, the latent class in which government is supposed to be responsible for "ideal" matters, but not for "material" matters appeared to be almost empty: $\hat{\pi}_{12}^{YZ} = .018$. A two latent variable model in

TABLE 3.6 Religiousness, Income, Party Preference and Responsibilities of Government Toward Issues: Germany[a]

A	B	C	D	1. Rel. / 1. Low / 1. L-W	1. Rel. / 1. Low / 2. R-W	1. Rel. / 2. High / 1. L-W	1. Rel. / 2. High / 2. R-W	2. N-R / 1. Low / 1. L-W	2. N-R / 1. Low / 2. R-W	2. N-R / 2. High / 1. L-W	2. N-R / 2. High / 2. R-W	Total
1	1	1	1	6	15	5	14	18	7	13	16	94
1	1	1	2	5	21	7	17	22	13	25	18	128
1	1	2	1	2	0	1	2	0	1	4	2	12
1	1	2	2	0	5	1	2	3	1	3	5	20
1	2	1	1	2	1	2	1	1	2	2	0	7
1	2	1	2	0	8	1	4	4	0	7	5	34
1	2	2	1	3	2	3	2	0	1	1	0	6
1	2	2	2	2	3	0	6	8	8	8	6	38
2	1	1	1	15	9	11	4	16	8	15	10	64
2	1	1	2	1	59	0	32	53	41	33	39	283
2	1	2	1	4	3	0	5	1	0	4	2	16
2	1	2	2	4	12	2	20	15	13	19	12	97
2	2	1	1	10	4	0	1	4	2	2	3	20
2	2	1	2	0	33	6	17	27	22	23	25	163
2	2	2	1	10	0	1	2	2	1	2	2	10
2	2	2	2	10	44	14	31	44	26	39	41	249
			Total	65	219	54	160	218	139	200	186	1241

[a] R. Religiousness
- 1. Rel. = very or somewhat religious
- 2. N-R = a little or not very religious

I. Monthly Family Income
- 1. Low = <1,500 DM
- 2. High = ≥1,500 DM

P. Party Preference
- 1. L-W = left wing party (SPD, DKP)
- 2. R-W = center, right wing party (CDU/CSU, FDP)

A. Guaranteeing equal rights for men and women (see A)
- 1 = an essential responsibility of government
- 2 = not an essential responsibility of government

B. Providing good education (see A)
C. Providing good medical care (see A)
D. Providing equal rights for guest (foreign) workers (see A)
NOTE: See note for Table 3.1.

Barnes and Kaase (1979)

TABLE 3.7 Two Latent Variable Model For Data On Government's Responsibilities: Germany

Latent Class				A. Equal Rights Men/Women $\hat{\pi}_{lrs}^{AYZ}$		B. Good Education $\hat{\pi}_{jrs}^{BYZ}$		C. Good Medical Care $\hat{\pi}_{krs}^{CYZ}$		D. Equal Rights Guest Workers $\hat{\pi}_{lrs}^{DYZ}$	
X_t	Y_r	Z_s	$\hat{\pi}_{rs}^{YZ}$	1. Yes	2. No	1. Yes	2. No	1. Yes	2. No	1. Yes	2. No
1	1	1	.288	.622	.378	.914	.086	.886	.114	.498	.502
—	1	2	0	—	—	—	—	—	—	—	—
2	2	1	.231	.132	.868	.914	.086	.886	.114	.057	.943
3	2	2	.481	.132	.868	.209	.791	.372	.628	.057	.942

$L^2 = 9.14$, df = 5, $p = .10$, Pearson-$\chi^2 = 9.31$
NOTE: The estimated probability of belonging to latent class YZ = 12 has been fixed to zero.
SOURCE: Table 3.6, Column Total

137

which this class has been fixed to zero yielded a reasonable fit, as reported under Table 3.7.

As can be seen in Table 3.7, almost 50% of the respondents in Germany feel government has no essential responsibilities with regard to "ideal" or "material" matters ($X = 3$); almost 30% thinks that government has an essential task in both areas ($X = 1$); and 23% is of the opinion that government has an essential task with regard to "material" but not to "ideal" matters ($X = 2$).

Given these results and the fact that political parties vary in their opinion about the responsibilities of government, it may be interesting to investigate whether the different perspectives people have on the government task leads to different political party preferences. Moreover, one may want to know how this outlook on government's task is rooted in certain background characteristics, and perhaps can help to interpret theoretically the relations between background characteristics and party preferences.

The relevant data to answer these questions for Germany are presented in Table 3.6, taking religiousness and income as background characteristics and making a distinction between a left wing and a center/right wing party preference. As discussed in Section 3.3, there are essentially two ways in which the relations between the latent and the external variables can be investigated: Assigning each individual a latent class score or extending the latent class model by incorporating the external variables. The latter possibility is to be preferred and will be discussed first. The (extended) model ultimately selected is presented in Figure 3.9.

The first step to arrive at the model in Figure 3.9 was to verify whether the latent variable $X (\equiv YZ)$ is the underlying fundamental variable, not only in the sense of explaining the relations among the indicators A through D themselves, but also in the sense of explaining the relations A through D have with the external variables religiousness, income, and party preference. The empirical validity of model $\{RIPX, XA, XB, XC, XD\}$ has to be investigated when the relations of X with A through D are restricted, as in Table 3.7, to conform to the latent Guttman model. The parameters of this model can be estimated by means of the Goodman algorithm. Therefore, the two latent variable analysis similar to the one reported in Table 3.7 has to be carried out, but now with one additional "indicator" of X, the joint variable RIP having $2 \times 2 \times 2 = 8$ categories.

Model $\{RIPX, XA, XB, XC, XD\}$ fits the data in Table 3.6 excellently as appears from the test results mentioned under Figure 3.9 ($p = .30$). It

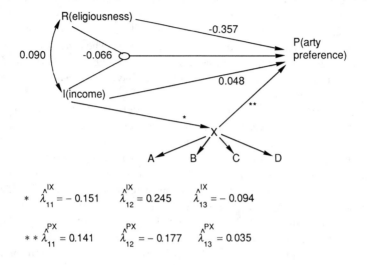

$$* \quad \hat{\lambda}^{IX}_{11} = -0.151 \qquad \hat{\lambda}^{IX}_{12} = 0.245 \qquad \hat{\lambda}^{IX}_{13} = -0.094$$

$$** \quad \hat{\lambda}^{PX}_{11} = 0.141 \qquad \hat{\lambda}^{PX}_{12} = -0.177 \qquad \hat{\lambda}^{PX}_{13} = 0.035$$

τ - parameters:

$$\hat{t}^{IR}_{11} = 1.094 \qquad \hat{t}^{IX}_{11} = 0.860 \qquad \hat{t}^{IX}_{12} = 1.278 \qquad \hat{t}^{IX}_{13} = 0.910$$

$$\hat{t}^{RP}_{11} = 0.700 \qquad \hat{t}^{IP}_{11} = 1.049 \qquad \hat{t}^{RIP}_{111} = 0.936$$

$$\hat{t}^{PX}_{11} = 1.152 \qquad \hat{t}^{PX}_{12} = 0.838 \qquad \hat{t}^{PX}_{13} = 1.036$$

Test Statistics:

Above model: {RI, IX}{RIX, RIP, XP, XA, XB, XC, XD}: $L^2 = 104.39$ df=106 p=.53
Pearson-$\chi^2 = 100.63$

Other models: {RIPX, XA, XB, XC, XD}: $L^2 = 102.88$ df= 96 p=.30
Pearson-$\chi^2 = 99.51$

{RI,IX}{RIX,RP, XP, XA,XB,XC,XD}: $L^2 = 113.17$ df=108 p=.35
Pearson-$\chi^2 = 109.17$

Figure 3.9. Causal Model with Latent Variable; Data from Table 3.6

is reassuring that, despite the many cells with very small frequencies, L^2 and Pearson-χ^2 have about the same value. The relations between X and the indicators A through D for this extended model are the same as obtained before (the largest difference being .013). The other parameter estimates, those that represent the relations among the latent variable X, religiousness, income, and party preference, have been used to find a more parsimonious model.

In searching for a more parsimonious model, the theoretical causal order of the variables was taken into account: Religiousness and income were considered to be causally prior to the latent variable X (to the latent variable Y and Z); R, I, and X were regarded as causes of party preference. Making use of the extra features of LCAG, the Goodman algorithm can be employed to evaluate the relevant, more restricted, models. Beside the latent variable X ($\equiv YZ$), three additional (quasi-)latent variables R', I', and P' are introduced. Not counting the (structurally empty) latent classes for which $YZ = 12$, there are now $3 \times 2 \times 2 \times 2 = 24$ latent classes. By setting the relevant conditional probabilities to 0 or 1, the conditional response probabilities are constrained in such a way that R' is perfectly related to R (religiousness), I' perfectly to I (income), and P' perfectly to P (party preference). In addition, restrictions on the conditional response probabilities are needed to ensure that the scores on A and D depend only on Y, and the scores on B and C only on Z.

Without imposing restrictions other than the ones above, the outcomes of a latent class analysis applied to the data in Table 3.6 correspond to model $\{RIPX, XA, XB, XC, XD\}$ discussed previously. Stepwise restrictions on the parameters $\pi_{m'n'o't}^{R'I'P'X}$ following the principles of the modified path analysis approach eventually resulted in the model in Figure 3.9. The estimates $\hat{\pi}_{m'n'o't}^{R'I'P'X}$ (or equivalently, $\hat{\pi}_{mnot}^{RIPX}$) for this latter model are such that the probabilities in the marginal (latent) table $R'I'X$ are in agreement with model $\{R'I', I'X\}$, and the probabilities in the complete (latent) table $R'I'XP'$ correspond to model $\{R'I'X, R'I'P, XP'\}$. As mentioned under Figure 3.9, this model also fits the data excellently ($p = .53$) and is not significantly worse than the previous model $\{RIPX, XA, XB, XC, XD\}$.

A competing model was a model in which all relations between income and party preference were left out. However, on the basis of conditional testing, the model in Figure 3.9 has to be preferred. The parameter estimates presented in Figure 3.9 are of course computed taking the causal order of the variables into account. For example, $\hat{\lambda}_{nt}^{IX}$ is computed by means of the marginal table $R'I'X$ holding religiousness constant, and not by means of the complete table $R'I'P'X$, holding religiousness and party preference constant.

The relations between X and the indicators A through D are not presented in Figure 3.9, but they are almost exactly the same as in Table 3.7. According to the $\hat{\lambda}$ effects presented in Figure 3.9, the two background characteristics religiousness and income are only weakly correlated. In the low income group, there are relatively few more persons who are very or somewhat religious than in the high income group.

Religiousness is not directly related to X, in contrast to family income. The lower income category has a relatively strong preference for the position that the government does have an essential responsibility with respect to "material," but not with respect to "ideal" matters $(X = 2)$. The higher income group is divided, in that preference is expressed for the position that the government has an essential responsibility with regard to both "ideal" and "material" matters, but also for the position that the government has no responsibilities at all in these respects.

As to the direct effects on party preference, those are strongest for the variable religiousness; people who are very or somewhat religious have a much stronger preference for a center/right wing party than those who feel less religious. From the three-variable interaction RIP we infer that the relation R-P is somewhat stronger for the lower income group than for the higher income category ($\hat{\lambda}^{RP/I}_{11\,1} = -.423$ and $\hat{\tau}^{RP/I}_{11\,1} = .665$ against $\hat{\lambda}^{RP/I}_{11\,2} = -.291$ and $\hat{\tau}^{RP/I}_{11\,2} = .748$).

The effects of income on party preference are very small. Among those who feel very or somewhat religious, the effect I-P is absent ($\hat{\lambda}^{IP/R}_{11\,1} = -.018$, $\hat{\tau}^{IP/R}_{11\,1} = .982$). The relation is stronger among those who feel less religious: The lower income group expresses a greater preference for the left wing party than the higher income category ($\hat{\lambda}^{IP/R}_{11\,2} = .114$, $\hat{\tau}^{IP/R}_{11\,2} = 1.121$).

It appears from the effects of X on P that those who support greater government responsibility with respect to "material" but not to "ideal" matters $(X = 2)$ have a relatively strong preference for a center or right wing party. The greater preference for the left wing with the latent classes 1 and 3 is strongest with class 1, so among the advocates of government responsibility with regard to both "material" and "ideal" matters.

Thus, latent class 2, the supporters of the traditional task of government with regard to "material" matters, but not to "ideal" matters, consists to a relatively large extent of people who have a low income and a center/right wing party preference. In latent class 1, the advocates of a large task for the government in both "ideal" and "material" matters, there are relatively many persons with a preference for a left wing party and relatively many persons from the higher income class. With respect to party preference and income, latent class 3 takes up an intermediate position though more in the direction of class 1 than of class 2.

Further speculations about the theoretical implications of the above results, for example, in the line of Inglehart's theory on post-materialism, would probably lead to the introduction of additional variables such as educational level and generation and to even larger tables than Table 3.6. These larger tables will probably contain many empty cells making it almost impossible to carry out the necessary extended latent class analyses such as the one above. One might then perhaps be forced to substitute the individual scores $\tilde{X}'$ for the latent variable X.

The consequences of the use of $\tilde{X}'$ for the model in Figure 3.9 have been examined. The point of departure was the latent class analysis for the column Total in Table 3.6, the outcomes of which were presented in Table 3.7. Given their score on the observed variables A through D, people were assigned to the conditional modal latent class t^*. The probability of a misclassification $\hat{E}$ was .20 and $\hat{\lambda}_{X.ABCD} = .62$. By means of these individual latent class scores, the observed table $RIP\tilde{X}'$ was set up.

To make the similarity with the previous (extended latent class) analysis as great as possible, the model in Figure 3.9 was also applied to the data in table $RIPX'$, that is, the modified path model $\{RI, I\tilde{X}'\}$ $\{RI\tilde{X}', RIP, \tilde{X}'P\}$. The general pattern of the outcomes was not different from the results presented in Figure 3.9, albeit that the strengths of the two- and three-variable effects $\hat{\lambda}$ have at least been halved. The marginal distributions of X and $\tilde{X}'$ were rather similar in spite of the expected 20% misclassifications.

It should be explicitly mentioned that the fact that the relevant effects were halved is not caused by a process similar to attenuation of correlation due to unreliability. Using the individual predicted latent scores can likewise inflate the associations. Because these predicted latent scores are needed in case of a large number of external variables, the problems connected with them deserve more attention.

3.8 Ordinal Latent Variables

In Section 2.8, a description was given of how the ordinal character of variables could be taken into account in the log-linear model by applying inequality restrictions on the log-linear effects and by fitting log-bilinear and log-linear association models. These methods extend to log-linear models with latent variables. In principle, the relations among the latent variables themselves and the relations between the latent and the manifest variables may be subjected to ordinal constraints.

Croon (1988), among others, used modified isotonic regression procedures to impose ordinal constraints on the relations between an underlying latent variable and the ordinal polytomous (manifest) indicators of this latent variable analogously to the inequality constraints in equation 2.44. The estimation procedures Goodman developed for the log-bilinear association model can (in principle) also be applied on the latent level. For the log-linear association approach in the case of a model with latent variables, Haberman's program NEWTON can be used.

An example of the latter approach can be given using the data in Table 3.1 on government's responsibilities in the Netherlands. Let us assume that the underlying latent variable X is an interval variable with three categories: -1, 0, 1. The relations between the latent and the manifest variables A through D are such that the odds of scoring Yes rather than No on each manifest item change by a constant factor as X increases by a score. The log-linear two variable effects of model $\{AX, BX, CX, DX\}$ are subjected to linear constraints, similar to the ones in equation 2.46. In terms of the parameters of the corresponding effect model, these constraints are:

$$\gamma_t^{X\overline{A}} = (\gamma^{*A})^t \quad \gamma_t^{X\overline{B}} = (\gamma^{*B})^t \quad \gamma_t^{X\overline{C}} = (\gamma^{*C})^t \quad \gamma_t^{X\overline{D}} = (\gamma^{*D})^t$$
$$\beta_t^{X\overline{A}} = (\beta^{*A})t \quad \beta_t^{X\overline{B}} = (\beta^{*B})t \quad \beta_t^{X\overline{C}} = (\beta^{*C})t \quad \beta_t^{X\overline{D}} = (\beta^{*D})t \tag{3.21}$$

This model does not fit the data in Table 3.1 very well, as appears from the test statistics mentioned under Table 3.8.[8] It definitely does not fit as well as the previous three latent class model where the latent variables Y and Z formed a perfect Guttman scale and which also had five degrees of freedom.

If we nevertheless accept the model, we may look at the parameter estimates in Table 3.8, which have been presented both in terms of the log-linear parameterization and in terms of the latent class parametrization. The higher the score on the latent variable, the more people are

TABLE 3.8 "Ordinal" Latent Variable Model for Data on Government's Responsibilities: The Netherlands

Latent Class X_t	$\hat{\pi}^X_t$	A. Equal Rights Men/Women $\hat{\pi}^{\bar{A}x}_{i|t}$		B. Good Education $\hat{\pi}^{\bar{B}x}_{j|t}$		C. Good Medical Care $\hat{\pi}^{\bar{C}x}_{k|t}$		D. Equal Rights Guest Workers $\hat{\pi}^{\bar{D}x}_{l|t}$	
		1. Yes	2. No	1. Yes	2. No	1. Yes	2. No	1. Yes	2. No
−1	.131	.531	.469	.988	.012	.940	.060	.656	.344
0	.376	.314	.686	.884	.116	.731	.269	.323	.677
+1	.493	.157	.843	.413	.587	.320	.680	.108	.892
		$\hat{\gamma}^{*A} = 0.405$		$\hat{\gamma}^{*B} = 0.092$		$\hat{\gamma}^{*C} = 0.173$		$\hat{\gamma}^{*D} = 0.252$	
		$\hat{\beta}^{*A} = -0.904$		$\hat{\beta}^{*B} = -2.386$		$\hat{\beta}^{*C} = -1.754$		$\hat{\beta}^{*D} = -1.378$	

$L^2 = 11.68$, df = 5, $p = .04$, Pearson-$\chi^2 = 11.72$
SOURCE: Table 3.1

opposed to an ample task for government. This latent variable is most strongly related to B (good education): each time the score on X increases by 1, the odds of responding yes/no on this item decrease by .09 (as can be checked by computing these odds from the conditional response probabilities presented). The weakest relation exists between X and A: With a unit increase of X, the odds of scoring yes/no on A decrease by a factor 0.405.

Latent class models with ordinal restrictions are directly related to Guttman, Rasch, and Likert scales (Andrich, 1985; Clogg, 1979b; Clogg & Sawyer, 1981; Langeheine and Rost, 1988). In general, they can be conceived of and formulated as latent trait models in which the underlying continuous latent trait has been divided into a few discrete categories (Heinen, Hagenaars & Croon, 1988). In this way, two important traditions in measurement theory can be brought together.

Notes

1. Actually, the latent class and the factor analysis model, as well as the latent trait models (Reiser, 1981) have much in common — they are all variants of the general latent structure model (Lazarsfeld & Henry, 1968; Mooijaart, 1978, 1982). Nevertheless, the problems of unreliability and invalidity in categorical data have seldom been explicitly attacked in a general and systematic way by means of the latent class model. For instance, Sutcliffe (1965a, 1965b) proposed a general approach to handle unreliability in the case of categorical data without using the latent class model, although his ideas fit perfectly well into this framework. Recent attempts which do make use of latent class analysis have been made by Hagenaars (1985) and Schwartz (1985, 1986).

2. All data from the Political Action Study used in this book have been obtained with the aid of Felix Heunks. Philip Stouthard, Felix Heunks, and Cees de Graaf, all members of the Sociology Department at Tilburg University, were the principal investigators for the Netherlands.

3. The EM-algorithm is a very general algorithm to obtain maximum likelihood estimates in the case of missing data (the scores on X are missing here). Dempster, Laird, and Rubin (1977) presented a general exposition of the EM-algorithm. Everitt (1984a, 1984b) and Kiiveri and Speed (1982) and also discussed the use of this algorithm for (log-linear) models with latent variables.

4. If one wishes to indicate that A and D are equivalent indicators in the sense that the relationships between A and X and between D and X are the same in terms of log-linear effects, the only restriction needed is $\tau_{it}^{AX} = \tau_{it}^{DX}$, without imposing: $\tau_i^A = \tau_i^D$. This restriction of equivalent relationship cannot be simply rendered in terms of equalities between conditional response probabilities, but has to be made in terms of equalities between particular products of response probabilities.

Sometimes, restrictions on the set of latent class parameters cannot be expressed in terms of restrictions on the set of log-linear parameters. For instance, a model in which the relations of A and X and of D and X are equally strong in terms of "percentage differences" $(\pi_{11}^{\overline{AX}} - \pi_{12}^{\overline{AX}}) = (\pi_{11}^{\overline{DX}} - \pi_{12}^{\overline{DX}})$ is not a log-linear model and cannot be handled by the algorithms discussed here.

5. Assuming that all models are identifiable, one could try out a number of latent class models, each having one latent variable, but a varying number of latent classes. The model providing the highest significance level might then be chosen as the most appropriate one. Although these models seem to be nested into each other — for example, the three latent class model reduces to a two latent class model by fixing all parameters related to the third latent class to zero — conditional tests cannot be used. In this case, the conditional test statistic $L_{r/u}^2$ does not approximate the χ^2-distribution because certain regularity conditions are not satisfied (the parameter set of the restricted model lies on the boundary of the parameter space of the unrestricted model (Bishop et al., 1975, Section 14.8.1).

6. The idea to consider the categories of two or more latent variables as the latent classes of one joint latent variable has already been brought forward by Lazarsfeld (1950a) and Wiggins (1955).

7. Actually, in so doing, Lazarsfeld's version of the probabilistic Guttman scale, the latent distance model, will be carried out, albeit in another formulation (Lazarsfeld & Henry, 1968, chapter 5). One can (re)parametrize Lazarsfeld's latent distance model as a model where each dichotomous manifest variable is directly connected to one and only one dichotomous latent variable (as in Figure 3.5) and where all latent variables together form a perfect Guttman scale. Although the model is presented here with two latent variables, extensions to any number of latent variables is possible.

It is known that if all latent variables are directly related to only one manifest variable, identifiability problems arise for the parameters referring to the easiest item and to the most difficult one. Lazarsfeld solves this problem by introducing equality constraints for particular conditional probabilities referring to the extreme items. If Y is the most difficult latent item, with indicator A, and Z the easiest latent item, with indicator F, Lazarsfeld makes the assumption: $\pi_{11}^{AY} = \pi_{22}^{AX}$ and $\pi_{11}^{FX} = \pi_{22}^{FX}$, where the scores on the latent and the manifest variables have the same substantive meaning, for example, 1 means "Yes." This solves the identifiability problem, but so does selecting at least two indicators for both the easiest and the most difficult latent item, as has been done here. Both approaches (and several others) are identical to each other in the sense that they yield the same estimated expected frequencies and the same values for the test statistics. Clogg and Sawyer (1981) presented an extensive survey of several (other) variants of the latent distance model.

8. Estimates were obtained by means of an ad hoc program developed and written by M.A. Croon from Tilburg University which employs a variant of the EM-algorithm. At that time, Haberman's program NEWTON was not available and no suitable initial estimates for LAT could be found to prevent the algorithm in LAT from breaking down. Later on, NEWTON was used without difficulties.

4. Panel Analysis: Answering Typical Panel Questions

4.1 Introduction

The systematic development of panel analysis techniques started with the work of Lazarsfeld and his associates. Lazarsfeld was one of the first to perform a large-scale panel study (Lazarsfeld, Berelson, & Gaudet, 1944) and to explain the basic potentialities and problems of the panel design (Lazarsfeld, 1948; Lazarsfeld and Fiske, 1938; Rosenberg, Thielens, & Lazarsfeld, 1951).[1] Lazarsfeld compared the possibilities of panel studies with those of cross-sectional surveys and trend studies and concluded that the panel design had to be preferred in all respects.

Lazarsfeld's work had a particular influence on sociologists who became very enthusiastic about the possibilities of panel analysis. As Davis (1978) remarked: "If the sociologist's attitude toward longitudinal research (i.e., trend study — J.H.) is one of religious faith, our feeling toward panels . . . approaches superstitious reverence. If longitudinal studies are Good, panel studies must be Wonderful" (p .173). Methodologists in psychology, however, have been more critical emphasizing the deficiencies of the panel design compared with a truly experimental design. In this and the following chapter both the possibilities and the limitations of panel studies will be elucidated.

The most striking feature of the panel design is that it offers the possibility to analyze (gross) changes at the individual level. With a panel study, it is possible to identify the persons who have changed their opinions and attitudes from one wave to another, to establish the nature of the individual changes, and to determine in what way certain kinds of changers differ from each other and from nonchangers.

147

In a panel study, it is possible to set up a turnover table in which the scores on a particular characteristic obtained in a certain wave are related to the scores on this same characteristic obtained in a previous wave. The log-linear analysis of such turnover tables will be dealt with in Section 4.2. In Section 4.3, these analyses of a turnover table will be extended to answer typical panel questions like: Are the observed changes in a particular characteristic different among older and younger people? Are the changes in one period different from those in another period? Are the changes in one characteristic different from those in another characteristic?

Studying and comparing changes is a precarious undertaking when the measurements are not quite reliable. In panel studies, unreliability of measurements perhaps plays an even more confounding role than in cross-sectional surveys or trend studies. At the same time, panel data offer unique possibilities to investigate the extent and the nature of the unreliability. This will be shown in Section 4.4 where panel analyses by means of log-linear models with latent variables will be dealt with.

In the next chapter, the panel design will be regarded as a quasi-experimental design, that is, the panel design will be examined as to its capabilities for detecting and testing causal relationships. Furthermore, within the same causal framework, some typical panel problems will be discussed in Chapter 5, especially the problems caused by nonresponse (panel attrition) and by reinterview effects and correlated error terms.

4.2 Changes in One Characteristic:
Log-linear Analysis of a Turnover Table

4.2.1 Introduction

Turnover tables are an instance of square tables in which the row and column variables are identically coded. Square tables arise, for example, when husbands and wives respond to the same questions or when the same items are repeated at two or more interviews. In the latter case, the square table is called a turnover table. Turnover tables show what changes in a discrete characteristic have taken place between two interviews or, more precisely, how an individual's score at time t_1 differs from the score at time t_2 (see Section 1.2). As such, the turnover table forms the nucleus of the panel analysis of discrete data.

Turnover tables and, in general, square tables have received much attention in the literature on log-linear modeling (Bishop et al., 1975, Chapter 8; Haberman, 1979, Chapter 8; Hout, 1983; Hout, Duncan & Sobel, 1987; Sobel, 1988). In fact the log-linear analysis of mobility tables in which father's occupation or former occupation is cross-classified with present occupation has almost become an independent, specialized branch of applied statistics. The most relevant approaches will be discussed in Sections 4.2.2 through 4.2.5.

By way of illustration, the data in Table 4.1 will be used. These data are taken from the Political Action Study (Barnes & Kaase, 1979) mentioned in the previous chapter and refer to the Dutch respondents who were eligible to vote in 1972 and participated in both the 1974 wave and the 1979 wave. In the first wave, the respondents were asked which party they had voted for in the Parliamentary Elections of 1972 and in the second wave this question was repeated for the Parliamentary Elections of 1977. Table 4.1 presents the cross-classification of the 1972 and the 1977 votes.

Table 4.1 contains a large amount of information (and is almost as complicated as the Dutch Party System itself). To explore this information systematically, Lazarsfeld and his associates worked out several basic approaches in which the distinction between gross and net change is fundamental.

A possible measure of the amount of gross change is the proportion of respondents who did not belong to the same category in both waves, that is, the proportion of respondents outside the main diagonal in Table 4.1: $1 - \Sigma_i p_{ii} = 0.395$. No fewer than 40% of all respondents voted differently in 1972 and 1977. Apparently, (reported) voting behavior is not a stable characteristic.

Gross change does not necessarily lead to net change (trend), that is, to differences between the two marginal distributions of the turnover table. The gross changes may compensate each other. In principle, the marginal distributions of voting in 1972 and 1977 could have been the same, despite the presence of a large number of people that switched parties. A possible measure for the size of the net change is the sum of the absolute differences between the marginal proportions in the corresponding categories, divided by two: $\Sigma_i |p_{i+}^{AB} - p_{+i}^{AB}|/2 = 0.132$. In this case, the net change is about one third of the gross change.

This amount of net change implies that in Table 4.1 at least 13% of the respondents should have had a different score on one of the two

TABLE 4.1 Voting Behavior at the Parliamentary Elections 1972 and 1977: Dutch Respondents Eligible to Vote in 1972 and 1977

	B. 1977							
A. 1972	1. did not vote	2. Labour Party	3. Christ. Democ. Party	4. Cons. Party	5. Lib. Democ. Party	6. Minor Right	7. Minor Left	Total
1. did not vote	38.2% (39)	20.6% (21)	18.6% (19)	9.8% (10)	4.9% (5)	3.9% (4)	3.9% (4)	15.7% (102)
2. Labour Party	8.9% (15)	67.9% (114)	6.0% (10)	3.6% (6)	10.1% (17)	0.6% (1)	3.0% (5)	25.8% (168)
3. Christian Democratic Party	2.2% (4)	2.2% (4)	86.9% (159)	6.6% (12)	1.6% (3)	0.5% (1)	0.0% (0)	28.2% (183)
4. Conservative Party	5.8% (5)	2.3% (2)	20.9% (18)	59.3% (51)	11.6% (10)	0.0% (0)	0.0% (0)	13.2% (86)
5. Liberal Democratic Party	4.2% (1)	12.5% (3)	20.8% (5)	12.5% (3)	45.8% (11)	0.0% (0)	4.2% (1)	3.7% (24)
6. Minor Right Wing Parties	5.1% (2)	10.3% (4)	30.8% (12)	23.1% (9)	12.8% (5)	15.4% (6)	2.6% (1)	6.0% (39)
7. Minor Left Wing Parties	6.3% (3)	36.5% (18)	10.4% (5)	10.4% (5)	8.3% (4)	0.0% (0)	27.1% (13)	7.4% (48)
Total	10.6% (69)	25.5% (166)	35.1% (228)	14.8% (96)	8.5% (55)	1.8% (12)	3.7% (24)	100% (650)

NOTE: In the Cells: horizontal percentages; in the Column Total: vertical percentages; in the Row Total: horizontal percentages; between brackets: absolute figures; Dutch party labels: 2.PvdA, 3.CDA, 4.VVD, 5.D66.
SOURCE: Political Action Study 1974 and 1979 — The Netherlands; see Barnes and Kaase (1979)

variables to make both marginal distributions equal to each other. In other words: 13% is the minimally necessary gross change in Table 4.1 given the observed (differences between the) marginal distributions.

Besides the minimum amount of gross change, it is also possible to calculate the maximum amount of gross change that could have been observed given the marginal distributions, or the expected gross change in the case of statistical independence given the marginal distributions. The amount of gross change actually found may then be evaluated against the minimal, maximum, or "independence" gross change.

Net and gross change can also be studied in greater detail. With regard to the net changes, it is possible to inspect the net gains and losses between all pairs of categories, that is, all pairs of parties. By comparing cell (i, j) with cell (j, i) for all values of i and j, it is possible to discover how much each party has lost or gained from all other parties. The parties in Table 4.1 can then be ordered in such a way that each party has a net gain from all parties occurring later in the rank order, but loses to all parties occurring earlier, the rank order being: Christian Democratic Party — Liberal Democratic Party — Conservative Party — Labour Party — Minor Left Wing Parties — Minor Right Wing Parties — Nonvoters. Compared with the 1972 Elections, the Christian Democratic Party is the great winner of the 1977 Elections, and the Labour Party along with the small parties the main losers.

Such an ordering based on the comparisons of the cells (i, j) and (j, i) need not correspond with the ordering on the basis of the total net gain based on the marginal distributions. But it is true in this case. All pairwise relations between the parties perfectly reflect the overall trend.

Gross change may be specified by means of horizontal and vertical percentages. Vertical percentages show where the votes for a particular party in 1977 came from and the horizontal percentages indicate where the votes for a particular party in 1972 went to in 1977. Among other things, the horizontal percentages make it clear that a large majority of those who voted for the minor left and right wing parties in 1972 made a different choice in 1977, in contrast to the Christian Democrats, the large majority of which remained loyal to their first choice. When the 1972 Christian Democrats changed their vote, most of them voted for the Conservatives.

As will be shown in the next sections, log-linear analyses of turnover tables supplement these elementary procedures, making it possible to provide better, more unequivocal, and detailed answers to more complex research questions.

At the same time, many of the problems connected with the use of these elementary procedures also apply to log-linear analyses. For example, it is known (although often forgotten) that the statements made above about the relative stability of the votes for the Christian Democratic Party compared with the votes for the minor parties may be very misleading. Given the differences in size of these parties, such differences in relative stability may be totally due to even a slight degree of unreliability of the measurements. Maccoby had already shown this conclusively in the 1950s (Maccoby, 1956). As will be explained in Section 4.4, this problem also plagues log-linear analyses.

Moreover, the horizontal percentages presented in Table 4.1 are based on very small numbers for some categories, leading to very large confidence intervals for the percentages. As was discussed in Chapter 2, small cell frequencies also cause serious problems in log-linear analysis.

4.2.2 (Quasi-)Independence

The log-linear analysis of a turnover table is often started with the independence model. Usually, this model is not seriously expected to be valid in the population, but the independence model is useful as a baseline model against which the existing patterns of association may be evaluated.

Not surprisingly, for the data in Table 4.1, the independence model $\{A, B\}$ has to be rejected: $L^2 = 660.09$, df = 36, $p < .001$. Because of the many small cell frequencies (13 of the 49 estimated expected cell frequencies are smaller than 5), L^2 may not be χ^2-distributed. The deviating value of Pearson-χ^2 underlines this possibility: Pearson-$\chi^2 = 802.53$. Nevertheless, given what we know about the patterns of voting behavior and given the large values of both test statistics, it is safe to conclude that the voting behavior in 1977 was not independent of the voting behavior in 1972.

As indicated in Section 2.5, the search for a better fitting model can benefit from a close inspection of the standardized or adjusted residual frequencies. The by far largest residual frequencies (not presented here) appear in the cells on the main diagonal: the observed frequencies f_{ii}^{AB} are systematically underestimated by model $\{A, B\}$. Voting behavior is much more stable than indicated by the independence model. People are inclined to vote for the same parties in 1972 and 1977 and our log-linear model should take this into account.

The logical next step away from the independence model is a model in which the tendency of people to vote for the same party is taken into account, but in which the votes at the two elections are otherwise independent of each other. This model can be used to determine whether or not those who originally supported a particular party, for example, the Labour Party, but changed their vote have a special preference for particular other parties, for example, the Minor Left. The question to be investigated is whether, when we leave the main diagonal cells aside, the remaining cells show particular systematic patterns of association or whether there is independence in this (truncated) table.

This model is an instance of a *quasi-independence model* (Goodman, 1968). In quasi-independence models, certain cells in a table are left aside and the independence model is postulated for the remaining cells. In this case, the question is whether or not the independence model is valid for those who changed their vote, leaving aside the cells on the main diagonal. The parameters of this quasi-independence model can be estimated in two ways with the same results. One may completely ignore the observed frequencies on the main diagonal by assigning the fixed value 0 both to all observed frequencies f_{ii} and to all estimated expected frequencies $\hat{F}_{ii}$ and postulate the independence model $\{A, B\}$ for the remaining cells. Most standard programs for log-linear models have the capacity to handle structural zero cells.

It is also possible by using a design matrix and the Newton-Raphson procedure to define the independence model for the whole table and add to this model the parameters τ_{ij}^{AB} for all $i = j$, which is the same as starting with the saturated model $\{AB\}$ but imposing the restrictions $\tau_{ij}^{AB} = 1$ if $i \neq j$. In this approach, the cell frequencies on the main diagonal are exactly reproduced: $\hat{F}_{ii}^{AB} = f_{ii}^{AB}$. The parameter estimates $\hat{\tau}_{ii}$ indicate the stability of party i.

The test statistics for this model are computed on the off diagonal cells comparing the observed and the estimated expected frequencies. The degrees of freedom are determined as follows. For a square $I \times I$ table, the ordinary independence model has $(I - 1)(I - 1) = (I - 1)^2$ degrees of freedom. In the quasi-independence model discussed above, I extra parameters τ_{ii}^{AB} must be estimated or I observed cell frequencies f_{ii}^{AB} are "missing." Thus the number of degrees of freedom is $(I - 1)^2 - I$.

Applied to Table 4.1, this quasi-independence model yields the following test outcomes: $L^2 = 77.91$, df = 29, $p < .0001$, Pearson-$\chi^2 = 74.92$.

Though this quasi-independence model fits much better than the ordinary independence model, it still has to be rejected. Those who change their vote in 1977 compared with 1972 follow specific patterns and do not change randomly within the boundaries set by the marginal distributions.

As before, inspection of the residual frequencies guided by theoretical notions may suggest models that can be meaningfully interpreted and are more parsimonious than the saturated model. The application of the quasi-independence model to Table 4.1 produced very large absolute values (≥ 3.00) of the adjusted residual frequencies (not reported here) for the cells (7, 2), (3, 4), (4, 3), (4, 2), and, to a lesser extent, cell (2, 4), where the first three residuals are positive and the latter two negative. In substantive terms, this means that the number of people who changed their votes from Labour to Conservative and vice versa (cells [2, 4] and [4, 2]) is smaller than expected on the basis of the quasi-independence model. Left (Labour) and Right (Conservative) represent opposite political views and people do not easily switch from the one to the other. Following the same line of reasoning, the Christian Democrats and the Conservatives appear to be kindred parties (cells [3, 4] and [4, 3]). Finally, those who originally supported one of the minor left wing parties voted more than expected for the Labour Party (cell [7, 2]) in 1977.

One could improve the quasi-independence model defined above by adding to the ordinary independence model not only the parameters for the main diagonal cells but also the parameters referring to the off diagonal cells with the large residuals mentioned. Eventually, repeated application of this procedure will most certainly yield a well-fitting model. If this final model has a clear substantive theoretical interpretation, there is not much wrong with applying this strategy, although it is necessary to test the final model on new data because there is the risk that sampling fluctuation has been taken for systematic variation (see Section 2.5).

In this case, for Table 4.1, we will not pursue this any further but in the next section present parameter estimates of a well-fitting model derived from a different perspective.

In order to find a well-fitting quasi-independence model one could also follow the reverse route and take the *saturated model* as the starting point. Application of the saturated model to Table 4.1 presents problems as some cells are empty. As discussed in Sections 2.4 and 2.9, there is no easy solution for this and the choice of a particular solution may have a great influence on the results. For example, replacing in Table 4.1 all observed cell frequencies equal to zero by 0.5 produced parameter estimates for the saturated model that were very different to those produced by replacing the empty cells by 0.1 (Hagenaars, 1985).

Seldom or never will there be a clear theoretical and substantive reason to choose between replacement by 0.1 and replacement by 0.5. Probably the best thing to do is follow the advice given in Section 2.9: Try out several solutions; compare the results; be careful with attaching meaning to those parameter estimates that vary widely among the different solutions.

If the empty cell problem is somehow solved, the saturated model can be applied. Inspection of the parameter estimates of the saturated model may suggest a more parsimonious model. Hauser (1979) has developed an interesting procedure to do this. The essence of his approach is to identify the τ_{ij}^{AB} parameters which are approximately equal to each other, after which a new model is constructed in which these parameters are set exactly equal to each other.

Cells for which the τ_{ij}^{AB} parameters are set equal to each other are considered to belong to the same *level*. A level contains at least one cell, but preferably many cells. The various levels which are thus distinguished within a table must be theoretically interpretable. Therefore, Hauser prefers a symmetric assignment of cells to levels: If cell (j, i) belongs to level k, cell (j, i) must, if possible, also be assigned to level k.

Because the τ_{ij}^{AB} parameters for the cells within a particular level are equal to each other, statistical independence applies to the cells within this level. If we denote the original variables by A and B and create a new variable L (Level) containing as many categories as levels distinguished, Hauser's proposal amounts to turning the original table AB into a three dimensional table ABL. This three dimensional table contains a number of structural zero cells: Within each particular category of L, all cells which do not belong to that particular level are empty. Hauser then assumes the quasi-independence model $\{A, B, L\}$ to be valid for this table, taking into account the structural zero cells (Hauser, 1979: 417).

There has been some discussion about Hauser's strategy particularly its alleged strongly empiric nature (Breiger, 1981, 1982; Goldthorpe, 1980; Macdonald, 1983). However, his method can be applied in a theoretical fashion. For example, in the case of Table 4.1, one could a priori indicate on theoretical grounds particular parties between which respondents choose more or less randomly and particular parties between which no exchange of voters takes place. Such theoretical suppositions may very well be translated in terms of Hauser's level analysis.

Breiger (1981; 1982) and especially Goodman (1981b) have elaborated Hauser's procedures by formulating in general and working out in detail how to arrive at homogeneous categories or groups of cells within a particular table. Marsden (1985) considered the possibilities of latent levels or classes (Section 4.4.2).

4.2.3 (Quasi-)symmetry and Marginal Homogeneity

In addition to (quasi-)independence, (quasi-)symmetry is a second important entry for the log-linear analysis of turnover tables. Central to it is the question to what extent the changes in a turnover table compensate each other. The most complete compensation occurs in the *symmetry model* which implies

$$F_{ij}^{AB} = F_{ji}^{AB} \qquad \text{for all } i \neq j \qquad (4.1)$$

In the symmetry model a possible loss of, for example, the Labour Party to the Christian Democratic Party is compensated by an equally great loss of the Christian Democratic Party to the Labour Party. Because the gross changes are symmetric, the marginal distributions of the table will be the same and there will be no net change. The log-linear model corresponding with equation 4.1 is in multiplicative form

$$F_{ij}^{AB} = \eta \tau_i^A \tau_j^B \tau_{ij}^{AB} \qquad (4.2)$$

with the restrictions

$$\tau_i^A = \tau_j^B \qquad \text{for all } i = j \qquad (4.3)$$

and

$$\tau_{ij}^{AB} = \tau_{ji}^{AB} \qquad \text{for all } i \neq j \qquad (4.4)$$

Because of restrictions 4.3 and 4.4, the effect parameters which determine the size of F_{ij}^{AB} are equal to the effects for F_{ji}^{AB}. The estimated expected frequencies for the symmetry model are as follows (Bishop et al., 1975: 283):

$$
\begin{aligned}
\hat{F}_{ij}^{AB} &= \tfrac{1}{2}(f_{ij}^{AB} + f_{ji}^{AB}) \qquad &\text{for } i \neq j \\
\hat{F}_{ij}^{AB} &= f_{ij}^{AB} \qquad &\text{for } i = j
\end{aligned}
\qquad (4.5)
$$

The estimated expected frequencies $\hat{F}_{ij}^{AB}$ of the symmetry model exactly reproduce the observed frequencies f_{ii}^{AB} and the sums $(f_{ij}^{AB} + f_{ji}^{AB})$. Furthermore, it follows that the frequencies $\hat{F}_{ij}^{AB}$ and $\hat{F}_{ji}^{AB}$ are equal to each other.

Parameter estimates and the values of the χ^2-test statistics of the symmetry model can be obtained in the usual manner by means of the estimated expected frequencies in equation 4.5. Given an $I \times I$ turnover table, the number of degrees of freedom df equals $I(I - 1)/2$ as follows from comparison of the number of cell frequencies and the number of parameters to be estimated.

The estimated expected frequencies can also be obtained by means of standard computer programs for log-linear models using the Newton-Raphson or Iterative Proportional Fitting (IPF) procedures (Bishop et al., 1975, Chapter 8; Haberman, 1979, Chapter 8). If the Newton-Raphson algorithm is being used, the restrictions 4.3 and 4.4 may be directly imposed through the design matrix.

In order to use the IPF procedure, the observed table AB has first to be converted into an "extended" table ABZ, where all cell frequencies f_{ij1}^{ABZ} for which $Z = 1$ correspond to the cell frequencies of the original table AB and all cell frequencies f_{ij2}^{ABZ} for which $Z = 2$ are identical to the cell frequencies of the transposed table AB. In the transposed table AB, the rows and columns of the original table AB are interchanged, such that each cell frequency (i, j) of this transposed table corresponds to the cell frequency (j, i) of the original table. Consequently, $f_{ij1}^{ABZ} = f_{ji2}^{ABZ}$. The symmetry model is obtained by applying model $\{AB\}$ to this extended table ABZ.[2]

The parameter estimates of model $\{AB\}$ for the extended table ABZ are the estimates for the parameters of the symmetry model. The values obtained for the test statistics L^2 and Pearson-χ^2 must be halved, as table ABZ contains twice the number of observations of the original table. The number of degrees of freedom has to be (re)set at $I(I - 1)/2$.

Application of the symmetry model to Table 4.1 yields the following test results: $L^2 = 95.61$, df $= 21$, $p < .001$ (Pearson-$\chi^2 = 81.02$). Evidently, Table 4.1 is not a symmetrical table and some political parties get more votes from another party than they lose to that party. This is in line with the results of the provisional analyses reported in Section 4.2.1 and with the fact that the observed marginal distributions of Table 4.1 differ from each other.

This raises the question whether the lack of symmetry in Table 4.1 might be explained by the differences in the marginal distributions, that is, by the net changes. Is Table 4.1 symmetrical as far as the marginal distributions permit, or are there special asymmetric relations between particular parties which do not follow completely from the observed overall trend?

We have seen in Section 4.2.1 that the political parties can be unequivocally ordered according to the sizes of their net gains and losses. The order based on the comparisons of all pairs of political parties was the same as the order based on comparison of the marginal distributions. This suggests that except for the trend, Table 4.1 is symmetrical. A formal test of this suggestion is possible by means of the *quasi-symmetry* model.

The quasi-symmetry model can be represented by equation 4.2 with the restrictions in equation 4.4 applied to the two-variable effect parameters. However, in contrast to the symmetry model, the restrictions in equation 4.3 for the one-variable effects are not imposed in the quasi-symmetry model. As in the symmetry model, the estimated expected frequencies of the quasi-symmetry model reproduce exactly the observed frequencies f_{ii}^{AB} and $(f_{ij}^{AB} + f_{ji}^{AB})$ but, unlike the symmetry model, the quasi-symmetry model also reproduces the observed marginal frequencies f_{i+}^{AB} and f_{+j}^{AB}.

If there is marginal homogeneity, that is, if the two marginal distributions are the same, the quasi-symmetry model gives results which are identical to those of the symmetry model. Quasi-symmetry together with marginal homogeneity implies and is implied by symmetry (Bishop et al., 1975: 286, 287):

$$\text{Quasi-symmetry} \cap \text{Marginal Homogeneity} \leftrightarrow \text{Symmetry} \quad (4.6)$$

So quasi-symmetry may be interpreted as "symmetry as far as the marginal distributions permit."

The estimated expected frequencies and the parameter estimates for the quasi-symmetry model can be obtained by imposing the restrictions in equation 4.4 directly through the design matrix.

If the standard IPF procedure is employed, model $\{AB, AZ, BZ\}$ has to be applied to the extended table ABZ defined above (Bishop et al., 1975: 289). The values of the test statistics must again be halved. The number of degrees of freedom is $(I - 1)(I - 2)/2$ as the quasi-symmetry model contains $(I - 1)$ parameters more than the symmetry model. The estimates for the parameters τ_{ij}^{AB} obtained with this extended table are identical to the estimates for τ_{ij}^{AB} in the original quasi-symmetry model. The estimates for the one-variable effects τ_i^A and τ_j^B of the original quasi-symmetry model are obtained by multiplying particular parameter estimates of the "extended model" {AB, AZ, BZ} with each other, namely, the products $\hat{\tau}_i^A \hat{\tau}_{i1}^{AZ}$ and $\hat{\tau}_j^B \hat{\tau}_{j1}^{BZ}$ respectively.

The quasi-symmetry model fits the data in Table 4.1 very well: $L^2 = 15.61$, df = 15, $p = .41$ (Pearson-$\chi^2 = 13.28$). So it may be concluded that the failure of the symmetry model was due to the dissimilarity of the marginal distributions and that, aside from the overall trend, there are no specific asymmetric relationships between the parties.

The exact nature of the relation between the voting behavior in 1972 and 1977 can be determined from the parameter estimates for the quasi-symmetry model presented in Table 4.2. Not surprisingly, the largest two-variable effects are to be found on the main diagonal. There are many more people who voted for the same party twice than expected on the basis of the one-variable effects alone. This tendency to remain loyal to a particular party is greatest for the Christian Democratic Party and for the minor left wing parties. This last result was not expected in view of the percentages presented in Table 4.1 from which it was concluded that the minor left wing parties were among the least stable. This analysis suggests that taking into account the general decline of the minor left wing parties, there remains a hard core of supporters of minor left wing parties.

Confining our attention to the off diagonal two-variable effects that are significant, it is seen that the combinations Labour-Christian Democrats (cells [2, 3] and [3, 2]) and Labour-Conservatives (cells [2, 4] and [4, 2]) occur far less often than expected on the basis of the one-variable effects. Likewise, the combination Christian Democrats and Minor Left occurs less often (cells [3, 7] and [7, 3]). The combination Labour-Minor Left (cells [2, 7] and [7, 2]) occurs more often than expected on the basis of the one-variable effects. Finally, nonvoting and voting for the Liberals (cells [1, 5] and [5, 1]) occur less often.

A more parsimonious well-fitting model might perhaps be obtained by setting all nonsignificant parameters to zero. Such a model would look very much like the quasi-independence models and Hauser's level models described above. As a matter of fact, the quasi-symmetry model is closely related to the quasi-independence model in which the observed frequencies on the main diagonal are exactly reproduced (or are structural zeros). Quasi-independence implies quasi-symmetry. The reverse

TABLE 4.2 Log-linear Effects ($\hat{\lambda}$) of the Quasi-symmetry Model for Table 4.1

			B. 1977					
	1. did not vote	2. Labour Party	3. Christ. Democ. Party	4. Cons. Party	5. Lib. Democ. Party	6. Minor Right	7. Minor Left	Total
A. 1972								
1. did not vote	1.42*	0.28	-0.31	-0.38	-0.90*	-0.07	-0.04	0.62
2. Labour Party	0.28	1.81*	-1.00*	-1.24*	0.12	-0.69	0.72*	0.73
3. Christian Democratic Party	-0.31	-1.00*	2.48*	0.29	-0.38	-0.02	-1.06*	-0.07
4. Conservative Party	-0.38	-1.24*	0.29	1.75*	0.25	0.01	-0.68	-0.03
5. Liberal Democratic Party	-0.90*	0.12	-0.38	0.25	1.47*	-0.24	-0.32	-1.01
6. Minor Right Wing Parties	-0.07	-0.69	-0.02	0.01	-0.24	1.84*	-0.83	-0.16
7. Minor Left Wing Parties	0.04	0.72*	-1.06*	-0.68	-0.32	-0.83	2.21*	-0.08
Total	0.03	0.60	1.07	0.62	0.35	-1.48	-1.20	1.59

NOTE: The two-variable effects that are significant at the .05 level are denoted by *.

is not true, except for a 3 × 3 table, in which case the two models have the same expected frequencies (Bishop et al., 1975: 287; Haberman, 1979: 429).

As Bishop et al. (1975) and Haberman (1979) indicate, the quasi-symmetry model is important not only in its own right, but has still another function: Together with the symmetry model, a test on *marginal homogeneity* can be derived from it, making it possible to test hypotheses about net changes. The hypothesis of marginal homogeneity:

$$F^{AB}_{i+} = F^{AB}_{+j} \qquad \text{for all } i = j \qquad (4.7)$$

cannot be formulated directly as a log-linear model. One might perhaps think that model $\{AB\}$ with the restrictions on the one-variable effects in equation 4.3 imposed on it implies marginal homogeneity. However, this is true only if the following equation is satisfied:

$$\sum_i \tau^{AB}_{ij} = \sum_i \tau^{AB}_{ji} \qquad (4.8)$$

This is the case in a 2 × 2 table and for the independence model, but is certainly not true in general. Recalling that quasi-symmetry plus marginal homogeneity imply symmetry (equation 4.6), the hypothesis of marginal homogeneity (equation 4.7) can be tested by conditionally testing the symmetry model against the quasi-symmetry model:

Marginal Homogeneity ≡ Symmetry minus Quasi-symmetry

$$L^2_{MH} \equiv L^2_{S/QS} \qquad = \qquad L^2_S \qquad - \qquad L^2_{QS} \qquad (4.9)$$

df: $(I - 1)$ $= I(I - 1)/2$ $- (I - 1)(I - 2)/2$

Because of its conditional nature, this test for marginal homogeneity is most adequate if the unrestricted model, that is, the quasi-symmetry model, need not be rejected. This is the case for Table 4.1. Applying the marginal homogeneity test to Table 4.1, the results are $L^2 = 95.61 - 15.61 = 80.00$, df $= 21 - 15 = 6$, $p < .001$. As was to be expected given earlier results, the hypothesis of marginal homogeneity has to be rejected and there is significant net change in Table 4.1.

A disadvantage of this conditional testing procedure for marginal homogeneity is that the unrestricted (quasi-symmetry) model has to be valid in the population and that it does not provide us with the estimated expected frequencies in the event the model is valid. In the remainder of this chapter, (quasi-)symmetry and marginal homogeneity models will be put to uses that are seemingly totally unrelated to questions about (quasi-)symmetry and marginal homogeneity. The lack of estimated expected frequencies in particular will then prove to be a disadvantage.

There are alternative unconditional procedures for testing marginal homogeneity which produce estimated expected frequencies. Most of these alternatives have been developed outside the log-linear framework (Bishop et al., 1975, Section 8.2.4; Haberman, 1979: 499). More recently, Haber and Brown (1986) introduced methods for defining log-linear models with linear constraints on the expected frequencies, including constraints that involve homogeneous marginal distributions. For many purposes when the interest lies in questions of marginal homogeneity as such, these (unconditional) tests may be useful, even more useful than the conditional testing procedure outlined above. However, as will be shown in Section 4.3, the flexibility of the conditional testing procedure is very great in some respects that are very relevant for the analysis of panel data. In these respects, unconditional tests of marginal homogeneity are less versatile.

The conditional testing procedure for marginal homogeneity and the log-linear formulation of (quasi-)symmetry models are the more useful because they can easily be extended to test hypotheses about (quasi-)symmetrical relations between three or more variables (Bishop et al., 1975, Section 8.3; Haberman, 1979, Section 8.3).

4.2.4 A Turnover Table with Ordered Categories

A very large part of the literature on ordinal categorical variables mentioned in Section 2.8 actually deals with the analysis of turnover and other square tables with ordered categories. An overwhelming number of models and methods have been suggested to investigate what is going on in such tables. Readers interested in "how to ransack social mobility tables and other kinds of [turnover] . . . tables" (Goodman, 1969) are well advised to study this literature thoroughly. Here only the general strategies proposed in the literature will be briefly discussed without working out in detail how these models are formally defined.

By way of example, let us take Table 4.1, leaving out the category Did Not Vote and rearranging the table by ordering the categories from politically left to right: 1. Minor Left Wing Parties, 2. Labour Party, 3. Liberal Democratic Party, 4. Christian Democratic Party, 5. Conservative Party, 6. Minor Right Wing Parties.

If the category scores 1, 2, . . . , 6 in the rearranged Table 4.1 are considered as strictly ordinal, inequality restrictions on the two-variable parameters make sense. Inequality restrictions are implied, for example, by the following hypothesis: The propensity of people to vote for the same party is greater than the propensity to change, but if they change, people will be more inclined to change toward parties that are close to their original party on the left/right dimension than toward parties that are further away. This kind of expectation can be tested by imposing on the two-variable parameters of the saturated or the quasi-symmetry model restrictions like $\tau_{11} \geq \tau_{12} \geq \tau_{13} \geq \ldots \geq \tau_{16}, \tau_{21} \leq \tau_{22}, \tau_{22} \geq \tau_{23} \ldots \geq \tau_{26}$, etc.

The distances between all successive category scores may also be assumed to be equal. In the case of (assumed) interval measurement it makes sense to consider the possibility of a linear relationship between the two variables A and B. The strictest model in this respect is the linear-by-linear interaction model in which the log-linear two-variable effects λ_{ij}^{AB} are subjected to the linear constraints: $\lambda_{ij}^{AB} = \lambda^* ij$. This log-linear association model has just one more parameter than the independence model $\{A, B\}$ and does not fit the data in the rearranged table at all ($L^2 = 273.23$, df = 24, $p < .001$).

A possible cause of the failure of this linear-by-linear interaction model might be that the distances between the successive categories are not equal. A log-bilinear association model in which a linear-by-linear interaction is also assumed but where the category scores are parameters to be estimated may be more appropriate.

Less strict variants of the class of association models are diagonal and distance models (Goodman, 1972c; Hout, 1983, Section 3). If one expects that the propensity of people to remain loyal to a particular party is the same for all parties, it makes sense to set all λ_{ii}^{AB} parameters referring to the cells on the main diagonal equal to each other. If it is expected that the propensity to choose a party one category to the right of the first party the second time is the same for all parties, all parameters λ_{ij}^{AB} referring to the cells on the diagonal for which $i - j = -1$ have to be set equal to each other. Similar equality restrictions can be applied to the parameters for which $i - j = -2$, etc. If the direction of the change does not matter, all parameters referring to the cells for which $|i - j|$ has the same value may be set equal to each other.

These diagonal models are called distance models if the two-variable effects referring to the off diagonal cells are a linear function of the distances between i and j, for example, $\lambda_{ij}^{AB} = (\lambda^*)(|i - j|)$. The category scores in these distance models are either fixed (log-linear association models) or parameters that have to be estimated (log-bilinear association models).

4.2.5 "Difference" Scores

In principle, turnover tables such as Table 4.1 offer the possibility to construct difference scores "A minus B" to indicate to what extent and in what way the scores on a particular characteristic deviate in the second wave from those in the first wave. These difference scores may be related to other variables to determine the causes and consequences of the changes. An often used "other variable" is variable A, the scores at t_1.

The use of difference scores is controversial, mainly because they are very sensitive to measurement unreliability and because they are logically related to the initial and final positions, in this case, to A and B (Bohrnstedt, 1969; Greenberg and Kessler, 1982; Kessler & Greenberg, 1981). It is not the purpose of this subsection to discuss these controversies extensively; in particular the discussion of the consequences of unreliability is postponed to Section 4.4. The main concern here is to show how difference scores if wanted could be employed when the variables have been measured at nominal level.

As difference scores essentially express in what way each individual score at t_2 deviates from the score at t_1, in the case of nominal level variables these difference scores consist of all combinations of all categories of the measurements at t_1 and t_2, and the difference variable actually is the joint variable AB having $I \times I$ categories.

Just as any categorical variable, this joint variable AB can be related to other variables using ordinary log-linear models. However, investigating the relation between AB and the initial position A or controlling for A when investigating the relationship between AB and other variables is obviously impossible: A and AB are perfectly related to each other in the sense that each category of AB occurs only in combination with one category of A. Meaningful analyses of the relations between the difference variable AB and the initial position A are possible only when it is theoretically meaningful to collapse particular categories of AB. To mention a few possibilities: One may collapse all those categories of AB where people did not change their vote from wave 1 to wave 2; the categories of AB may be combined in such a way that the turnover from each small party to each large party falls into the same category; one may use "absolute" difference scores where the turnover from party 1 to party 2 falls in the same category as the turnover from 2 to 1; etc.

Although the original difference variable AB and the initial position A are perfectly associated, this is no longer the case for the relations between A and these collapsed difference variables. Nevertheless, the relationships may still be very strong as it is possible that several combinations of categories of A and a collapsed AB cannot logically occur. However, by introducing structural zero cells for these logically impossible combinations, analyses can be carried out in which both A and AB (in collapsed form) appear.

An often used difference variable which does not give rise to tables with structural zeros is the stability variable S, consisting of the categories 1. Stable: the scores on A and B are the same, and 2. Unstable: the scores on A and B are different. It is easy to construct the 7×2 table AS from Table 4.1. Independence in table AS implies that the percentages of stable people or the ratios stable/unstable are the same for all categories of A. This independence model must be rejected: $L^2 = 144.63$, df = 6, $p < .001$ (Pearson-$\chi^2 = 136.09$).

The parameter estimates $\hat{\gamma}_i^{AS}$ for the saturated logit model indicate what the differences in stabilities are between the categories of A. The categories of A may then be ordered according to their relative stability (between parentheses the parameter estimates $\hat{\gamma}_i^{AS}$): Christian Democrats (6.94), Labour Party (2.21), Conservatives (1.53), Liberals (0.89), non-voters (0.65), minor left (0.39), minor right (0.19).

This order of stability is different in many respects from the "stability" or "loyalty" order based on the $\hat{\tau}_{ii}^{AB}$ parameters in Table 4.2, but corresponds exactly to the order that appears from the horizontal percentages in Table 3.1. We have to be aware of the fact that we are using different variables and posing different questions when working with turnover table AB and table AS. Moreover, the warning issued when discussing the differences in horizontal percentages that they could have been caused by an even slight degree of measurement unreliability also applies to the log-linear analysis of table AS.

4.3 Analyzing Tables with More Than Two Variables

However interesting the analysis of a single turnover table may be, the typical questions we like to have answered by our panel data usually involve more variables. We are generally interested in questions like the following. Are the changes in voting behavior the same among older and younger people? Are the preferences for political parties more or less stable than the preferences for political candidates? Does voting behavior change more rapidly during one period than during another? As will be shown below, these typical panel questions can be answered excellently by means of log-linear models. This section recapitulates and extends the work of Duncan who applied in a very ingenious way (quasi-)symmetry models to these problems that seemingly have nothing to do with (quasi-)symmetry (Duncan, 1979a, 1981; see Landis & Koch, 1979, and Kritzer, 1977, for similar contributions employing the GSK approach and weighted least squares procedures.)

4.3.1 Comparing Changes in Different Subgroups

Political theory suggests that newcomers into the Electoral System more readily change their vote over time than those who have been part of it for some time (Converse, 1976: 12, 13). Age may be taken as an indicator for the length of time people participate in the electoral system. The hypothesis about the instability of newcomers can then be tested by inspecting the conditional turnover tables AB on the 1972-1977 voting behavior (see Table 4.1) for all age categories. Some possible log-linear models for this table ABC (where C refers to Age) will be presented below, without data, however, as it involves a standard application of log-linear models.

If all conditional turnover tables AB corrected for possible differences in the number of persons belonging to the age categories are the same, the hypothesis about the instability of newcomers must be rejected. This similarity of all conditional turnover tables AB implies that the log-linear model $\{AB, C\}$ is valid.

If model $\{AB, C\}$ must be rejected, a less restrictive model which still implies rejection of the hypothesis about the instability of newcomers is model $\{AB, AC, BC\}$. This latter model exactly reproduces the observed marginal distributions of all conditional turnover tables AB. According to this less restrictive model, the conditional turnover tables AB differ from each other not only because of the varying number of persons belonging to the age categories, but also because the (relative) conditional marginal distributions of A are not the same for all age categories, neither are the (relative) conditional marginal distributions of B. At the same time model $\{AB, AC, BC\}$ implies that, in terms of the conditional parameters $\tau_{ijk}^{AB/C}$ there are no differences between the conditional turnover tables. Although their votes may be distributed differently over the parties, the age groups show the same kinds of changes in voting behavior. There are no three-variable interaction effects and the conditional turnover tables are equal to each other as far as their differences with regard to the marginal distributions permit. If this model need not be rejected, the parameter $\hat{\tau}_{ij}^{AB}$ tells us how voting behavior has changed between 1972 and 1977 in a similar way for all age groups.

Rejection of model $\{AB, AC, BC\}$ leads to the saturated model $\{ABC\}$ where the conditional turnover tables are assumed to be different from each other in every respect. Analyzing the turnover in vote by means of the saturated model $\{ABC\}$ for table ABC is identical to the analysis of the turnover table AB for each age category separately. From comparisons of the conditional two-variable effects $\tau_{ij\,1}^{AB/C}, \tau_{ij\,2}^{AB/C}, \tau_{ij\,3}^{AB/C}$, etc, it can be inferred whether newcomers are indeed more unstable with regard to their voting behavior than established members of the Electorate System.

It is easy to extend these analyses to higher dimensional tables and to analyze the conditional turnover tables AB within subgroups formed by the combinations of two or more variables. For example, one might be interested in the effects of Age (C) and Gender (G) on changes in voting behavior. Points of departure are table $ABCG$ and some assumptions about the effects of age and gender. For example, it may be supposed that at both times the votes vary according to gender and age, even allowing the possibility of a three-variable effect such that the differences in voting behavior between men and women may vary among the age categories. However, with regard to the changes in voting behavior between 1972 and 1977, it is hypothesized that these changes depend directly only on age but not on gender. These hypotheses imply the validity of the log-linear model $\{ABC, ACG, BCG\}$. In this model, the observed marginal distributions of the conditional turnover tables AB are exactly reproduced for each subgroup, that is, for each category of the joint variable CG. Furthermore, the turnover A-B as expressed by the parameters $\tau_{ij\,kl}^{AB/CG}$ varies according to age, but is the same for men and women within each particular age category.

In the above, no assumptions were made about the nature of the change itself, for example, the turnover being symmetrical. However, such extra restrictions can be applied fairly easily. Let us take table ABC and models $\{AB, C\}$, $\{AB, AC, BC\}$, and $\{ABC\}$ discussed above. The compound hypothesis that all conditional turnover tables are quasi-symmetrical and are exactly the same for all age groups can be tested by means of log-linear model $\{AB, C\}$ in which restriction 4.4 is applied to the two-variable effects τ_{ij}^{AB}. This restriction can be imposed by means of a design matrix or, when the standard IPF procedure is used, by means of an extended table $ABCZ$. In this extended table there exists for each age category an original $(Z = 1)$ and a transposed table AB $(Z = 2)$. To test the above compound hypothesis, model $\{AB, CZ, AZ, BZ\}$ has to be fitted to the extended table $ABCZ$. As usual, the values of the χ^2-test statistics have to be divided by 2.

The expectation that the marginal distributions of the conditional turnover tables will differ from age group to age group, but that there are otherwise no differences between the quasi-symmetrical conditional turnover tables, leads to model $\{AB, AC, BC\}$ in which restriction 4.4 is again imposed on the two-variable parameters τ_{ij}^{AB}. In terms of the extended table $ABCZ$, model $\{AB, ACZ, BCZ\}$ must be applied.

The least parsimonious quasi-symmetrical model is model $\{ABC\}$ with restrictions:

$$\tau_{ij}^{AB} = \tau_{ji}^{AB} \qquad \tau_{ijk}^{ABC} = \tau_{jik}^{ABC} \qquad (4.10)$$

Each conditional turnover table AB is supposed to be quasi-symmetrical, but otherwise different in all respects for all age groups. Parameter estimates and the test statistic L^2 may be found by applying the quasi-symmetry model to each conditional turnover table AB separately and summing the separate L^2-values and degrees of freedom. Alternatively, a design matrix can be used to impose restrictions 4.10 in model $\{ABC\}$, or model $\{ABC, ACZ, BCZ\}$ can be applied to the extended table $ABCZ$.

Perfect symmetry instead of quasi-symmetry can be imposed in similar ways. Comparisons of symmetry and quasi-symmetry models yield conditional tests on marginal homogeneity and make it possible to investigate whether the net changes are significantly affected by particular variables like age. For example, comparison of the test results of the least restrictive symmetry model $\{ABC\}$ with the least restrictive quasi-symmetry model $\{ABC\}$ provides a test for the hypothesis that all conditional turnover tables AB within all age groups have homogeneous marginals, that is, that there is no net change in any of the conditional turnover tables. If this hypothesis is accepted, it may be concluded that age has no influence on net change in voting behavior. If the hypothesis has to be rejected, partial symmetry models discussed in the next subsection can be used to investigate whether or not the nature of the conditional net change is the same in all age categories and a mere reflection of the overall trend (see Section 4.3.3, note 3).

4.3.2 Comparing Changes in Related Characteristics

Panel studies usually register changes in different kinds of phenomena and investigators are often interested in comparing these changes with each other. For example, Hagenaars (1983) examined whether preferences for particular political parties are less vulnerable to short term changes than preferences for particular political leaders, the underlying question being whether people identify more strongly with parties or with political leaders. The relevant data obtained from a national sample of the Dutch electorate and pertaining to the turnover in vote intention and preference for Prime Minister between February 1977 and March 1977 are presented in Table 4.3.

If the two subtables (a) and (b) of Table 4.3 came from two different subpopulations, say $C = 1$ and $C = 2$, the methods and significance tests described in the previous subsection could be used to investigate the differences and similarities of the two turnover tables. However, both turnover tables contain the same persons, and significance tests must take this dependence of the observations into account.

Duncan (1979a, 1981) has suggested a testing procedure that takes this dependence into consideration. His point of departure is the full cross-classification of all variables concerned, in this case the two measurements of vote intention and of preference for Prime Minister. Table 4.4 renders the relevant cross-classification. The marginals of Table 4.4 are identical to the original Tables 4.3a and 4.3b. Therefore, tests of the marginal homogeneity of Table 4.4 are also tests of the similarity of Tables 4.3a and 4.3b.

By regarding A and B as one joint variable AB and C and D as one joint variable CD, (quasi-)symmetry models can be fitted and tests on marginal homogeneity can be carried out for Table 4.4 by straightforward application of the methods discussed in Section 4.2.3. In terms of the standard IPF procedure, use is made of the extended table $ABCDZ$ with $Z = 1$ referring to the original Table 4.4 and $Z = 2$ to the transposed Table 4.4.

First, the most restrictive hypothesis, that the marginals of Table 4.4 are completely identical, is tested. This hypothesis is equivalent to the hypothesis that the turnover tables for vote intention (Table 4.3a) and preference for Prime Minister (Table 4.3b) are exactly the same. The

TABLE 4.3   Turnover in Political Preference in The Netherlands:
            February 1977 - March 1977

(a) Vote Intention: February 1977 - March 1977

	B. March			
	1. Christ. Democ.	2. Left Wing	3. Other	Total
A. February				
1. Christian Democratic	242	17	25	284
2. Left Wing	9	350	26	385
3. Other	3	51	337	431
Total	294	418	388	1100

(b) Preference for Prime Minister: February 1977 - March 1977

	D. March			
	1. Christ. Democ./ Van Agt	2. Left Wing/ Den Uyl	3. Other	Total
C. February				
1. Christian Democratic/Van Agt	111	19	42	172
2. Left Wing/Den Uyl	16	410	49	475
3. Other	53	66	334	453
Total	180	495	425	1100

SOURCE: Hagenaars (1983)

symmetry model may be obtained by applying model $\{ABCD\}$ to the
extended table $ABCDZ$. In this case, the symmetry model has $9(9 - 1)/2$
$= 36$ degrees of freedom. The test results are reported in Table 4.5,
symmetry model 1. From these test results, it is clear that Table 4.4 is
not symmetrical, but that is not of much direct relevance here.

The quasi-symmetry model may be obtained by applying model $\{ABCD,$
$ABZ, CDZ\}$ to the extended table $ABCDZ$. This model has $(9 - 1) = 8$
degrees of freedom less than the symmetry model and according to the
test results in Table 4.5, quasi-symmetry model 2, fits the data very well.
So there are no obstacles to performing the conditional test on marginal
homogeneity by testing the validity of the symmetry model against the
validity of the quasi-symmetry model. The outcomes of this conditional
test, reported in row Model 6 of Table 4.5, are such that the hypothesis
of marginal homogeneity of Table 4.4 has to be rejected and the conclu-

TABLE 4.4 Vote Intention and Preference for Prime Minister in The Netherlands: February 1977 - March 1977

A	B	$C=1$ $D=1$	$C=1$ $D=2$	$C=1$ $D=3$	$C=2$ $D=1$	$C=2$ $D=2$	$C=2$ $D=3$	$C=3$ $D=1$	$C=3$ $D=2$	$C=3$ $D=3$	Total
1	1	84	9	23	6	13	7	24	8	68	242
1	2	0	1	0	0	8	1	2	2	3	17
1	3	3	1	2	0	2	3	2	3	9	25
2	1	1	1	0	1	2	2	1	0	1	9
2	2	2	4	0	1	293	6	1	22	21	350
2	2	1	0	0	1	8	7	0	0	9	26
3	1	6	1	1	4	5	0	9	1	16	43
3	2	0	1	1	0	31	0	2	9	7	51
3	3	14	1	15	3	48	23	12	21	200	337
Total		111	19	42	16	410	49	53	66	334	1100

NOTE: The symbols A, B, C, D and 1, 2, 3 have the same meaning as in Table 4.3.
SOURCE: Hagenaars (1983a).

TABLE 4.5   Test Results for (Quasi-)symmetry and Marginal Homogeneity
            Models For Table 4.4

Model	L^2	df	p	Pearson-χ^2
1. Symmetry	160.23	36	.00	141.26
2. Quasi-symmetry AB, CD	21.38	28	.81	17.51
3. Partial symmetry A, C	74.61	34	.00	75.79
4. Partial symmetry B, D	75.92	34	.00	71.46
5. Partial symmetry A, C, B, D	43.89	32	.08	39.56
6. (1) – (2)	138.85	8	.00	
7. (3) – (2)	53.33	6	.00	
8. (4) – (2)	54.54	6	.00	
9. (5) – (2)	22.51	4	.00	

sion must be that the turnover tables of vote intention and preference for
Prime Minister (Tables 4.3a and 4.3b) are not identical.

   However, rejection of the hypothesis of complete homogeneity does
not necessarily imply that one of our phenomena, party preference or
candidate preference, is the more volatile. Weaker forms of marginal
homogeneity in Table 4.4 and weaker forms of similarity between Tables
4.3a and 4.3b can be defined that are more relevant from our theoretical
point of view.

   A conspicuous difference between Tables 4.3a and 4.3b is that the
marginal distributions are different. The marginal distribution of A differs
from C, and B differs from D. Both in February and in March, the number
of people who preferred the Socialist candidate Den Uyl is larger than
the number of people who preferred a left wing party, whereas the
opposite is true for the Christian Democratic candidate Van Agt and the
Christian Democratic Party. In this period, Den Uyl actually was Prime
Minister and this greater preference for the socialist candidate Den Uyl
may be a reflection of the "empirical regularity" that the Prime Minister
in office is more popular than his or her political opponents, often even
among those who do not subscribe to the Prime Minister's politics. When
comparing the turnover tables for vote intention and party preference
with regard to which characteristic is the most volatile, we would like to
disregard the differences between the tables that are caused by such
differences between the marginal distributions. The relevant null hypothesis to be tested is that Tables 4.3a and 4.3b are similar as far as the
differences in the marginal distributions permit.

   Such hypotheses can be tested by means of what will be called *partial
symmetry models*, models halfway between the symmetry and the quasi-

symmetry model. The symmetry model for Table 4.4 implies perfect symmetry and equal marginal distributions. Its estimated expected frequencies reproduce exactly the sums $(f^{ABCD}_{ijkl} + f^{ABCD}_{klij})$ and the observed frequencies f^{ABCD}_{ijij} on the main diagonal. The quasi-symmetry model for Table 4.4 implies symmetry as far as the differences in the marginal distributions of the joint variables AB and CD permit. Its estimated expected frequencies reproduce exactly not only the sums $(f^{ABCD}_{ijkl} + f^{ABCD}_{klij})$ and the observed frequencies f^{ABCD}_{ijij} on the main diagonal, but also the marginal distributions f^{ABCD}_{ij++} and f^{ABCD}_{++kl}. The quasi-symmetry model takes all differences between the marginal distributions AB and CD into account.

Partial symmetry models for Table 4.4 are used to examine to what extent there is symmetry as far as particular partial dissimilarities of the marginal distributions of AB and CD permit. The estimated expected frequencies of partial symmetry models reproduce exactly the sums $(f^{ABCD}_{ijkl} + f^{ABCD}_{klij})$, the observed frequencies f^{ABCD}_{ijij} on the main diagonal, and some or all of the marginal single-variable distributions of A, B, C, and D, but, in contrast to the quasi-symmetry model, not the marginal joint variable distributions AB and CD.

Table 4.5 contains the test outcomes for three partial symmetry models. The partial symmetry model 3 implies that Table 4.4 is symmetrical as far as the dissimilarity of the distributions of A and C make this possible. The estimated expected frequencies of this particular partial symmetry model reproduce exactly the sums $(f^{ABCD}_{ijkl} + f^{ABCD}_{klij})$, the observed frequencies f^{ABCD}_{ijij} on the main diagonal, and, moreover, the marginal distribution of A, f^{ABCD}_{i+++}, and of C, f^{ABCD}_{++k+} (but not f^{ABCD}_{ij++} and f^{ABCD}_{++kl}). Compared with the symmetry model for Table 4.4, this partial symmetry model loosens the equality restrictions on the τ^A_i and τ^C_i parameters.

The estimated expected frequencies can most easily be obtained by means of the extended table $ABCDZ$ defined above. To obtain the estimates for the partial model 3 in Table 4.5, model $\{ABCD, AZ, CZ\}$ has to be applied to the extended table $ABCDZ$. As usual, the test statistics for this extended table have to be halved. The number of degrees of freedom is $(3 - 1) = 2$ less than for the symmetry model, because A and C are trichotomies and the partial symmetry model loosens the equality restrictions on τ^A_i and τ^C_i.

Partial symmetry models are particularly interesting because they can be compared with the quasi-symmetry model and in this way yield conditional tests of "marginal homogeneity under certain conditions."

In the same way that quasi-symmetry in combination with marginal homogeneity implies and is implied by symmetry (equation 4.6), quasi-symmetry in combination with "marginal homogeneity as far as the marginal distributions of A and C permit" implies and is implied by the partial symmetry model 3 in Table 4.5. By comparing the quasi-symmetry model with this particular partial symmetry model, a conditional test is carried out of the hypothesis that the marginal distributions of AB and CD in Table 4.4 are similar to each other as far as this is possible given the dissimilarity of the marginal distributions of A and C. In terms of our original Tables 4.3a and 4.3b, this conditional test provides a test of the hypothesis that the turnover Tables 4.3a and 4.3b are similar as far as this is possible given the differences between the marginal distribution of A (Vote Intention in February) and the marginal distribution of C (Preference for Prime Minister in February).

As appears from the test result in Table 4.5, model 7, this hypothesis has to be rejected. The same negative result is obtained when the similarity of the two turnover tables of party preference and candidate preference is tested allowing for different marginal distributions in March (Table 4.5, models 4 and 8).

The weakest but in this case theoretically the most interesting test of the similarity of the turnover Tables 4.3a and 4.3b takes into account that the marginal distribution of vote intention and preference for Prime Minister may differ from each other both in February and in March. Formulated in terms of the parameters of the log-linear model, the hypothesis is tested that model $\{AB\}$ is valid for table AB (Table 4.3a) and model $\{CD\}$ for table CD (Table 4.3b) with the restriction that

$$\tau_{ij}^{AB} = \tau_{kl}^{CD} \qquad \text{if } i = k \text{ and } j = l \qquad (4.11)$$

In order to test hypothesis 4.11, a partial symmetry model for Table 4.4 has to be defined in which all four observed marginal distributions A, B, C, and D are exactly reproduced. Using the extended table $ABCDZ$, model $\{ABCD, AZ, BZ, CZ, DZ\}$ has to be applied. This partial symmetry model, indicated as model 5 in Table 4.5, has $2(3 - 1) = 4$ degrees of freedom less than the symmetry model. Comparison of the test outcomes of this partial symmetry model with those of the quasi-symmetry model indicates that hypothesis 4.11 has to be rejected (Table 4.5, model 9). Even taking into account their differences with respect to the marginal distributions in February and March, the turnover Tables 4.3a and 4.3b are not similar. The changes in party preference are different from those in candidate preference.

What the relevant differences are can be seen by fitting models $\{AB\}$ and $\{CD\}$ to tables AB and CD in Table 4.3. From a comparison of the $\hat{\tau}_{ij}^{AB}$ with the corresponding $\hat{\tau}_{ij}^{CD}$ parameters it appears that vote intention (party preference) is more stable than preference for Prime Minister (candidate preference): The parameters $\hat{\tau}_{ii}^{AB}$ are about 1.5 times as large as the $\hat{\tau}_{ii}^{CD}$ parameters, the (geometrical) mean value for table AB being 5.37 and for table CD being 3.61. Over time changes between the Christian Democratic candidate Van Agt and the other candidates (Den Uyl and Other) occur more readily than the corresponding changes between the Christian Democratic Party and the other parties (Left wing and Other): The negative off-diagonal effects $\hat{\tau}_{1j}^{AB}$ and $\hat{\tau}_{i1}^{AB}$ are stronger than $\hat{\tau}_{1l}^{CD}$ and $\hat{\tau}_{k1}^{CD}$ (for $j = l$ and $i = k$).

As is clear from the above, partial and (quasi-)symmetry models provide flexible tools for testing out the changes in related characteristics. Nevertheless, several problems remain.

In principle, these procedures can be extended to test hypotheses about changes in more than two characteristics or time periods. Bishop et al. (1975, Chapter 8) and Haberman (1979, Chapter 8) discussed the necessary extensions of the (quasi-)symmetry models. However, the full cross-classification of all variables concerned will then yield a very sparsely distributed multi-dimensional table with all the problems caused by (nearly) empty cells discussed in Chapter 2.

Furthermore, the conditional testing procedures do not provide the estimated expected frequencies by which the estimated effect parameters can be computed. For example, if model 9 in Table 4.5 and therefore the hypothesis in equation 4.11 could have been accepted, the conclusion would have been that the changes in vote intention and preference for Prime Minister were the same. But what do the (similar) patterns of change look like? Because of the lack of estimated expected frequencies, it is not directly clear how the "change" parameters $\hat{\tau}_{ij}^{AB}$ and $\hat{\tau}_{kl}^{CD}$ should be computed taking the equality restriction 4.11 into account.

An obvious solution is to consider the two turnover tables of vote intention and preference for Prime Minister in Table 4.3 as two sub-tables FG cross-classified by means of the variable H. F then refers to the measurements at t_1 (February), G to the measurements at t_2 (March), $H = 1$ to the cell frequencies of the turnover table of vote intention, and

$H = 2$ to the turnover table of preference for Prime Minister. The parameter estimates $\hat{\tau}_{ij}^{FG}$ in model $\{FG, FH, GH\}$ correspond with the two-variable "change" parameter estimates $\hat{\tau}_{ij}^{AB}$ and $\hat{\tau}_{kl}^{CD}$ for Tables 4.3a and 4.3b subjected to the equality restrictions in equation 4.11.

If other hypotheses about the equality of the turnover Tables 4.3a and 4.3b had been accepted, parameter estimates in agreement with these hypotheses could have been obtained in a similar manner by means of table FGH. For example, if the hypothesis that the two turnover tables in Table 4.3 are completely identical had been accepted, model $\{FG, H\}$ would have been the appropriate model whose parameter estimates satisfy all intended restrictions.

Another shortcoming of the conditional testing procedures is that they can be used to test only hypotheses about whether or not the turnover tables are similar under various restrictions on the marginal distributions, and not to test whether the similar turnover tables show particular patterns of change, for example, a symmetrical pattern. There is no standard testing procedure available to perform this latter task. The three dimensional table FGH in which the two turnover tables are conceived as parts of a three dimensional table cannot be used because of the dependence of the observations. Nevertheless, this table FGH can be used for estimating the parameters of models in which particular patterns of change are postulated. Whether or not the postulated model fits the data in table FGH has to be judged by means of descriptive measures of goodness-of-fit (Section 2.5).

4.3.3 Testing Changes in Association

In line with the example in Section 4.3.2, it might be expected that the link between a candidate and a particular party becomes stronger as the election date comes closer and that therefore the association between party and candidate preference at the beginning of an election campaign will be weaker than just before the elections. In a trend study in which a sample is drawn at the beginning and a new sample at the end of the campaign, such hypotheses about changing associations are easily tested (Chapter 6). Panel studies present more difficulties in this respect, because in a panel study, the table for the association between party and candidate preference at the beginning of the campaign refers to the same groups of respondents as the corresponding table at the end of the campaign.

The solution to this problem of dependent observations is essentially the same as proposed above using partial and (quasi-)symmetry models. If we would like to know whether the association between vote intention and preference for Prime Minister in February is different from this association in March, Table 4.4 has to be rearranged in such a way that the marginals of this four dimensional table represent the distributions of the joint variables *AC* and *BD*, that is, the relevant table for the association between vote intention and preference for Prime Minister in February and the table for the association in March respectively.

Quasi-symmetry and (partial) symmetry models can be used to test hypotheses about the similarity of tables *AC* and *BD* under different conditions. As these procedures have been extensively discussed, we will not go into this any further, but just mention the main result: The association between candidate and party preference has been strengthened from February to March, at least with regard to the link between the Socialist candidate Den Uyl and the left wing parties and between the Other candidates and Other parties.

In the above example, both characteristics showed variation over time. To answer the question whether the association between two variables remains the same over time when one of the characteristics involved is stable, the same procedures are applicable albeit that the multidimensional table has a number of a priori zero cells (Duncan, 1981).

Data pertaining to the relation between age (measured in February) and vote intention (measured in February and March) coming from the same panel study as the data in Table 4.4 may serve as an example. A table whose marginals correspond with the tables age-vote intention February and age-vote intention March is needed. This is the case for Table 4.6.

The variables *A* and *C* in Table 4.6 refer to age as measured in February (where Young means less than 40 years old) and *B* and *D* refer to vote intention as measured in February and March respectively. So *A* and *C* are actually identical variables. People who score Young on *A* by definition also belong to the category Young on *C*, and vice versa. This means that the cells Young-Old and Old-Young in table *ABCD* (Table 4.6) contain no observations and are structurally empty. The log-linear analyses have to be carried out with the restriction that the estimated expected frequencies pertaining to these cells are zero. Otherwise, everything remains the same.

TABLE 4.6   Age and Vote Intention in The Netherlands: February 1977 - March
            1977

| | | March | | | | | | |
| February | A | C = 1 | C = 1 | C = 1 | C = 2 | C = 2 | C = 2 | |
	B	D = 1	D = 2	D = 3	D = 1	D = 2	D = 3	Total
1	1	90	10	10	0	0	0	110
1	2	4	196	15	0	0	0	215
1	3	20	24	220	0	0	0	264
2	1	0	0	0	152	7	15	174
2	2	0	0	0	5	154	11	170
2	3	0	0	0	23	27	117	167
Total		114	230	245	180	188	143	1100

NOTE: Variables A and C refer to age, where 1 = young and 2 = old. Variables B and D
refer to vote intention, where 1 = Christian Democratic Party, 2 = Left Wing, and 3 =
Other.
SOURCE: Hagenaars (1983)

The estimated expected frequencies for the symmetry model may be
obtained by imposing model $\{ABCD\}$ on the extended table $ABCDZ$,
where both the original table ($Z = 1$) and the transposed table ($Z = 2$)
contain a number of structural zero cells. In this case, because of the
special structure of table $ABCD$, the estimated expected frequencies of
the symmetry model not only reproduce exactly the sums (f_{ijkl}^{ABCD} +
f_{klij}^{ABCD}) and the observed frequencies on the main diagonal (f_{ijij}^{ABCD}), but
also the number of older and younger people, that is, the frequencies
f_{i+++}^{ABCD} and f_{++k+}^{ABCD} .

Given the test outcomes presented in Table 4.7, model 1, the symmetry model has to be rejected. The number of degrees of freedom is six.
This is most easily seen by noting that symmetry in Table 4.6 implies
that both the upper left subtable (the turnover table of vote intention
among younger people) and the lower right subtable (the turnover table
for vote intention among older people) are symmetrical and that the
symmetry model in a 3 × 3 table has three degrees of freedom.

This points to another way of obtaining the test statistics in Table 4.7.
Using the methods of Section 4.3.1, we might test whether the conditional turnover tables of vote intention among younger and older people
are symmetrical without imposing further restrictions as to the equality
of the parameters for the younger and older people. This would yield the
same results as mentioned in Table 4.7, model 1.

TABLE 4.7   Test Results for (Quasi-)symmetry and Marginal Homogeneity Models for Table 4.6

Model	L^2	df	p	Pearson-χ^2
1. Symmetry (A, C)	17.14	6	.01	16.74
2. Quasi-symmetry AB, CD	2.27	2	.32	2.22
3. Partial symmetry B, D (A, C)	2.43	4	.66	2.41
4. (1) – (2)	14.87	4	.005	
5. (3) – (2)	0.16	2	.92	

The quasi-symmetry model for Table 4.6 has two degrees of freedom, one degree of freedom for each subtable. The test statistics may be obtained by applying model $\{ABCD, ABZ, CDZ\}$ to the extended table $ABCDZ$ or by postulating quasi-symmetry for each of the two conditional turnover tables of vote intention among younger and older people. As appears from Table 4.7, the quasi-symmetry model fits the data rather well and provides a good starting point for the conditional test on marginal homogeneity.

Comparing the symmetry and quasi-symmetry model (Table 4.7, model 4), the conclusion must be that the hypothesis of marginal homogeneity has to be rejected for Table 4.6, meaning that the marginal tables AB and CD, age-vote intention, are not exactly the same in both waves.

Weaker forms of similarity between tables AB and CD are possible by taking the differences between the marginal distributions of B and D (vote intention February, March, respectively) into account. The marginal distributions of A and C (age) are by definition equal for both periods and are already exactly reproduced by the symmetry model. The relevant partial symmetry model for Table 4.6 in which the observed marginal distributions of B and D (and also A and C) are reproduced has two degrees of freedom less than the symmetry model. Its test results are reported in Table 4.7, model 3.

The same test results could have been obtained by means of a model in which it is postulated that both conditional turnover tables vote intention among younger and older people are symmetric as far as the overall differences in the marginal distributions of B and D permit.[3]

Conditionally testing the empirical validity of the partial symmetry model against the quasi-symmetry model provides a test of the hypothesis that the marginal distributions AB and CD in Table 4.6 are similar as far as the differences between the marginal distributions of B and D permit.

This conditional test determines whether the marginal tables AB and CD are similar in the sense that the association between age and vote intention in February, indicated by the two-variable parameters $\hat{\tau}_{ij}^{AB}$ estimated for the marginal table AB, is equal to this association in March, that is, to the two-variable parameters $\hat{\tau}_{ij}^{CD}$ estimated for the marginal table CD.

The test outcomes (Table 4.7, model 5) show that there is no reason to reject this hypothesis of equal association. Once we allow for the fact that the marginal distribution of vote intention has changed between February and March, the tables vote intention-age are similar at both points in time.

Because the conditional test procedure does not provide estimated expected frequencies corresponding with the hypothesis of equal association, the strength of the relationship vote intention-age must be estimated by the procedure indicated in Section 4.3.2. (The other evaluating remarks made there apply here as well.) First, the tables Age (F) by Vote Intention (G) for February and for March are set up. Next, we consider these two tables as two subtables within the categories of H, where $H = 1$ refers to the measurements in February and $H = 2$ to the data obtained in March. Our confirmed hypothesis implies that the two-variable effects FG are the same in February and March. Therefore, model $\{FG, FH, GH\}$ has to be fitted for table FGH. The parameters $\hat{\tau}_{ij}^{FG}$ in this model provide the estimates of the stable association between age and vote intention. In this case, it appears that the preference of younger people for the left wing parties is about the same as that of older people ($\hat{\lambda}_{12}^{FG} = 0.066$), but younger people are more inclined to vote for the Other parties, and older people for the Christian Democrats ($\hat{\lambda}_{13}^{FG} = 0.206$, $\hat{\lambda}_{11}^{FG} = -0.272$).

4.3.4 Comparing Changes in Different Periods

Probably most characteristics do not vary regularly at a constant rate over time. For example, party preferences may be rather stable during long periods only to change rapidly because of an economic crisis. If a panel study has more than two waves and encompasses the relevant periods, it may be investigated whether the rate of change did indeed increase in some periods compared with others.

Suppose that the variable vote intention has been measured in four consecutive months January (J), February (F), March (M), and April (A).

If we want to know whether the turnover in vote intention from January to February, table *JF*, differs from the turnover from Mach to April, table *MA*, there are four non-overlapping months and variables *J*, *F*, *M*, and *A*. We can set up table *JFMA* with marginals *JF* and *MA* and follow the procedures outlined in Section 4.3.2 to investigate the differences between the two turnover tables *JF* and *MA*.

The months and variables may also overlap, for example, the turnover from January to February (table *JF*) is compared with the turnover from February to March (table *FM*). In this case, the procedures must be used that are mentioned in Section 4.3.3 with regard to testing changes in association if one of the variables is stable over time; the overlapping variable (here: *F*) plays the role of the stable characteristic. No new techniques are required, although determination of the correct number of degrees of freedom may turn out to be difficult, as is often the case when tables contain a priori zero cell frequencies (Bishop et al., 1975, Chapter 5; Haberman, 1979, Chapter 7).

Simultaneously comparing changes among three or more periods is possible by the extensions of the (quasi-)symmetry models provided by Haberman and Bishop et al. mentioned before.

Although not without difficulties (see the end of Section 4.3.2), log-linear partial symmetry and (quasi-)symmetry models provide flexible tools for the testing of many key hypotheses in panel research (Duncan, 1979a). On the other hand, exactly because of these difficulties, the further development of unconditional tests of marginal homogeneity under different restrictions is still desirable.

4.4 Latent Variable Models

4.4.1 Introduction: Change and Unreliability

In all analyses discussed so far in this chapter, it is assumed that the observed changes indicate true changes and are not just reflections of measurement unreliability. However, it is very well possible that some people's vote intentions remained the same during February and March, but that measurement errors were made when recording their intentions, as a result of which they seemingly changed their opinion. And of course the opposite may also occur: True changes take place which are not observed because of measurement error.

Lazarsfeld (1972a), referring to the work of Kendall (1954), made a very strong stand against interpreting observed changes as originating from measurement unreliability:

> But why adopt this idea from a psychometric tradition? Certainly if a man changes his vote intention it does not mean that the interview is ambiguous; much more likely a multiplicity of influences make him oscillate around his "true" position from one interview to the next. And in the case of mood . . . every reader knows only too well how realistic change is. Mood shifts would manifest themselves irrespective of the way they are "measured." (p. 360)

In this quotation, Lazarsfeld actually indicated three sources of instability in the observed score. In the first place, the observed scores may reflect real changes, that is, changes in a person's true score. The true score may be operationally defined as the score an investigator wants to measure, for example, the party preference a person has on average during a certain period of time or a person's voting behavior in a particular election.[4] Second, the observed scores may mirror a person's casual fluctuations around the true score. People who truly have a left wing political orientation may as a result of circumstances which happen to play a role at that particular moment of observation express themselves as belonging to the political middle or right wing. Their moods draw them to the right although they are basically left wing people. Third, observed changes may occur because the questions are incorrectly or ambiguously formulated, because the respondents make mistakes when answering the questions, because interviewers, observers, and coders make errors during the data recording and processing phases, etc.

Lazarsfeld assigned a minor role to these ("psychometric") mistakes and recording and processing errors as sources of observed change. In this, he may be mistaken. In Hagenaars and Bijnen (1974) and in the references given there, it is shown that the observed scores of background characteristics such as age, education, church attendance, and income show rather sizeable changes over time that can be explained only in terms of measurement errors in the strict sense of the word, that is, as response errors, interviewer errors, coder errors, etc. The measurement of these background characteristics is generally considered to be easy, at least much easier than the measurement of most attitudes, desires, preferences, beliefs, etc. The fact that measurement error in the strict sense plays an important role in the measurement of most of the characteristics we are interested in must therefore be reckoned with.

Moreover, Lazarsfeld implicitly uses a rather narrow and in many circumstances not very useful definition of measurement error and unreliability. From the viewpoint of wanting to discover the changes in the true score, it does not matter whether the observed scores are confounded by "psychometric" measurement error or by "real" but incidental fluctuations around the true score caused by factors like the mood a person is in. Therefore, both unsystematic deviations from the true score will be regarded here as unreliability and measurement error.

As has been explained in Chapter 3, for variables measured at nominal level, unreliability is defined as the conditional probability of a misclassification, given the true category one belongs to. For example, if A is a manifest variable, X a latent variable, and there is a one-to-one correspondence between the categories of A and X, the unreliability of variable A is determined by means of $\pi_{i\,t}^{AX}$, for $i \neq t$. Latent class models and log-linear models with latent variables are excellent tools for taking unreliability in the form of misclassification into account. Wiggins had already recognized this in the 1950s and he used latent class models ("latent probability models" as he called them) to distinguish between observed manifest change and underlying latent change, although, like Lazarsfeld, he objected to interpreting the relation between manifest and latent change in terms of measurement error (Beck, 1975; Wiggins, 1955, 1973: 7-11).

In Section 4.4.2, it will be shown how latent class models can be used for the analysis of a turnover table, in line with the analyses reported in Section 4.2. Extensions to more manifest variables and/or times of measurements will be discussed in Section 4.4.3. Special attention will be paid to latent (quasi-)symmetry models by means of which the typical panel questions posed in Section 4.3 can be answered taking measurement error into account.

4.4.2 Analyzing a Turnover Table
by Means of Latent Variable Models

Most of the essential features of the latent variable approach to turnover tables can be illuminated by using a simple 2 × 2 turnover table showing the changes in a dichotomous characteristic between two points of time. Table 4.8 presents the changes in the intentions of Dutch respondents to vote or not to vote between November and December 1971 (Hagenaars, 1978). According to Table 4.8, 16.5% (100 × (43 + 95)/ 837) of the respondents changed their intention between November and

TABLE 4.8   Vote Intention in The Netherlands: November 1971 - December
            1971

| | B. December 1971 | | Total |
	1. Voter	2. Nonvoter	
A. November 1971			
1. Voter	93.8%	6.2%	100%
	(652)	(43)	(695)
2. Nonvoter	66.9%	33.1%	100%
	(95)	(47)	(142)
Total	89.2%	10.8%	100%
	(747)	(90)	(837)

SOURCE: Hagenaars (1978)

December. The horizontal percentages seem to indicate that especially the intentions of the nonvoters are unstable.

Once the possibility of measurement error is allowed, the question arises to what extent the manifest changes reflect unreliability instead of true change. The most extreme answer is that there is no true change at all and that the observed changes result solely from measurement error. In terms of the latent class model, each and every person belongs either to the latent class "voter" or to the latent class "nonvoter" without changing his or her latent position. Manifest changes occur only because the latent positions are not perfectly reliably measured. If all this is true, the data in Table 4.8 can be explained by a standard latent class model with one dichotomous latent variable X, the true vote intention, and two indicators A, the observed vote intention in November, and B, the observed vote intention in February:

$$\pi_{ijt}^{ABX} = \pi_t^X \pi_{it}^{\overline{AX}} \pi_{jt}^{\overline{BX}} \tag{4.12}$$

Formulated as a log-linear model, the equivalent equation in multiplicative form is

$$F_{ijt}^{ABX} = \eta \tau_i^A \tau_j^B \tau_t^X \tau_{it}^{AX} \tau_{jt}^{BX} \tag{4.13}$$

This standard latent class model is not identifiable for Table 4.8: Equation 4.13 contains five unknown nonredundant parameters and Table 4.8 provides only three "knowns." If there is reason to believe that

TABLE 4.9 Two latent Class Model for Data on Vote Intention in table 4.8: Equal Reliabilities for both Latent Classes

Latent Class		A. Vote Intention November 1971 $\hat{\pi}_{f\mid t}^{\bar{A}X}$		B. Vote Intention December 1971 $\hat{\pi}_{j\mid t}^{\bar{B}X}$	
X_t	$\hat{\pi}_t^X$	1. Voter	2. Nonvoter	1. Yes	2. No
1	.940	.876	.124	.946	.054
2	.060	.124	.876	.054	.946
		$\hat{\tau}_{11}^{AX} = 2.653$		$\hat{\tau}_{11}^{BX} = 4.184$	

$L^2 = 0$, df = 0

the reliabilities for the true voters and the true nonvoters do not differ from each other, the number of parameters can be reduced by two and the model becomes exactly identifiable.[5] The relevant restrictions are

$$\pi_{11}^{\bar{A}X} = \pi_{22}^{\bar{A}X} \qquad \pi_{11}^{\bar{B}X} = \pi_{22}^{\bar{B}X} \qquad (4.14)$$

or, in terms of the parameters in equation 4.13 (see Section 3.2.3 and equation 2.22)

$$\tau_i^A = \tau_j^B = 1 \qquad (4.15)$$

This restricted model has zero degrees of freedom and fits the observed frequencies in Table 4.8 perfectly ($L^2 = 0$). The parameter estimates are presented in Table 4.9.

According to the manifest data in Table 4.8, the percentage of voters was 83.0% in November and 89.2% in December. In the restricted latent class model in Table 4.9, the percentage of true voters is estimated at 94.0%. The distribution of the latent variable Vote Intention is more skewed and has smaller variance than the manifest distributions of vote intention, a result that resembles the consequences of the postulates of the classical error theory where the observed variance equals the sum of the true variance and the error variance (Bohrnstedt, 1983: 71).

The observed marginal distribution of B is closer to the true distribution of vote intention than the observed marginal distribution of A. This is a consequence of the greater reliability of the measurements in November (A) than in December (B), as can be seen from comparison of the estimates $\hat{\pi}_{11}^{\bar{A}X}$ with $\hat{\pi}_{11}^{\bar{B}X}$ or $\hat{\tau}_{11}^{AX}$ with $\hat{\tau}_{11}^{BX}$ in Table 4.9. It is possible to test whether the difference in reliability between A and B is significant. Therefore, in the model in Table 4.9, the extra restriction that the reliabilities of A and B are the same is introduced:

$$\pi_{it}^{\bar{A}X} = \pi_{jt}^{\bar{B}X} \qquad \tau_{it}^{AX} = \tau_{jt}^{BX} \qquad \text{for all } i = j \qquad (4.16)$$

Because of this extra restriction 4.16, there is now one degree of freedom. The test results are $L^2 = 20.09$, $p < .001$ (Pearson-$\chi^2 = 19.59$). So it may be concluded that the increase in reliability from November to December as estimated in Table 4.9 is significant.

In panel analysis, it is often seen that the reliabilities increase in successive waves and that the measured characteristics become more stable. It is possible that under the influence of the questions asked at the first wave the respondents start to think more seriously about the matters at hand. Consequently, during the next waves the opinions are more firmly established, show less incidental fluctuations, and are less likely to truly change. To deal effectively with these kinds of reinterview effects, models are needed in which true change is separated from manifest change due to unreliability and due to direct relations between the manifest variables (see Sections 4.4.3 and 5.5.2).

No such analyses are possible for Table 4.8. The model in Table 4.9 provides a reasonable explanation of the turnover in vote intention in the sense that the estimates of the latent class parameters are substantively plausible, but it is not possible to test the basic postulate of the model that there is no true change. Models in which latent change is allowed require more data: larger tables, two or more indicators for each latent variable, or measurements from more than two waves.

Before dealing with these more complicated models, the simple no-latent-change model for Table 4.8 may be used to illustrate a few other consequences of unreliability for the patterns of manifest change. In Section 4.2.5 the possible use of "difference" scores was pointed out. The turnover Table 4.8 can be rearranged into a table AS where A indicates the manifest vote intention in November and S the stability variable with $S = 1$ referring to those whose manifest vote intention did not change and $S = 2$ to those whose manifest vote intention did change from November to December. The saturated logit model for this table AS with S as the dependent variable yields $\gamma_1^{AS} = 5.536$ and $\gamma_2^{AS} = 0.181$. The same conclusion is drawn from these effects as from the horizontal percentages in the original turnover Table 4.8: The vote intention of those who voted in November is much more stable than that of those who did not vote in November.

However, assuming that the simple no-latent-change model in Table 4.9 is valid in the population, these conclusions about differences in stability between voters and nonvoters are completely misleading. In the first place, no true changes take place at all among either the voters or the nonvoters. In the second place, as far as unreliability reflects incidental but "real" fluctuations, these oscillations have the same sizes for voters and nonvoters, because the reliabilities in the model in Table 4.9 are the same for both latent classes. As Maccoby (1956) already observed, given the assumptions that there is no true change and that the unreliability is similar for all classes, a manifest turnover table arises in which the smallest category seems to be the most unsteady.

If this simple no-latent-change model with equal reliabilities for both latent classes is extended to three or more waves, manifest data result which may be very misleading in still other respects. An often encountered pattern in the manifest data is that those who changed their manifest scores in previous waves are more inclined to change in the following waves as well than those who obtained the same score in the previous waves. There seem to be two groups of people, changers and nonchangers, and investigators are inclined to look for systematic differences between the two groups.

However, exactly this pattern of manifest change is to be expected on the basis of the no-latent-change model. Those people who obtained the same manifest score "voter" ("nonvoter" respectively) in the previous waves, most probably all belong to the latent class "voter" ("nonvoter" respectively) and, given fairly reliable measurements, they all have a reasonably large probability of obtaining the same manifest score "voter" ("nonvoter" respectively) again in a following wave. However, those whose manifest scores changed during the first waves from "voter" to "nonvoter" (from "nonvoter" to "voter" respectively) are more equally divided between the latent classes "voter" and "nonvoter." Each of these manifest changers has a reasonably large probability of obtaining during the following waves a manifest score in agreement with the latent class to which he or she belongs; for the group changers as a whole, a lot of manifest change will then be observed during the following waves. As far as these differences between "changers" and "nonchangers" result solely from measurement error, it does not make much sense to look for systematic differences between these two groups.

As the turnover table becomes larger, a variety of latent class analyses may be performed. Clogg (1981b) and Marsden (1985) suggested several interesting latent class models for the analysis of turnover tables with nondichotomous variables that supplement the techniques discussed in Section 4.2. To illustrate these models, use will be made again of turnover Table 4.1 on the 1972 and 1977 voting behavior.

One of the "manifest" log-linear models discussed in Section 4.2.2 was Hauser's Levels Model. Hauser's ideas can also (and perhaps even more adequately) be formulated in terms of latent class parameters. Let us assume that, although there are many political parties to choose from, the vote preferences of the Dutch electorate can be reduced to two basic categories (two latent levels): People truly prefer either a left wing or a right wing party. If these latent preferences are stable, we have a standard latent class model with one dichotomous latent variable X, where $X = 1$ refers to the left wing preference and $X = 2$ to the right wing preference, and two manifest variables A, manifest voting behavior in 1972, and B, manifest voting behavior in 1977.

Although it is not immediately clear, (local) identifiability tests (Section 3.2.4) reveal that the parameters of this model are not identifiable for Table 4.1. The model can be made identifiable by introducing theoretically meaningful restrictions that are analogous to what Clogg (1981b) suggested in his analyses of social mobility tables. It is reasonable to assume that those people who belong to the left wing latent class ($X = 1$) will never vote for an extreme right wing party ($A = 6, B = 6$) and that, in a similar vein, people from the right wing latent class ($X = 2$) will never vote for an extreme left wing party ($A = 7, B = 7$). So we impose the following restrictions on the conditional response probabilities: $\pi_{61}^{AX} = \pi_{72}^{AX} = \pi_{61}^{BX} = \pi_{72}^{BX} = 0$.[6]

A consequence of these restrictions is that there are no people whose manifest party choices change from the "minor left" to "minor right" or vice versa. According to the postulated model: $F_{67}^{AB} = F_{76}^{AB} = 0$. The estimated manifest turnover table contains some structural zero cells. The program LCAG can take such a priori zero cells into account, although the number of degrees of freedom reported in the program output may be wrong. If we look at the observed Table 4.1, it appears that cell (7, 6) is indeed empty, but that cell (6, 7) contains one observation. It was decided to ignore this one observation and replace the cell frequency (6, 7) by zero.

Applying the two latent class model to Table 4.1 with the four response probabilities fixed to zero yielded a very bad fit: $L^2 = 286.87$, df = 25, $p < .001$ (Pearson-$\chi^2 = 332.10$).[7] The voting behavior manifested in Table 4.1 cannot be explained in terms of just two stable basic political preferences.

The reason for this may be that the number of latent "levels" (two) is too small. However, keeping the two (left and right wing) latent classes in the model along with the restrictions on the response probabilities and adding a third latent class (the political "center"), for which no extra restrictions have to be made to make the model identifiable, did not improve the fit. Several four latent class models with additional restrictions to make the model identifiable also failed. The conclusion must be that the various categories of A and B, the manifest preferences for particular political parties in Table 4.1, cannot be reduced to just a few fundamental levels or classes within which the voting behavior of 1972 is independent of the voting behavior in 1977.

Instead of having too few latent classes, the problem with the latent levels models used above may be the assumption that the basic political orientations are completely stable over time (Marsden, 1985). After all, the 1972 and 1977 elections were five years apart and some true changes may have to be expected. It is perhaps better to postulate two latent variables Y and Z where Y indicates the true vote preference in 1972 and Z the true vote preference in 1977.

Given the above results and the characteristics of the Dutch political system, both latent variables are assumed to be trichotomous, with categories 1 = left wing, 2 = center, 3 = right wing vote preference. The latent class model has nine latent classes, forming the 3×3 latent turnover table on true vote preference.

Several restrictions have to be imposed on the conditional response probabilities π_{irs}^{AYZ} and π_{jrs}^{BYZ}. First, equality restrictions are needed to ensure that the scores on A depend only on Y, and the scores on B only on Z (see Section 3.4.1). Second, to ensure identifiability, it is assumed that those with a truly left wing vote preference will never vote for a minor right wing party, and those with a truly right wing vote preference will never vote for a minor left wing party: $\pi_{61s}^{AYZ} = \pi_{73s}^{AYZ} = 0$, for $s = 1, 2, 3$, and $\pi_{6r1}^{BYZ} = \pi_{7r3}^{BYZ} = 0$, for $r = 1, 2, 3$.

This two latent variable model also has to be rejected: $L^2 = 140.11$, df = 10, $p < .001$ (Pearson-$\chi^2 = 162.65$) and it does not make much sense to inspect the latent turnover table YZ to see what are the true changes in vote preferences. Even allowing for true change over time, the manifest turnover in vote behavior cannot be satisfactorily explained in terms of a few basic classes. It is much more likely that many people really identify themselves with a particular party, and not just with a general political view.

Clogg (1981b) suggested a model that could cope with the existence of voters who really identify with particular parties. The starting point is the basic latent class model with one latent variable X and the two manifest variables A and B of Table 4.1. The latent variable X has many categories. The first seven latent classes represent the true supporters of the various parties. Latent class 1 contains the hard core of nonvoters. It is absolutely certain that the members of latent class $X = 1$ did not vote in 1972 and 1977 ($A = 1, B = 1$). The members of latent class $X = 2$ each time vote only for the Labour Party ($A = 2, B = 2$); the members of latent class $X = 3$ only for the Christian Democratic Party; and so on. This implies the following restrictions on the conditional response probabilities for the first seven classes:

$$\pi_{it}^{\bar{A}X} = \pi_{jt}^{\bar{B}X} = \begin{cases} 1 & \text{for all } i = t, \ j = t, \ t = 1, 2, \ldots, 7 \\[2mm] 0 & \text{for all } i \neq t, \ j \neq t, \ t = 1, 2, \ldots, 7 \end{cases} \qquad (4.17)$$

If these seven latent classes were the only ones, according to this latent class model, there would not be any manifest change. Observed changes may come from people who more or less randomly choose between the various parties at both times of measurement, most probably people who are not very interested in politics. So an eighth latent class is added without any extra restrictions on its conditional response probabilities.

This eight latent class model yields the same estimated expected frequencies $\hat{F}_{ij}^{AB}$ as the quasi-independence model discussed in Section 4.2.2 where, disregarding the observations on the main diagonal of Table 4.1, the independence model was postulated for those who changed their manifest vote (the off-diagonal cells). The eight latent class model and the quasi-independence model make the same kind of assumptions about the turnover in Table 4.1. In the latent class model, all observed changes come from the eighth latent class, and within this class, A and B are independent of each other. The first seven latent classes fulfil the same role as the seven parameters τ_{ii}^{AB} in the quasi-independence model and also in this latent class model, the observed frequencies on the main diagonal f_{ii}^{AB} are exactly reproduced. The test results are of course the same for both models, and so this eight latent class model must be rejected ($L^2 = 77.91$, df $= 29$, $p < .001$).

The weakest part of this eight latent class model is the assumption that all changers choose randomly among the parties within the boundaries set by the marginal distributions. It is probably more realistic to assume that there are at least two classes of manifest changers: those who basically have a preference for left wing parties in both elections and those who basically are more inclined to vote each time for a right wing party. It is better to add two latent classes to the first seven (leaving out the former eighth class). As above, to ensure identifiability, additional restrictions are needed for the conditional response probabilities. For latent class $X = 8$, the left wing latent class, the conditional response probabilities π_{68}^{AX} and π_{68}^{BX} of voting for a minor right wing party in 1972 or 1977 are set to zero. For latent class $X = 9$, the right wing latent class, the conditional probabilities π_{79}^{AX} and π_{79}^{BX} of voting for a minor left wing party in 1972 and 1977 are fixed at zero. This again leads to structural zero cells (6, 7) and (7, 6).

This nine latent class model in which it is assumed that some people are truly nonvoters, that some have stable identifications with a particular party, and that the others have a stable but more general left or right wing political orientation without a special attachment to a particular party fits the data in Table 4.1 rather well: $L^2 = 25.47$, df = 18, $p < = .11$ (Pearson-$\chi^2 = 24.73$). The parameter estimates are presented in Table 4.10.

From the estimates $\hat{\pi}_t^X$ in Table 4.10, it appears that 4.8% of the respondents are truly nonvoters; 26.4% has a general left wing and 26.6% a general right wing political orientation; 42.3% truly identifies with a particular party. The Christian Democratic Party has the most loyal supporters; no other party has as many true identifiers. Moreover, if the number of true identifiers with a particular party is expressed as a percentage of the total votes that this party received in 1972, according to the marginal distribution of A in Table 4.1, 72% of the total votes for the Christian Democratic Party in 1972 consisted of votes from those who truly identified with the Christian Democrats. The corresponding percentage for 1977 is 63%. The corresponding percentages for all other parties are much lower in both elections (at most 45%).

Inspection of the estimated conditional response probabilities in Table 4.10, first of all indicates that although they are not truly nonvoters, members of latent classes 8 and 9 still have a rather large propensity not to vote. Furthermore, in both elections, latent class 8, the general left wing political orientation, had a relatively large tendency to vote for the Labour Party and the minor left wing parties, while latent class 9, the right wing political orientation, prefers the Christian Democratic, the Conservative, and the minor right wing parties.

TABLE 4.10 Nine Latent Class Model for Data on Vote Intention in Table 4.1: Seven Quasi-Latent Classes.

	1. Did Not Vote	2. Labour Party	3. Christ. Democ. Party	4. Cons. Party	5. Liberal Party	6. Minor Right	7. Minor Left	8. "Left"	9. "Right"
$\hat{\pi}_t^X$	.048	.105	.222	.059	.014	.007	.016	.264	.266
$\hat{\pi}_{i\,8}^{\bar{A}X}$	.203	.541	.002	.000	.033	0*	.221		
$\hat{\pi}_{i\,9}^{\bar{A}X}$	.209	.043	.225	.275	.054	.194	0*		
$\hat{\pi}_{j\,8}^{\bar{B}X}$	.130	.849	.100	.071	.135	0*	.075		
$\hat{\pi}_{j\,9}^{\bar{B}X}$	.090	.083	.388	.263	.133	.043	0*		

$L^2 = 25.47$, df = 18, $p = .11$, Pearson-$\chi^2 = 24.73$

NOTE: Because the conditional response probabilities of obtaining a particular score on A and B for the latent classes 1 through 7 are all fixed to zero or one, they are not shown in this table and this table does not have the usual format in which the estimated latent class parameters are presented.

* Fixed values

In this nine latent class model, manifest net gains and losses of parties come about only because the conditional response probabilities for the people with a general political orientation (classes 8 and 9) change over time. The differences between the marginal distributions of Table 4.1 are explained by the fact that, for latent classes 8 and 9, the tendencies not to vote or to vote for one of the minor parties has declined over time, while the propensity to vote for the Christian Democratic Party or the Liberals has increased. If this model is valid, explanations of political change in this period should be given in terms of the causes that led people from latent classes 8 and 9, the people who are less involved in the party system, to change their preferences somewhat.

4.4.3 Analyzing Tables with More than Two Manifest Variables by Means of Latent Variable Models: Latent (Quasi-)symmetry

The latent class analyses of a particular turnover table can be extended in many ways in much the same vein the analyses on the manifest level of a particular turnover table (Section 4.2) have been extended to answer typical panel questions involving more than two manifest variables (Section 4.3). Actually, all models mentioned in Section 4.3 can in principle be carried out on the latent level, thereby taking measurement error into account.

"Comparing changes in different subgroups" (Section 4.3.1) on the latent level is possible by applying the latent class analyses of a turnover table outlined above to the relevant subgroups. Following the principles of the "Simultaneous analyses in several groups" explained in Section 3.6, it is possible to test whether or not the parameters of the postulated latent variable models vary among subgroups.

For "Comparing changes in related characteristics" (Section 4.3.2), "Testing changes in association" (Section 4.3.3), and "Comparing changes in different periods" (Section 4.3.4) on the latent level, use can be made of the latent (quasi-)symmetry models developed in Hagenaars (1986). The principles of latent class analyses with (quasi-)symmetric relations between the latent variables will be outlined below. But first some latent variable models will be discussed that have no direct parallels at the manifest level.

If a particular characteristic, for example, vote intention, has been measured at comparatively many waves, for example, three or more points in time, complicated patterns of manifest change arise which partly result from underlying true changes in vote intention, but also

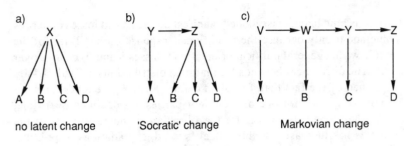

Figure 4.1. Latent Variable Models for Four Measurements in Time of One Characteristic; Manifest Variables A, B, C, and D; Latent Variables V, W, X, Y, and Z

follow from measurement error. Very often latent class models provide a simple and clear picture of the nature of the manifest changes. Three examples of possibly relevant models have been presented in Figure 4.1. The symbols A, B, C, and D in Figure 4.1 represent the recorded vote intentions at time t_1, t_2, t_3, and t_4 respectively, and X indicates the underlying true vote intention.

The simplest model in Figure 4.1, model a, is applicable whenever it is assumed that the true vote intention does not change over time and the manifest changes result completely from measurement error (Dayton & McReady, 1983). Essentially, this is the same kind of model that was postulated for the two-variable turnover Table 4.8 on vote intention. In the case of four manifest variables, it is generally not necessary to impose (extra) restrictions on the conditional response probabilities in order to achieve identifiability. Nevertheless, for substantive reasons, the investigator may want to test whether the reliabilities of the indicators are the same for all latent classes or stable over time and therefore imposes extra restrictions on the conditional response probabilities. As the model in Figure 4.1a is a standard latent class model, the estimation of the parameters with or without restrictions provides no special difficulties.

Models b and c in Figure 4.1 allow for latent change to take place. In these models, manifest changes result not only from measurement errors but also from particular patterns of underlying true changes. Model b constitutes one way to represent the "Socratic Effect" (Jagodzinski, Kühnel, & Schmidt, 1987; McGuire, 1960). As Socrates made the citizens of Athens conscious of their beliefs and opinions by questioning them, modern questionnaires may make respondents aware of their implicitly held opinions and attitudes. Because change is most likely to occur when opinions and attitudes remain implicit and less likely when an attitude or opinion has been consciously formed, latent change occurs more likely between the first and the second wave and not between later waves. This is exactly what model b represents. Model b is an ordinary two latent variable model, where A is the indicator of Y, the true score at the first wave, and B, C, and D are the indicators of Z, the stable true score during waves 2, 3, and 4.

Model c allows for a particular pattern of latent change during the whole period of investigation. The latent change is Markovian in the sense that the true score at time t is directly related only to the true score at $t - 1$, but not to the score at $t - 2$, $t - 3$, etc. The parameters of this model can be estimated by means of the LCAG program. Restrictions on the conditional response probabilities are necessary to ensure that each manifest variable is directly related to one and only one latent variable. The latent probabilities π_{pqrs}^{VWYZ} have to be restricted in such a way that the cell probabilities in the latent table VWY are in conformity with model $\{VW, WY\}$ and the cell probabilities in the latent table $VWYZ$ with model $\{VWY, YZ\}$, according to the principles of Goodman's modified path analysis approach (Sections 2.7.3 and 3.4.2). Making the direct effects between the latent variables equal to each other, as required by the standard Markov model, is not possible when using LCAG or comparable programs. Bye and Schechter (1986) and Van de Pol and De Leeuw (1986) provided algorithms to impose such equality restrictions.

Many variations of the models in Figure 4.1 can be defined in which all kinds of patterns of systematic true change are postulated. Because these models subtract as it were the change caused by measurement error from the observed change, the resulting patterns of true change are very often much simpler and easier to interpret than the patterns of change suggested by the manifest data as such.

This easier interpretation of latent change is certainly also true for those panel investigations in which particular theoretical concepts are operationalized by means of more than one indicator. In that case, the manifest changes will look very complex even if the panel investigation has only two waves. Table 4.4 which shows the simultaneous turnover in vote intention and preference for Prime Minister illustrates this point.

Latent class analysis can be used to investigate whether or not vote intention and preference for Prime Minister may be conceived as indicators of one underlying concept, political preference, and whether or not the phenomena underlying the manifest variables are stable over time. The composite hypothesis that vote intention and preference for

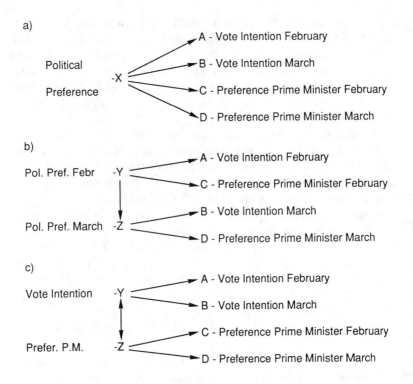

Figure 4.2. Latent Variable Models for Two Wave Panel Data on Vote Intention and Preference For Prime Minister; Data from Table 4.4

Prime Minister are indeed just expressions of one underlying concept, political preference, and that the true political preference did not change in February and March leads to the standard latent class model {XA, XB, XC, XD} in Figure 4.2a, where X represents the underlying stable political preference having three categories in correspondence with the categories of the manifest variables.

This model {XA, XB, XC, XD} fails to fit the data in Table 4.4: $L^2 = 362.71$, df = 54, $p < .001$ (Pearson-$\chi^2 = 491.23$). The reason that model a does not fit might be that true changes have taken place between February and March 1977. After all, there was a (Dutch) national election campaign going on at that time. Relaxing only the restriction of no latent change results in model {YZ, YA, YC, ZB, ZD} in Figure 4.2b, where the latent trichotomous variable Y denotes the true political preferences in

February and the latent trichotomy Z the true political preferences in March. The parameters π_{rs}^{YZ} of this model are the cell entries of the latent turnover table YZ on the true changes of political preference.

Model b does not fit the data better than model a: $L^2 = 351.14$, df = 48, $p < .001$ (and Pearson-$\chi^2 = 501.02$). Apparently, the assumption of "no latent change" was not the (sole) cause of the failure of the original model a. The weak point of model a might have been the assumption that vote intention and preference for Prime Minister are indicators of the same theoretical variable political preference. Abandoning this latter assumption but restoring the assumption of stable latent scores results in the two latent variable model c in Figure 4.2, where the trichotomous latent variable Y now indicates the true and stable vote intention, and the trichotomous variable Z the true and stable preference for Prime Minister. The latent table YZ can be used to estimate the underlying true association between the stable features vote intention and preference Prime Minister. This log-linear model $\{YZ, YA, YB, ZC, ZD\}$ fits the data much better than the previous models a and b, but the decrease in L^2 is still not enough to accept model c: $L^2 = 84.74$, df = 48, $p < = .001$ (Pearson-$\chi^2 = 94.33$).

The logical way to proceed would be to abandon both the assumption that vote intention and preference Prime Minister are indicators of the same theoretical concept and the assumption that there is no latent change. Such a model has four latent variables, one for each manifest variable. It will be discussed in Section 5.4. Less extreme models are also possible. For example, it might be hypothesized that the true vote intention is a stable phenomenon, but that the true preference for Prime Minister has changed. If W denotes the true vote intention and Y and Z the true preference for Prime Minister in February and March respectively, the relevant model is $\{WYZ, WA, WB, YC, ZD\}$.

If the latent variable models contain more than one latent variable, the relations between these latent variables can be restricted to conform with various unsaturated log-linear models, including the quasi-independence and (quasi-)symmetry models that appeared to be so important for the analysis of change at the manifest level (Sections 4.2 and 4.3). Quasi-independence models for the relations between latent variables can be fitted routinely by means of programs that have facilities to handle structural zero cells such as LAT or LCAG. (Quasi-)symmetrical restrictions on the relations between latent variables that have the same number of categories can be imposed directly by means of the design matrix when using programs in which Newton-Raphson procedures have been implemented, for example, NEWTON. If the EM-algorithm is used, as in LCAG, (quasi-)symmetry models can be handled by setting up the same kind of extended table, albeit now on the latent level, that was used to impose the (quasi-)symmetric restrictions on the log-linear parameters by means of the IPF procedure (Section 4.2.3). Because of the advantages of the EM-algorithm, the main principles of defining (quasi-)symmetry models and setting up the necessary extended table will be briefly discussed. Hagenaars (1986) provided a more extensive discussion.

The data in Table 4.4 will be used as an example, starting from model $\{YZ, YA, YB, ZC, ZD\}$ (Figure 4.2c) and forgetting for the time being that this model had to be rejected. It will be shown how to impose on model $\{YZ, YA, YB, ZC, ZD\}$ the (extra) restriction that the relation between the latent variables Y and Z is (quasi-)symmetrical. In this exposition, it is assumed that the parameter estimates are such that the categories of the latent and manifest variables have similar meanings. The meaning of $Y = 1$ is similar to $A = 1$ (Christian Democratic vote intention), the meaning of $Z = 1$ is similar to $C = 1$ (preference for the Christian Democratic candidate), etc. This similarity can be achieved by choosing appropriate initial estimates to start (Goodman's variant of) the EM-algorithm.

A symmetrical latent table YZ implies the following restrictions on the parameters π_{rs}^{YZ}:

$$\pi_{rs}^{YZ} = \pi_{sr}^{YZ} \qquad \text{for all } r \neq s \tag{4.18}$$

Because the restriction in equation 4.18 is an ordinary equality restriction, its implementation by means of Goodman's variant of the EM-algorithm is straightforward. As we saw above, the original model in Figure 4.2c yielded the following test statistics when applied to Table 4.4: $L^2 = 84.74$, df = 48. Imposing the extra restrictions of equation 4.18 yields three extra degrees of freedom (because the latent Table YZ is a 3×3 table — see Section 4.2.3). Applying the same model but with the extra symmetry restrictions to Table 4.4 yields $L^2 = 106.27$, df = 51, $p < .001$ (Pearson-$\chi^2 = 114.19$). Taking the validity of the initial model for granted, a conditional test of the latent symmetry hypothesis may be performed: $L^2 = 106.27 - 84.74 = 21.53$, df = $51 - 48 = 3$, $p < .001$. Introduction of restriction 4.18 results in a significantly worse fit and it must be concluded that the relation between the true vote intention and the true preference for Prime Minister is not symmetrical.

Whereas the application of latent symmetry models is rather straightforward, this is not the case for partial symmetry and quasi-symmetry models. These latter models cannot be rendered in terms of simple equality restrictions on the parameters π_{rs}^{YZ}. We need to set up an extended latent table (by means of which the symmetry model can also be handled). In Section 4.2.3, the extended table was indicated by means of the auxiliary variable Z. To avoid confusion with the latent variable Z (the true preference for Prime Minister), the auxiliary latent variable will now be indicated by H. For all cells for which $H = 1$, the cells of the extended latent table YZH refer to the original table YZ; for all cells for which $H = 2$, the cells of the extended table YZH refer to the transpose of the original table YZ.

Adding symmetrical restrictions to the model in Figure 4.2c by means of the extended latent table implies the following steps. First, instead of the original nine latent classes formed by the categories of the joint variable YZ, we now have to postulate 18 latent classes, the categories of the joint latent variable YZH. The manifest table $ABCD$ need not be extended. Second, the conditional response probabilities with subscript $H = 1$ ($\pi_{i\,rs1}^{AYZH}, \pi_{j\,rs1}^{BYZH}$, etc.) are subjected to the usual restrictions to ensure that with regard to the original table YZ the scores on A and B depend only on Y, and the scores on C and D are directly related only to Z. Third, a number of equality restrictions are needed to ensure that table YZH for $H = 2$ is indeed the transposed table YZH for $H = 1$ and that, for all values of r and s, latent class $YZH = rs1$ is identical in every respect with latent class $YZH = sr2$:

$$\pi_{rs1}^{YZH} = \pi_{sr2}^{YZH} \qquad \pi_{i\,rs1}^{\overline{A}YZH} = \pi_{i\,sr2}^{\overline{A}YZH} \qquad \pi_{j\,rs1}^{\overline{B}YZH} = \pi_{j\,sr2}^{\overline{B}YZH}$$

$$\pi_{k\,rs1}^{\overline{C}YZH} = \pi_{k\,sr2}^{\overline{C}YZH} \qquad \pi_{l\,rs1}^{\overline{D}YZH} = \pi_{l\,sr2}^{\overline{D}YZH} \tag{4.19}$$

Finally, the unsaturated model $\{YZ\}$ for the extended latent table YZH is postulated to ensure a symmetrical relationship between Y and Z.

LCAG will report the correct values of L^2 and Pearson-χ^2. They need not be divided by two as the manifest data are not doubled. However, the reported number of degrees of freedom is incorrect and must be calculated as indicated above for the symmetry model. The estimates of the conditional response probabilities are also correct, but the desired estimates of π_{rs}^{YZ} must be obtained by doubling $\hat{\pi}_{rs1}^{YZH}$.

To impose the quasi-symmetrical model on the relation Y-Z by means of LCAG, all steps described above remain the same, except the last one. The extended latent table must be constrained to conform to model $\{YZ, YH, ZH\}$ instead of to $\{YZ\}$. To find the correct number of degrees of freedom, it should be remembered that a quasi-symmetric $I \times I$ table has

$(I - 1)(I - 2)/2$ degrees of freedom (Section 4.2.3). So the total number of degrees of freedom for the model in Figure 4.2c with the latent quasi-symmetry restrictions is 48 + 1 when applied to Table 4.4. The test outcomes for Table 4.4 are $L^2 = 84.81$, df = 49, $p < .001$ (Pearson-$\chi^2 = 94.35$). The results of the conditional test of the latent quasi-symmetry hypothesis given the validity of the initial model in Figure 4.2c are $L^2 = 84.81 - 84.74 = 0.07$, df = 49 - 48 = 1, $p = .79$. There is no reason to reject the hypothesis that the latent table YZ is quasi-symmetrical.

The similarity of latent marginal distributions can be (conditionally) tested in the same manner that the similarity of manifest distributions can be (conditionally) tested, namely, by using symmetry and quasi-symmetry models. If one wants to know whether the true support for a party in the example used is the same as the true support for a candidate subscribing to the political views of that party, one has to test the hypothesis that the marginal distribution of Y is identical to the marginal distribution of Z: $\pi_{r+}^{YZ} = {}_{+s}^{YZ}$, for $r = s$. Conditional tests of this hypothesis can be carried out by testing the validity of the latent symmetry model given the validity of the latent quasi-symmetry model. The hypothesis has to be rejected: $L^2 = 106.27 - 84.81 = 21.46$, df = 51 - 49 = 2, $p < .001$. A comparison of the $\hat{\pi}_{r+}^{YZ}$ and the $\hat{\pi}_{+s}^{YZ}$ parameters in the latent quasi-symmetry model can identify the differences between the latent marginals.

If sufficient data are available, latent (quasi-)symmetry models can be used to answer the typical panel questions discussed in Section 4.3 at the latent level. In this way, these questions are answered taking measurement error into account. Defining the necessary partial symmetry models at the latent level follows rather simply from the principles outlined in this subsection and in Section 4.3.2. However, finding the relevant data and sufficient cases may turn out to be a major problem in applying these models.

Notes

1. Although Lazarsfeld and his associates have written a great many articles on panel analysis in many journals, the most comprehensive account of their work is to be found in many unpublished reports and papers of the Columbia Bureau of Applied Research. Examples are the Proceedings of the 1954 Dartmouth Seminar on Social Process, B. Levenson's "The status and prospects of panel analysis" and "Panel analysis" by J. A. Davis. Section IV on Panel Analysis in the book *Continuities in the Language of Social Research* (Lazarsfeld, Pasanella, & Rosenberg, 1972) gives an overall impression of these early contributions, presenting several unpublished papers from this period.

2. Bishop et al. (1975, Chapter 8) mentioned another way of constructing an "extended table" ABZ to get the symmetry model. $Z = 1$ then relates to the cell frequencies below the main diagonal for which $i > j$, and $Z = 2$ to the cell frequencies above the main diagonal for which $i < j$. The main diagonal cells themselves are deleted from the extended table. In this way, a log-linear model can be defined in which the sum of the observed frequencies above and the sum below the main diagonal are reproduced exactly, resulting in estimated expected frequencies which are symmetric as far as the difference between the two sums permits. Such a model might make sense in the analysis of social mobility tables when there is a uniform increase in social status. Bishop et al. present other examples.

3. If A indicates the variable Vote Intention in February, B Vote Intention in March, C indicates Age, and $Z = 1$ refers to the original table AB and $Z = 2$ to the transposed table, the test statistics in Table 4.7 may be obtained, as indicated in text, by means of the extended table $ABCZ$. The test statistics for the symmetry model are obtained by fitting model $\{ABC\}$, and those for the quasi-symmetry model by fitting model $\{ABC, ACZ, BCZ\}$. In the latter model, the conditional marginal distributions of A and B are exactly reproduced. The partial symmetry model (Table 4.7, model 3) is arrived at by means of fitting model $\{ABC, AZ, BZ\}$ to the extended table $ABCZ$. In this partial symmetry model, allowance is made for the differences between the conditional marginal distributions A and B, but only as far as they reflect the differences between the overall marginal distributions of A and B.

4. In the literature, a distinction has been made between platonic and operational true scores (Bailey, 1988; Bohrnstedt, 1983, Section 3.2; Sobel and Arminger, 1986; Sutcliffe, 1965a, 1965b). A platonic true score exists with regard to a person's weight at a particular moment, being married or not, the voting behavior at a particular election, that is, categories to which a person really belongs due to his or her essential characteristics. An operationally defined true score is the score a person obtains on average in a series of independently conducted experiments or observations. In some respects, this distinction is useful when investigating sources of unreliability because for some characteristics (age, marital status, etc.) we have the notion that "objective" measurements can be found that really reflect the true states of affairs, whereas for other characteristics, such as attitudes, no such "objective" measurement can be imagined. However, this distinction is not absolute. Although a platonic true score "weight" may seem to exist, only seldom will an investigator be interested in a person's exact weight at a particular point of time after finishing a particular meal. Much more often, the interest will be in how much a person weighs on average during a couple of days. The definition used here, "the score an investigator wants to measure," is intended to cover both types of true scores.

5. Another way of reducing the number of unknown parameters is to assume that the reliabilities are the same in November and December, although they may differ between the latent classes: $\pi_{11}^{AX} = \pi_{11}^{BX}$ and $\pi_{22}^{AX} = \pi_{22}^{BX}$. For this latent class model too, there seem to be as many unknown parameters as known frequencies. However, a consequence of stable reliabilities over time is a symmetrical turnover table AB with $F_{12}^{AB} = F_{21}^{AB}$. So with this restriction, there are just two independent knowns left to estimate the three remaining nonredundant parameters, leaving the model unidentifiable.

6. To achieve identifiability, it is not necessary to impose the restrictions on the conditional response probabilities of both A and B. A or B suffices. Clogg imposes the restrictions only for A, thereby avoiding particular structurally zero cells. However, it is

hard theoretically to justify imposing the restrictions only for A, and structurally zero cells do not pose a problem when using the program LCAG.

7. Taking into account the two a priori zero cells, Table 4.1 contains 46 independent cell frequencies. Given the restrictions on the parameters, five conditional response parameters for variable A have to be estimated for each latent class, resulting in 10 parameters to be estimated for A. The same applies for variable B. Together with one latent class probability, this makes for 21 nonredundant parameters, and for $46 - 21 = 25$ degrees of freedom.

5. Panel Analysis: Investigating Causal Hypotheses

5.1 Introduction: The Panel Design as a Quasi-experimental Design

In the previous chapter a large number of questions about social change were posed that could be answered by means of panel data making use of log-linear models. Very often these questions and answers implicitly or explicitly involve conjectures about causal relationships among certain variables. In this chapter, the potentialities and shortcomings of the panel design as a quasi-experimental design oriented toward the testing and detecting of causal relationships will be dealt with. Using the terminology of Campbell and associates, attention will be paid to the one-group pretest-posttest design (Section 5.2), to the nonequivalent control group design (Section 5.3), and to the path analytic approach (Section 5.4) (Campbell & Stanley, 1966; Cook & Campbell, 1979). No new log-linear techniques will be introduced, but the models presented in Chapter 4 will be applied within an explicitly causal setting. Chapter 5 concludes with a path analytic approach toward two important panel problems, namely, nonresponse and test-retest effects.

5.2 The One-Group Pretest-Posttest Design

Because in a panel study, the same characteristic, buying behavior, for instance, is measured at several points in time, changes in this characteristic may tell us something about the impact of events that have taken place between the measurements, for example, an advertisement campaign. The logic underlying this kind of "impact panel" analysis corresponds in its simplest form with the logic of the one-group pretest-posttest design. This design is symbolized in the work of Campbell and associates by

$$O_1 \qquad X_e \qquad O_2$$

In this diagram, the symbol O refers to the scores on the dependent characteristic, O_1 pertaining to the pretest, the measurements at t_1, O_2 to the posttest, the measurements at t_2. X_e is the value of the independent variable X in the experimental group. The placement of all symbols in one row means that they all pertain to the same group of cases. The ordering of the symbols one after another, the respective columns of this row, represents an order in time.

Thus O_1 might represent the buying behavior of a group of people registered in wave 1, O_2 the buying behavior of this same group of people registered at wave 2, and X_e might symbolize the fact that between wave 1 and 2 all these people were exposed to an advertisement campaign.

In principle, the effect of the campaign on buying behavior is determined in the one-group pretest-posttest design by comparing the buying behavior reported at wave 1 with the behavior at wave 2. The trend or net change in buying behavior is a measure of the effect of the advertisement campaign.

How the differences between the marginal distributions of the dependent characteristic at t_1 and t_2, the net changes, are evaluated within the context of log-linear modeling will be shown in the following paragraphs in conjunction with a discussion of the extent to which this variant of panel analysis permits valid conclusions about the effects of X and the extent to which all kinds of disturbance factors make this difficult. In this discussion, the panel design will be compared with the trend design presented in Chapter 6. After all, for establishing trends, net changes, we do not need panel data, trend data suffices. By comparing the trend design with the panel design, insight may be gained into the comparative potentialities and weaknesses of the two main designs discussed in this book.

Campbell and Stanley (1966) called the "trend" equivalent of the one-group pretest-posttest design, the "separate-sample pretest-posttest design" (p. 53), symbolized by

$$R \qquad O_1 \qquad (X_e)$$

$$R \qquad\qquad X_e \qquad O_2$$

This representation of the separate-sample pretest-posttest group design has two rows indicating that there are two groups of respondents. The first group receives the pretest of buying behavior at t_1 (O_1) and is subsequently exposed to the advertisement campaign (X_e). Because the exposure of this group to X_e is not essential, the symbol X_e has been placed between parentheses. The second row denotes the second group which is exposed to the advertisement campaign after t_1 (X_e) and subsequently receives the posttest of buying behavior at t_2 (O_2). So the first group receives only the pretest and the second group only the posttest. The symbol R in the beginning of each row indicates that each group constitutes a random sample from the same (statistical) population. The distribution of the total number of respondents to the two groups occurs randomly. As in the panel variant of the pretest-posttest design, the effect of X is established in the trend variant by means of the differences between posttest and pretest.

Most explanations of how to analyze both pretest-posttest designs deal with pretest and posttest scores that have been measured at interval level. Most researchers will be familiar with the required interval level techniques. Nevertheless, a brief exposition of the main principles of analyzing the pretest-posttest design with interval level data is presented below, mainly to provide a framework in which the log-linear approach can be discussed.

If the pretest and posttest scores are measured at interval level, the size of the effect of the intervening event X_e in both the one-group and the separate-sample pretest-posttest design is usually determined by subtracting the posttest (arithmetic) mean M_2 from the pretest (arithmetic) mean M_1. Testing the statistical significance of this difference between means is done with a standard t-test (Hays, 1981, Chapter 8):

$$t = \frac{M_1 - M_2}{\text{est. } \sigma_{\text{diff}}} \tag{5.1}$$

In the separate-sample pretest-posttest design, this t-test has $N_1 + N_2 - 2$ degrees of freedom, where N_1 refers to the sample size of the first (pretest) group and N_2 to the second (posttest) group. In the one-group pretest-posttest design, there are $N - 1$ degrees of freedom, where N is the (one) group size. If all groups have size N, the number of degrees of freedom are twice as large with trend than with panel data. Because of this, the t-test in equation 5.1 has more power, that is, the probability of detecting existing differences between means in the population is greater with trend than with panel data, other things being equal. However, the significance level corresponding to a particular t-value is seriously affected only by a doubling of the number of degrees of freedom if the number of degrees of freedom is rather small. Therefore, even for moderately large numbers of degrees of freedom, say ≥ 30, this gain in power of trend over panel data is negligible for most practical purposes.

Much more influence on the power of the t-test is exerted by the estimated standard error of the differences between means, denoted as est.σ_{diff} in the denominator of equation 5.1. Assuming, for convenience's sake, that all groups have N cases and that the population variances σ of the scores at t_1 and t_2 are equal to each other, the standard error σ_{diff} of the difference between means in the separate-sample pretest-posttest design is given by

$$\sigma_{\text{diff}}^2 = \frac{\sigma_1^2}{N_1} + \frac{\sigma_2^2}{N_2} = \frac{2\sigma^2}{N} \tag{5.2}$$

In the one-group pretest-posttest design σ_{diff} is given by

$$\sigma_{\text{diff}}^2 = \frac{\sigma_1^2 + \sigma_2^2 - 2r_{12}\sigma_1\sigma_2}{N} = \frac{2\sigma^2 - 2r_{12}\sigma^2}{N}$$

$$= (1 - r_{12})\frac{2\sigma^2}{N} \tag{5.3}$$

where r_{12} denotes the product-moment correlation between the pretest and posttest scores in the panel design. The estimated standard error est.σ_{diff} in equation 5.1 can routinely be obtained by substituting the unbiased (pooled) estimate s based on the sample data for σ in equations 5.2 and 5.3 (Hays, 1981, Chapter 8).

In the simple situation under study (for more complex cases, see Berger, 1985) it follows from comparison of equations 5.2 and 5.3 that the results of the t-test will be the same for panel and trend data if in the panel data the pretest and posttests are uncorrelated: $r_{12} = 0$. On the other hand, the standard errors σ_{diff} will be smaller, and therefore the t-values larger and the t-tests more powerful in the panel design compared with the trend design if pretest and posttest are positively correlated: $r_{12} > 0$. The reverse is true if $r_{12} < 0$. Because pretest and posttest are almost always positively correlated, the panel design generally has more power and is more suitable to detect small changes between means in the population than the trend design.

TABLE 5.1   Net Changes of Vote Intention in The Netherlands;
November 1971 - December 1971: Pretest-Posttest Design

	Campaign Exposure (X)	
Vote Intention (Y)	1. Before $(X_c — Nov.)$	2. After $(X_e — Dec.)$
1. Voter	83.0%	89.2%
2. Nonvoter	17.0%	10.8%
100% =	837	837
	$\hat{\gamma}_2^{x\bar{y}} = 1.302$	
	$\hat{\beta}_2^{x\bar{y}} = 0.264$	

SOURCE: Table 4.8

Similar conclusions about the two pretest-posttest designs can be drawn if the dependent variable has been measured at nominal level and log-linear analysis is being used. The essence of the log-linear approach can be exemplified by means of Table 5.1, derived from Table 4.8, on net changes in vote intention.

Let us assume that the Dutch Government, alarmed by the rising number of nonvoters in the last elections, launched a campaign to convince the people of the importance of going to vote. The campaign started in November 1971. The marginal distribution of the vote intention before the campaign (O_1) is as found in the turnover Table 4.8 and is identical to the first column in Table 5.1. The postcampaign measurements in December (O_2) result in the other marginal distribution of the turnover Table 4.8 and the second column of Table 5.1.

In this case, the data in Table 5.1 are obtained by means of a panel design and consequently the two columns in Table 5.1 contain the same people. However, a table such as Table 5.1 could have been obtained by means of a trend design, in which case the two columns would have contained two different groups of people. For descriptive purposes, however, to obtain an estimate of the effect of the campaign X_e, the distinction between trend and panel data is not important.

As argued before, the effect of X_e for both the one-group and the separate-sample pretest-posttest design is determined by means of the difference between O_1 and O_2. With nominal level data, it is not possible to express this difference as an ordinary difference between means. We could use the difference between the proportion of voters after and the proportion of voters before the campaign instead: 0.892 − 0.830 = 0.062. From this, it is concluded that the campaign has increased the percentage of (intended) voters and as such was successful.

The effect of X_e can also be described in log-linear terms. If the percentage of (intended) voters is exactly the same before and after exposure to the campaign, model $\{X, Y\}$ will exactly fit the data for the XY Table 5.1, or, what amounts to the same, the two variable parameter $\hat{\tau}_{ij}^{XY}$ in the saturated model $\{XY\}$ is 1. Parameter values different from 1 are indicative of an effect of the campaign. For Table 5.1, the relevant parameter for the saturated effect model is $\hat{\gamma}_2^{XY} = 1.302$. The odds of being a voter rather than a nonvoter are $1.302^2 = 1.696$ times greater after the campaign than before and it is concluded that the campaign has had the desired effect.

If the dependent variable has more than two categories, the analyses proceed in the same manner. Let us look at Table 5.2 in which data are presented on the kinds of cars people own. The net changes rendered in Table 5.2b might be used to evaluate the success of the advertisement campaigns each of the major automobile manufactures has conducted.

As the log-linear effects reported in Table 5.2b indicate, the "Other" brand's share of the market diminished, much more so than any other brand. Its advertisement campaign failed. The great winner with the most successful advertisement campaign was Chrysler Corporation; Ford and General Motors occupy a middle position.

Although the distinction between panel and trend design is not relevant for determining the strength of the relationship between X and Y as a measure of the effect of the intervening event, this distinction is important for determining the statistical significance of this relationship. If tables such as Tables 5.1 and 5.2b originate from a trend study, the usual χ^2-tests of statistical independence in Table XY are appropriate. In log-linear terms: The validity of the independence model $\{X, Y\}$ is tested. In the case of a panel design resulting in dependent observations, statistical independence in Table XY must be tested by means of tests of marginal homogeneity in turnover tables such as Tables 4.8 and 5.2a. Below we will show in more detail what kinds of χ^2-tests are available.

In general, it is important to notice that, contrary to the views of Davis (1978, p.175), the χ^2-tests used in the case of panel design usually have more power than those employed in the trend design, just as was indicated above for the t-tests between two means. This is easily seen if one remembers that there is a close relationship among the z, t and

TABLE 5.2 Brand Loyalty of Automobile Owners, Michigan 1966:
Gross and Net Changes

(a) Gross Changes

	Old Car (t_1)				
	General Motors	Ford Motor Company	Chrysler Corporation	Other	Total
New Car (t_2)					
General Motors	120,019	17,710	4,840	4,626	147,195
Ford Motor Company	16,912	48,933	3,122	3,648	72,615
Chrysler Corporation	10,517	6,460	19,640	2,384	39,001
Other	3,044	1,739	646	7,867	13,296
Total	150,492	74,842	28,248	18,525	272,107

(b) Net Changes

Brand (Y)	1. Before ($X_c - t_1$)	2. After ($X_e - t_2$)	
1. General Motors	55.3%	54.1%	$\hat{\tau}_{21}^{XY} = 0.997$
2. Ford Motor Company	27.5%	26.7%	$\hat{\tau}_{22}^{XY} = 0.993$
3. Chrysler Corporation	10.4%	14.3%	$\hat{\tau}_{23}^{XY} = 1.184$
4. Other	6.8%	4.9%	$\hat{\tau}_{24}^{XY} = 0.854$
100% =	272,107	272,107	

SOURCE: Adapted from Zeisel (1968: 214)

χ^2-distributions (Hays, 1981, Section 9.11) and that a polytomous variable with k categories can be represented in the form of $k - 1$ dichotomous dummy variables with categories 0 and 1 whose mean is p, the proportion that scores one, and variance is $p(1 - p)$. Therefore, χ^2-tests on the similarity of frequency distributions can be transformed into z-tests or t-tests on equality between means (proportions). It was shown above that these t-tests or z-tests were generally more powerful for panel data than trend data. In the same vein, if the pretest and posttest scores are "uncorrelated" with each other, that is, there is independence in the turnover table, trend and panel data will yield the same χ^2-test statistics. If the "correlation" between pretest and posttest is positive, that is, most observations are to be found on the main diagonal of the turnover table, panel data yield more powerful tests. If the pretest-posttest "correlation" is negative, that is, most observations are to be found outside the main diagonal, trend data yield the more powerful results (Hagenaars, 1985, Table 4.14).

As stated above, if the data in Tables 5.1 and 5.2b had come from a trend study, the usual χ^2-testing procedure to test the significance of the relation between X and Y in those tables could have been used. In that case, the test statistics for the log-linear independence model $\{X, Y\}$ applied to Table 5.1 would have been $L^2 = 13.63$, df = 1, $p = .00$ (Pearson-$\chi^2 = 13.53$) and for Table 5.2b: $L^2 = 2660.11$, df = 3, $p = .00$ (Pearson-$\chi^2 = 2648.79$). Thus, the hypothesis that the intervening event had no effect would have been rejected for both examples.

Extensions of the simple separate-sample pretest-posttest design to more than two waves or several subgroups can be handled in the same straightforward manner, as will be shown in the next chapter.

The testing of the hypotheses of "no net change" using panel data presents more difficulties. Because the data in Tables 5.1 and 5.2b do come from panel studies, the before measurements involve the same people as the after measurements and the ordinary χ^2-test of statistical independence cannot be used. The relevant tests have to be performed using the turnover Tables 4.8 and 5.2a.

If the dependent variable is dichotomous, the tests are comparatively simple. The hypothesis of "no effect" for the voting example is identical to the hypothesis that, in turnover Table 4.8, F_{1+}^{AB} equals F_{+1}^{AB}, or, what amounts to the same, $F_{12}^{AB} = F_{21}^{AB}$. McNemar (1947) has derived the Pearson χ^2-test statistic to test this hypothesis for 2 × 2 tables. Hays presented the formula in which a correction for continuity is included (Hays, 1981, Section 15.9). The outcome for Table 4.8 is Pearson-$\chi^2 = 18.84$, df = 1, $p = .00$. This is somewhat larger than the corresponding test for trend data (which yielded Pearson-$\chi^2 = 13.53$), exemplifying the greater power of the panel test.

Bhapkar (1973), Cochran (1950 — see also Hays, 1981, Section 16.4), Hamdan, Pirie, and Arnold (1975), and Marascuilo and Serlin (1979) have generalized McNemar's test for application in more complicated situations in which the dependent characteristic is still dichotomous but in which the number of waves is more than two and/or in which the analysis is carried out among several subgroups.

McNemar's test is actually a test of symmetry in a 2 × 2 table. Applying the log-linear symmetry model to Table 4.8 yields L^2 = 20.09, df = 1, p = .00 (Pearson-χ^2 without the continuity correction is 19.59 and with the continuity correction 18.84, identical to McNemar's test statistic). The extensions of the symmetry model given by Bishop et al. (1975, Section 8.3) and Haberman (1979, Section 8.3) can be used to test the effects of intervening events on dichotomous characteristics if the one-group pretest-posttest design contains more than one pretest and/or posttest. Extensions of the one-group pretest(s)-posttest(s) design to include several subgroups poses no special problems with dichotomous variables as it has been shown in Section 4.3.1 how to test symmetry models simultaneously for several subgroups.

If the dependent variable is not dichotomous, the tests are more complicated. In the general $I \times I$ turnover table, marginal homogeneity and symmetry are no longer identical. The "no effect" hypothesis for the automobile example implies that marginal homogeneity, but not necessarily symmetry, exists in turnover Table 5.2a. As mentioned above in Section 4.2.3, Bishop et al. (1975, Chapter 8), Haberman (1979, Chapter 8), and Haber and Brown (1986) presented several tests of marginal homogeneity for polytomous data.

One of the most attractive unconditional test statistics is Stuart's Q (Haberman, 1979: 499; Stuart, 1955) which has approximately a χ^2-distribution with $I - 1$ degrees of freedom. For Table 5.2a this test statistic is Q = 5183.13, df = 3, p = .00. The "no effect" hypothesis definitely has to be rejected. If the value of Q is compared with the corresponding test for trend data, which yielded Pearson-χ^2 = 2648.79, the greater power of the test using panel data is apparent.

A conditional test of marginal homogeneity is obtained by testing the log-linear symmetry model given the validity of the log-linear quasi-symmetry model (Section 4.2.3). This test is most attractive if the quasi-symmetry model approximately holds. Applied to Table 5.2a, the test result for the symmetry hypothesis is L^2 = 5435.65, df = 6, p = .00 (Pearson-χ^2 = 5279.54) and for the quasi-symmetry hypothesis: L^2 = 95.75, df = 3, p = .00 (Pearson-χ^2 = 95.24).

Although, on the basis of the p-value, the quasi-symmetry model has to be rejected, the size of the test statistic L^2 is very small given the number of observations in Table 5.2. There is not much reason to doubt that the quasi-symmetry model fits the data rather well. The outcomes for the conditional test of marginal homogeneity are L^2 = 5339.90, df = 3, p = .00. These values are rather close to those obtained for Q. If one uses

Pearson-χ^2 instead of L^2 for testing the marginal homogeneity hypothesis, the values are almost identical: Pearson-χ^2 = 5184.30. Comparing the conditional with the unconditional tests, the former appear to be more flexible. The extensions of the (log-linear) symmetry and quasi-symmetry model indicated before make it possible to use the conditional tests to analyze rather complex panel designs with more than two waves and including subgroup analyses. These possibilities are not readily available when using the unconditional tests.

The size and the statistical significance of the difference $O_1 - O_2$ can be determined nicely by means of log-linear models. What log-linear analysis or any other analysis technique cannot do is to ascertain to what extent the difference between pretest and posttest may indeed be attributed to the intervening event. Although it has been established that the number of nonvoters decreased, is this difference really caused by the Government's campaign? Is an increase in automobile sales of one particular brand compared with another really due to the superior advertisement campaign of the former?

All kinds of disturbing factors may be effective which actually are responsible for the difference $O_1 - O_2$ or which prevent the effects of X_e from being reflected in the data. Campbell and Stanley (1966) and Cook and Campbell (1979) presented a systematic survey of all kinds of threats to the internal and external validity of several common designs, including the pretest-posttest design. The most important ones will be dealt with briefly.

The dangers of testing (test-retest, or reinterviewing) effects have been traditionally recognized as one of the main weaknesses of the panel design. The occurrence of "testing" implies that the respondent's answers at the posttest are directly influenced by the fact that a pretest has been taken and that this influence is confounded with the influence of the intervening event. For example, in line with the Socratic effect mentioned in Section 4.4.3, because the respondent has been questioned about his or her vote intentions, he or she may start to think more seriously about these intentions, come to the conclusion that voting is important, and eventually change from nonvoter to voter. The intermediate government's campaign may have had no influence at all, but still the number of nonvoters will have decreased from pretest to posttest and we will wrongly conclude that the campaign was successful. Test-retest effects will be discussed at length in Section 5.5. Now it suffices to notice that test-retest effects occur only in the one-group pretest-posttest design and not in the separate-sample pretest-posttest design where different groups of respondents are questioned at pretest and posttest.

Another disturbing factor, which will also be discussed more thoroughly in Section 5.5, poses a threat to the validity of both panel and trend studies. Campbell and associates call it "mortality" and its most common form is nonresponse. The key difference $O_1 - O_2$ may be influenced by nonresponse if nonresponse has a different effect on O_1 than on O_2. In panel studies, this influence can be eliminated by restricting the analysis to those respondents who have participated in all waves. However, by doing this, the generalizability of the results will be seriously endangered.

The biggest problem in the one-group and separate-group pretest-posttest designs is what Campbell and associates call "history," the occurrence of events that may directly influence the post-test scores and whose effects are confounded with those of X_e. In the classic experiment, the investigator determines exactly what X_e is. Care is taken that everybody in a particular experimental group is exposed to the experimental stimulus in the same manner and that there are no other relevant events occurring between pretest and posttest.

If the pretest-posttest design has come about by making use of existing surveys repeated over time, such control over the intervening events is impossible. In terms of our voting example, although the government has launched a nationwide campaign to enhance the number of voters, not everybody in the country will have noticed this to the same extent and in the same manner, and beside the campaign, many other things may have happened causing people to change their vote intention. It may be that the political issues that have come to the fore between pretest and posttest are particulary relevant for those social groups from which the nonvoters originate and that therefore, and not because of the government campaign, a number of nonvoters have become voters.

Between pretest and posttest, many relevant events may take place whose effects show complicated patterns of interaction, both on the long term and on the short term. Using the one-group or separate-sample pretest-posttest design, it will often be difficult to establish exactly and indisputably what caused the difference $O_1 - O_2$. A thorough knowledge of what intervening events took place in combination with theoretically and empirically based estimates of the possible influence of all these events provides the only way out.

Another very important and potentially disturbing factor, be it of a somewhat different nature, is unreliability of the measurements. The consequences of unreliability and the possibilities to correct for it have been discussed in more general terms in Chapter 3 and Section 4.4. The more specific consequences of unreliability for the analysis of the pretestposttest design will be briefly indicated below.

From the analysis of Table 4.8 on vote intention in Section 4.4.2, it appeared that a simple latent class model in which no latent change takes place can account for the observed data (see Table 4.9). In this model, the observed net change of vote intention results completely from the unreliability of the measurements, in particular from the fact that the unreliability of the pretest was somewhat lower than of the posttest. However, in the discussion of Table 5.1 in which the net changes were presented, it was concluded that the net changes were significant and were indicative of the success of the campaign. As this example shows, manifest net changes may not point to true net changes which are caused by the intervening event but result solely from measurement error.

Unreliability may also distort the conclusions in the case of true latent change. In most cases, unreliability will influence the marginal distributions of the manifest variables in such a way that the manifest distributions are less skewed, more uniformly distributed than the corresponding latent (true) distributions. Very often this will imply that the true differences between the marginal distributions of the pretest and posttest "scores" are underestimated by the corresponding manifest differences. Therefore, the effect of the intervening event will be underestimated and may not even be judged statistically significant on the basis of the χ^2-tests. Other outcomes are possible, but in general, as far as unreliability influences the marginal distributions of variables, the conclusions about the effect of the intervening event will be distorted.

Another consequence of unreliability is that unreliability in many cases attenuates the true association. Mostly, the positive association between the manifest pretest and posttest scores will be less than the underlying true positive association. This has consequences for the marginal homogeneity tests performed in the one-group pretest-posttest design. After all, as has been discussed above, the size of the χ^2-test statistic and the power of the χ^2-test in a panel design depend on the strength of the association in the turnover table. The lower this association is and the fewer people are to be found on the main diagonal, the lower is the value of the χ^2-statistics (other things being equal).

So unreliability will generally lower the values of the χ^2-statistics in a panel design in two ways: by diminishing the true differences and by diminishing the association between pretest and posttest scores. Because of unreliability, existing differences pointing to existing effects of the intervening event may not be detected. In more complicated designs, other kinds of consequences must be reckoned with. The solution is to take unreliability explicitly into account by building models with latent variables and testing the net changes on the latent level. In Section 4.4 it was pointed out how to perform such analyses. The techniques are available, the necessary data will often be missing.

5.3 The Nonequivalent Control Group Design

The major weaknesses of the panel design as a one-group pretest-posttest design are the confounding of the effect of X_e with the effects of other intervening events (history) and the uncertainty about the degree to which the respondents have been exposed to the "experimental stimulus" X_e.

Both major shortcomings can be remedied by dividing the total sample into several groups according to the degree or the manner in which they were affected by the intervening event. This of course is possible only if the exposure to the "experimental stimulus" has been explicitly measured. In terms of the examples used in the previous section, it is necessary to measure explicitly to what degree the respondents have noticed the Government's campaign urging people to vote or to what extent they have been exposed to the advertisement campaigns of the car manufacturers.

The resulting design is known as the nonequivalent control group design (Campbell & Stanley, 1966; Cook & Campbell, 1979). In its most elementary form this design may be represented as follows:

$$O_1 \quad X_e \quad O_2$$
$$\overline{}$$
$$O_3 \quad X_c \quad O_4$$

In this nonequivalent control group design, there are two groups of respondents, each of which receives a pretest and a posttest. The values of the "experimental" variable X are different for the two groups. For example, the first group has noticed the Governments's campaign to promote voting (X_e), the other group has not (X_c). The frequency distribution of the variable Voting at t_1 for the "experimental" and the "control" group is symbolized by O_1 and O_3 respectively; O_2 and O_4 refer to the measurements of voting for the two groups at t_2.

The assignment of the respondents to the two groups has not occurred randomly. The respondents themselves have "chosen" whether or not to notice the campaign. Because of this, the symbol R was not placed in the beginning of each row of the diagram, but the two rows are separated by a dashed line. In the notation used by Campbell and associates, this indicates nonrandom assignment of respondents to the groups.

The effect of X is determined by comparing the posttest difference $O_2 - O_4$ with the pretest difference $O_1 - O_3$, or, equivalently, by comparing the net change $O_2 - O_1$ in the "experimental" group with the net change $O_4 - O_3$ in the control group. (In algebraic terms: $[O_2 - O_4] - [O_1 - O_3] = [O_2 - O_1] - [O_4 - O_3]$.)

The strong and weak points of this variant of the panel design can be clarified by comparing it with the cross-sectional one-shot survey, with the separate-group and the one-group pretest-posttest designs, and with the classic experiment.

The one-shot survey may be considered as an "after only" or "posttest only" design. In terms of the above diagram, all respondents' scores on the posttest and on X (X_e or X_c) are known, but no pretest was conducted. In the one-shot survey, the posttest difference $O_4 - O_2$ is used as a measure of the effect of X. This is not without risks as can be clearly seen from the nonequivalent control group diagram: The posttest differences may not come from X but may merely reflect group differences that already existed at the pretest. The nonequivalent control group design is a much stronger design: The pretest differences can be taken into account when evaluating the effect of X by means of the posttest differences.

The nonequivalent control group design also has some advantages over the pretest-posttest design, especially the one-group pretest-posttest design. As discussed above, the pretest-posttest design assumes two forms, that is, the separate-sample and the one-group pretest-posttest design, depending on whether the data were gathered by means of a panel or a trend study. The nonequivalent control group design, a panel design, seldom or never has a "trend counterpart." Setting up such a counterpart is possible only if it is known at t_1 which of the people interviewed at t_1 will later be exposed to X_e (e.g., the campaign) and which to X_c (did not notice the campaign). After all, in a trend study this cannot be established afterwards, as, by definition, the people interviewed at t_1 will not be reinterviewed. In most cases, this kind of information will not be available. The shortcomings of the separate-sample pretest-posttest design have to be overcome by increasing the number of pretest and posttests, as will be shown in the next chapter.

However, most of the confounding factors operating in the one-group variant of the pretest-posttest design are eliminated by the nonequivalent control group design. The main effects of "mortality" (nonresponse), of "testing" (reinterview effects), and, especially, of "history" (intervening events other than X_e and X_c) play a confounding role in each group but as such cancel out in the total nonequivalent control group design because there the effect of X is determined by comparing the net changes in the "experimental" group with the net changes in the "control" group. "History, " for example, may determine the net changes in the "experimental" and in the "control" group, but, as far as these effects are the same in both groups, the difference between the net changes is not influenced by "history."

In many respects, the nonequivalent control group design resembles the design of the classic experiment. There is, however, one overriding difference. In a true experiment, the individuals are randomly assigned to the experimental and the control groups. In the nonequivalent control group design, the assignment of the respondents to the groups does not occur randomly, and even worse: the "assignment rule" is unknown. The unknown "assignment variable" (Judd & Kenny, 1981) may cause the "experimental" and "control" groups to be different in many unknown respects. Using Campbell and associates' terminology: The confounding factor "Selection" is not completely under control (Campbell & Stanley, 1966; Cook & Campbell, 1979).

Because of the importance of this distorting factor, it warrants a closer look. When the nonequivalent control group design was compared above with the one-shot survey, it was remarked that since the one-shot survey was a posttest-only design, the effects of X were determined only by means of the posttest differences and therefore may be confounded with the unknown pretest differences. "Selection" is an important confounding factor in the one-shot survey. As far as selection has the same effects on the pretest and the posttest, the distorting effects of selection are eliminated in the nonequivalent group design: To determine the effects of X, the posttest differences are compared with and adjusted for the known pretest differences. The main effects of selection are under control in the nonequivalent control group design, just as the main effects of mortality, testing, and history.

However, excluding the occurrence of distorting main effects does not guarantee the absence of distorting interaction effects. The (main) effects of mortality, history, and testing are under control as far as they are the same in both groups. This is less likely to the extent that the groups differ from each other, that is, to the extent that the assignment variable has created differences between the groups. Interaction effects of testing and selection and especially of history and selection must therefore be reckoned with in the nonequivalent control group design. It is also possible that selection interacts with the causal factor X in the sense that different results might have been obtained if the present "control group" had been exposed to X_e, and the present "experimental group" to X_c.

Nevertheless, although many disturbing interaction effects may occur, the control of the main effects of several important potentially confounding factors is an important asset of the nonequivalent control group design compared with most other quasi-experimental designs.

A great deal has been written about how to analyze data from the nonequivalent control group design.[1] Not surprisingly in the light of the above, the discussion mostly concerns the question how to correct or adjust the posttest differences for the pretest differences in order to eliminate the main effects of selection and to get an unbiased estimate of the effect of X.

There are two main approaches. One is the unconditional (Goldstein, 1979; Plewis, 1985) or analysis of variance approach in which the effect of X is estimated by simply subtracting the pretest differences from the posttest differences. The basic rationale behind this approach is that if the intervening event X_e has no influence, the pretest differences remain the same over time. The other is the conditional or analysis of covariance approach. In this latter approach, one predicts the posttest differences on the basis of the relation between pretest and posttest scores and estimates the effect of X as the difference between observed posttest differences and predicted posttest differences. In both approaches, the consequences of measurement unreliability have to be taken into account.

Most of the literature deals with pretest and posttest scores measured at interval level. (Somewhat of an exception are Campbell & Clayton, 1961, and Plewis, 1981, 1985.) The essence of this literature will be briefly discussed to make clear what exact form the unconditional and conditional analyses assume in the case of interval level data. Next, conditional and unconditional analyses techniques for categorical data will be dealt with, followed by a discussion of the appropriateness of the conditional and unconditional approaches. Finally, attention will be paid to the role of unreliability in these approaches.

The unconditional analysis amounts to performing an analysis of variance on the differences between the posttest scores $Y(t_2)$ and the pretest scores $Y(t_1)$. The dependent variable is $Y(t_2) - Y(t_1)$ and the treatment variable is X with two categories X_e and X_c. The changes in Y are attributed to the causal factor X. In terms of regression analysis, this implies the following equation:

$$Y(t_2) - Y(t_1) = a + b_x X + e \tag{5.4}$$

where X assumes the value 0 if the respondent belongs to the "control" group (has been exposed to X_c) and the value 1 if the respondent belongs to the "experimental" group (has been exposed to X_e).

If the symbols O_i in the above diagram of the nonequivalent control group design are replaced by the mean values $\overline{Y}_i$, it can be shown that the effect of the causal factor b_x in equation 5.4 is a straightforward function of the pretest and posttest group mean differences (Judd & Kenny, 1981: 107):

$$b_x = (\overline{Y}_2 - \overline{Y}_4) - (\overline{Y}_1 - \overline{Y}_3) \tag{5.5}$$

In the conditional approach, an analysis of covariance is performed with $Y(t_2)$ as the dependent variable, X as the causal factor, and $Y(t_1)$ as the covariate:

$$Y(t_2) = a + b_x X + b_y Y(t_1) + e \tag{5.6}$$

According to equation 5.6, the effect of X is estimated as (Judd & Kenny, 1981: 107)

$$b_x = (\overline{Y}_2 - \overline{Y}_4) - b_y (\overline{Y}_1 - \overline{Y}_3) \tag{5.7}$$

In the conditional approach, the effect of the intervening event is not simply calculated by subtracting the pretest differences between the group means from the posttest differences between the group means. Rather, the expected differences at the posttest are determined on the basis of the linear relationship between pretest and posttest scores and these expected differences $b_y(\overline{Y}_1 - \overline{Y}_3)$ are subtracted from the mean posttest differences $(\overline{Y}_2 - \overline{Y}_4)$.

As can be seen from these equations, in the unconditional approach, the pretest and posttest differences are assumed to be stable if there is no effect of X, whereas in the conditional approach, even in the absence of an effect of X, the posttest differences deviate from the pretest differences to an extent that depends on the nature of the linear relationship between pretest and posttest. It should be clear that the two ap-

TABLE 5.3 Anti-Semitism in November (*A*. Pretest) and May (*B*. Posttest) and Seeing the Film "Gentleman's Agreement" (*X*)

Anti-Semitism				Anti-Semitism		
A. November	*B. May*	*Frequencies*		*A. November*	*B. May*	*Frequencies*
X. 1. Respondent saw film				X. 2. Respondent did not see film		
1. High	1. High	20		1. High	1. High	92
1. High	2. Moderate	6		1. High	2. Moderate	20
1. High	3. Low	6		1. High	3. Low	20
2. Moderate	1. High	4		2. Moderate	1. High	34
2. Moderate	2. Moderate	12		2. Moderate	2. Moderate	15
2. Moderate	3. Low	10		2. Moderate	3. Low	27
3. Low	1. High	3		3. Low	1. High	24
3. Low	2. Moderate	5		3. Low	2. Moderate	28
3. Low	3. Low	49		3. Low	3. Low	121
Subtotal		115		Total		496

SOURCE: Campbell and Clayton (1961) and Glock (1955)

proaches will not lead to the same conclusions about the effect of X, unless $b_y = 1$. A discussion of which approach to choose under what circumstances will be postponed until the log-linear analysis of the nonequivalent control group design has been addressed.

The data in Table 5.3 will be used to explain how to employ log-linear models in the case of categorical pretest and posttest scores.

The data in Table 5.3 pertain to the effects on anti-Semitism of (voluntarily) seeing the movie "Gentleman's Agreement" between November and May. Because of the contents of the film, it is expected that seeing the film will reduce the degree of anti-Semitism. These data were analyzed before by Campbell and Clayton (1961) and Glock (1955). A rearrangement of the data in Table 5.3 according to the diagram of the nonequivalent control group design is presented in Table 5.4.

The percentages in Table 5.4 indicate that the degree of anti-Semitism has declined somewhat between November and May among those who saw the movie, but increased slightly among those who did not see the movie. This suggests a small effect of the movie on anti-Semitism. The same follows from comparing the pretest and posttest differences. The people who have seen the movie are somewhat less anti-Semitic than those who did not see the movie, but this difference is somewhat larger at the posttest than at the pretest.

A more formal conditional (covariance) analysis confirms this. The essential point of a covariance analysis of the nonequivalent control group design is that the posttest scores are influenced both by the pretest scores and by the score on the causal factor. In terms of the log-linear effect model, this means that the odds of obtaining a particular score on B (the degree of anti-Semitism in May) depend directly on the score on X (whether one saw the movie or not) and on the pretest score A (the degree of anti-Semitism in May).

Three log-linear models are therefore relevant for the data in Table 5.3. First of all, there is model $\{AX, AB\}$ which represents the null hypothesis that, in the log-linear effect model, seeing the film or not has no direct effect on the degree of anti-Semitism in May when the pretest scores B are held constant. This model has to be rejected employing a .05 significance level: $L^2 = 17.02$, df = 6, $p = .01$ (Pearson-$\chi^2 = 16.09$). Seeing the movie apparently did influence the degree of anti-Semitism.

This brings us to the second relevant model, model $\{AX, AB, BX\}$, in which the posttest scores (B) are directly influenced by both the causal factor X and the pretest scores (A), but in which the three-variable interaction ABX is absent. This model can be accepted: $L^2 = 7.90$, df = 4,

TABLE 5.4 "Nonequivalent Control Group Design": Representation of the Effects of Seeing the Film "Gentleman's Agreement" on Degree of Anti-Semitism

A. *Anti-Semitism November*

O_1:			
	1. High	32	(27.8%)
	2. Moder.	26	(22.6%)
	3. Low	57	(49.6%)
	Total	115	(100%)

X_e: 1. Saw film

B. *Anti-Semitism May*

O_2:			
	1. High	27	(23.5%)
	2. Moder.	23	(20.0%)
	3. Low	65	(56.5%)
	Total	115	(100%)

X_c: 2. Did not see film

O_3:			
	1. High	132	(34.6%)
	2. Moder.	76	(19.9%)
	3. Low	173	(45.4%)
	Total	381	(100%)

O_4:			
	1. High	150	(39.4%)
	2. Moder.	63	(16.5%)
	3. Low	168	(44.1%)
	Total	381	(100%)

SOURCE: Table 5.3

$p = .10$ (Pearson-$\chi^2 = 7.82$). Conditionally testing the validity of the first model, in which no effect of the film appeared, against this model, which includes the main effects of seeing the film, yields $L^2 = 9.12$, df $= 2$, $p = .01$. Employing the .05 significance level, the second model has to be preferred. It is concluded that seeing the film has directly influenced the degree of anti-Semitism in May and this effect cannot be explained by means of the differences in anti-Semitism already present at the pretest in November between those who eventually went to see the movie and those who did not.

The nature of the effects of the film on the degree of anti-Semitism can be computed from the parameter estimates in the log-linear covariance model $\{AX, AB, BX\}$. The results are

$$\hat{\beta}^{X\bar{B}}_{11/2} = -0.367 \qquad \hat{\beta}^{X\bar{B}}_{11/3} = -0.431 \qquad \hat{\beta}^{X\bar{B}}_{12/3} = -0.064$$

$$\hat{\gamma}^{X\bar{B}}_{11/2} = 0.693 \qquad \hat{\gamma}^{X\bar{B}}_{11/3} = 0.649 \qquad \hat{\gamma}^{X\bar{B}}_{12/3} = 0.937$$

In agreement with the expectations, the odds of scoring high on anti-Semitism in May ($B = 1$) rather than middle ($B = 2$) or low ($B = 3$) are (somewhat) greater among those who did not see the movie ($X = 2$) than among those who did ($X = 1$). The odds moderate/low are hardly influenced by seeing the movie.

There is a third log-linear covariance model possible which also includes a direct effect of the film on the posttest scores, but an effect that depends on the pretest scores. This is the saturated model $\{ABX\}$. The main disadvantage of this model is the presence of the three-variable interaction term γ^{AXB}_{ik}. This term may be looked upon as indicating to what extent the effect of seeing the film on the posttest scores varies among those who scored differently at the pretest. The three-variable effect may also be interpreted as indicating to what extent the effect of the pretest on the posttest is different for those who saw the film and those who did not. This three-variable interaction effect cannot be attributed unequivocally to either the causal factor "film" or the pretest. Therefore, this saturated model will not be preferred if the simpler model without the three-variable effect fits the data.

Although somewhat more complicated, it is also possible to define a log-linear analogue to the analysis of variance of the pretest and posttest differences. As was indicated above, if seeing the film does not have any effect on the degree of anti-Semitism, the analysis of variance approach implies that the posttest differences ($O_2 - O_4$) are equal to the pretest differences ($O_1 - O_3$), or, equivalently, the net changes in the "experimental" group ($O_2 - O_1$) are equal to the net changes in the "control" group ($O_4 - O_3$). With categorical pretest and posttest scores, these equalities are expressed in terms of similarities among the "experimental" and "control" group pretest and posttest distributions.

The relevant distributions are presented in Table 5.4. Hypotheses about similarities among these distributions cannot be tested directly by means of the ordinary χ^2-tests applied to Table 5.4 as we have panel data here and the pretest and posttest distributions pertain to the same people. We have to use the tests of marginal homogeneity discussed in Section 4.2.3.

First, it will be shown how the comparison of the posttest differences ($O_2 - O_4$) with the pretest differences ($O_1 - O_3$) can be made in log-linear terms using (quasi-)symmetry models. After that, the comparison of the net change ($O_2 - O_1$) in the "experimental" group with the net change ($O_4 - O_3$) in the "control" group will be discussed in log-linear terms.

The left part of Table 5.4 renders the cell frequencies of the marginal table AX, where A refers to the pretest scores in November and X to having seen the film or not. If the relative distribution of anti-Semitism in November is the same for the "experimental" and the "control" group, A and X will be independent of each other and all parameters τ_{ik}^{AX} of the saturated model $\{AX\}$ applied to the marginal table AX will be equal to one. If, as might be expected here, the relative distribution of A is different for both groups, the parameter estimates $\hat{\tau}_{ik}^{AX}$ indicate in what respects the two relative distributions differ. In this sense, the parameters τ_{ik}^{AX} measure the pretest differences ($O_1 - O_3$).

In the same manner, with the aid of the right part of Table 5.4, model $\{BX\}$ for the marginal table BX can be defined and its parameters τ_{jk}^{BX} interpreted as measures of the posttest differences ($O_2 - O_4$). Consequently, absence of any effect of X and equality of pretest and posttest differences can be expressed as

$$\tau_{ik}^{AX} = \tau_{jk}^{BX} \qquad \text{for all } i = j \qquad (5.8)$$

An extensive discussion of how to test hypotheses about stable or changing associations was presented in Section 4.3.3. What is necessary is a rearrangement of Table 5.3 so that the marginals of the rearranged table coincide with table AX and BX respectively. Table 5.5 is that table.

The test results are reported in Table 5.6. Model 1 in Table 5.6 is the symmetry model by means of which the symmetry of Table 5.5 is tested. As has been explained in Section 4.3.3, in this case, the symmetry model

TABLE 5.5 Seeing the Film (X) and Degree of Anti-Semitism in November (A) and May (B)

		May[a]						
X. Seen Film	A. Anti-Semitism November	1. Yes 1. High	1. Yes 2. Mod.	1. Yes 3. Low	2. No 1. High	2. No 2. Mod.	2. No 3. Low	Total
1. Yes	1. High	20	6	6	0	0	0	32
1. Yes	2. Moderate	4	12	10	0	0	0	26
1. Yes	3. Low	3	5	49	0	0	0	57
2. No	1. High	0	0	0	92	20	20	132
2. No	2. Moderate	0	0	0	34	15	27	76
2. No	3. Low	0	0	0	24	28	121	173
Total		27	23	65	150	63	168	496

a. First row refers to X (Seen Film), the second row to B (Anti-Semitism May).
SOURCE: Table 5.3

TABLE 5.6   Test Statistics for (Quasi-)symmetry and Marginal Homogeneity in
            Table 5.5

Model	L^2	df	p	Pearson-χ^2
1. Symmetry $(X—t_1, X—t_2)$	7.18	6	.30	7.08
2. Quasi-symmetry XA, XB	0.74	2	.69	0.75
3. Partial symmetry A, B $(X—t_1, X—t_2)$	4.93	4	.29	4.90
4. (1) – (2)	6.44	4	.17	
5. (3) – (2)	4.19	2	.12	

reproduces exactly the observed marginal distribution of X. Because the symmetry model need not rejected, it is concluded that the marginals of Table 5.5 are equal to each other, that is, tables AX and BX are identical to each other in the population. The "no effects" hypothesis in equation 5.8 can be accepted. Seeing the film did not significantly influence the degree of anti-Semitism, a conclusion rather different from the results of the log-linear covariance analysis.

In this case, no further tests are necessary to test the "no effects" hypothesis. But if the symmetry model had been rejected, it would not have followed automatically that the "no effect" hypothesis also had to be rejected. Therefore, the implications of the other models mentioned in Table 5.6 will be briefly discussed.

Model 2 in Table 5.6 is the familiar quasi-symmetry model which reproduces the observed marginal distributions AX and BX. Model 3 is the partial symmetry model in which it is postulated that Table 5.5 is symmetric as far as the differences between the marginal distributions of A and B permit.

Model 4 in which the symmetry model is compared with the quasi-symmetry model is used to test whether there is marginal homogeneity in Table 5.5. Acceptance of this model implies acceptance of the "no effect" hypothesis in equation 5.8. Rejection of this hypothesis still does not necessarily imply rejection of the "no effect" hypothesis. Tables AX and BX may differ from each other only because the distribution of A differs from the distribution of B. If these differences between the marginal distributions A and B are the same for the "experimental" and the "control" group, there still is no reason to assume an effect of X.

Model 5, which compares the partial symmetry model with the quasi-symmetry model, is used to test whether there is marginal homogeneity in Table 5.6 as far as the differences in the marginal distributions of A and B permit. This is the final and conclusive test of the "no effect" hypothesis in equation 5.8. Accepting model 5 implies definitely accepting the "no effect" hypothesis and rejection of model 5 implies definitely accepting the existence of an effect of X.

Exactly the same results as presented in Table 5.6 are obtained if the net changes in the "experimental" group ($O_2 - O_1$) are compared with the net changes in the "control" group ($O_4 - O_3$). (See Table 5.4.) If the net changes of the degree of anti-Semitism are the same for both groups, it is concluded that the film did not influence the degree of anti-Semitism. To test the equality of the net changes, the conditional turnover tables AB for both "experimental" group and "control" group have to be used. The turnover table for the "experimental" group can be found in the first column of Table 5.3 or in the upper left corner of Table 5.5 and for the "control" group in the second column of Table 5.3 or in the lower right corner of Table 5.5.

In section 4.3.1, it has been shown how to define (quasi-)symmetry and marginal homogeneity models for several subgroups simultaneously. As has been explained in Section 4.3.3 and should be clear from inspection of Table 5.5, symmetry in both conditional turnover tables AB is identical to symmetry in Table 5.5. Thus the test statistics for the symmetry model in Table 5.6 are identical to the test statistics for the hypothesis that both conditional turnover tables are symmetrical.[2] From the test results in the first row of Table 5.6, it follows that both conditional turnover tables are symmetrical in the population and that neither in the "experimental" group nor in the "control" group has any significant net change in anti-Semitism taken place. Therefore, the film did not have any effect.

The quasi-symmetry and the marginal homogeneity models for both conditional turnover tables are related in a similar vein to the models in Table 5.6. Model 2 in Table 5.6 can be conceived as representing the hypothesis that both conditional turnover tables AB are quasi-symmetrical. Model 3 can be interpreted as the partial symmetry model in which it is postulated that both conditional turnover tables AB are symmetrical as far as the general trend in anti-Semitism permits; this trend is supposed to be identical in the "experimental" and the "control" group. Model 4 provides a simultaneous conditional test of the hypothesis of marginal homogeneity in both conditional turnover tables. Finally, by means of model 5 the hypothesis is tested that there is marginal homogeneity in both conditional turnover tables as far as the general trend in anti-Semitism permits. According to the partial marginal homogeneity hypothesis model 5, the marginal distributions A and B may differ from each other within each conditional turnover table — there may be net change in each group — but the differences are the same for both tables AB — the net changes in anti-Semitism are similar in the "experimental" and the "control" groups. Acceptance of this hypothesis definitely leads to accepting the "no effect" hypothesis; rejection to definitely accepting the existence of an effect of seeing the film on anti-Semitism.

No significant effects of X have been found by means of these unconditional analyses. But if a significant effect of X had been found, the next step would have been to determine the size and nature of these effects. In Section 4.3, it has been shown in principle how to obtain the desired effect estimates. Use has to be made of Table 5.4, which is now conceived as a three-dimensional table OXT, where O refers to the degree of anti-Semitism, X to seeing the film or not, and T to time of measurement ($T = 1$: pretest, $T = 2$: posttest).

The saturated model $\{OXT\}$ is applied to this table OXT. The two-variable parameters $\tau_{i\,k}^{OX}$ of the saturated model indicate how strong the average relationship at pretest and posttest is between the degree of anti-Semitism (O) and seeing the film or not (X), that is, how large the average differences at both times of measurement are between the relative distributions of anti-Semitism in the "experimental" and the "control" group. The three-variable parameters $\tau_{i\,kt}^{OXT}$ show in what respects the differences between "experimental" and "control" groups with regard to degree of anti-Semitism vary at pretest and posttest. These three-variable parameters $\tau_{i\,kt}^{OXT}$ actually measure the differences between $\tau_{i\,k}^{AX}$ and $\tau_{i\,k}^{BX}$ in equation 5.8 and therefore can be interpreted as measures of the effects of seeing the film.

From the other point of view: The parameters $\tau_{i\,t}^{OT}$ indicate what the average net changes of anti-Semitism are in the "experimental" and "control" groups whereas the interaction effects $\tau_{i\,kt}^{OXT}$ show how different the net changes are in the "experimental" group from the "control" group, differences that are attributed to seeing the film.

Different conclusions about the existence of the effects of X may be drawn depending on whether one performs a conditional "covariance" analysis or an unconditional "analysis of variance" of the pretest-posttest differences. In both approaches, one tries to correct for the confounding factor selection by adjusting the posttest differences for the pretest differences, but in a different way. This raises the question which approach is the right one.

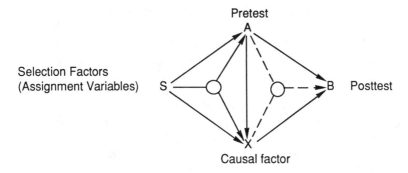

Figure 5.1. The Modified Mediational Model; Adapted from Judd and Kenny (1981: 111)

Judd and Kenny (1981, Chapter 6) answer this question in a very illuminating way by pointing out that it all depends on the way the selection factor operates. They recommend setting up causal models in which not only the causal relations among the causal factor (X), pretest (A) and posttest scores (B) are included but also the unknown assignment variable (S). Achen (1986) makes similar suggestions, but assumes that the assignment variables are partly measured. The function of these models is to make explicit what kind of hypothesis the investigator entertains about the way the respondents are assigned to and have become members of the "experimental" and the "control" groups and to examine what the possible distorting influence is of the assignment variable S on the relations among A, B, and X.

An interesting model Judd and Kenny (1981: 110) proposed is the mediational model the essence of which is that the assignment variables have a direct relation only with the pretest scores and the causal factor X but not with the posttest scores. Their mediational model is depicted in Figure 5.1 in somewhat modified form.

According to the modified mediational model {SAX, AB, BX} (or {SAX, ABX}), the decision to see the film or not is made on the basis of the respondent's degree of anti-Semitism at t_1 and on the basis of the scores on the unknown and unmeasured assignment variables S. In the mediational model, these selection factors S may be related to the initial degree of anti-Semitism, but not directly to the degree of anti-Semitism at t_2. This mediational model is especially attractive when the groups are formed on the basis of the pretest scores.

Judd and Kenny proved that when the pretest and posttest scores are measured at interval level, standard covariance analysis with B as the dependent variable, X as the treatment variable, and A as the covariate provides an unbiased estimate of the effect of X if the mediational model is valid in the population. The same is true when the pretest and posttest scores are categorical and the log-linear "covariance" model $\{AX, AB, BX\}$ or model $\{ABX\}$ is applied to table ABX. This follows from the collapsibility theorem (Section 2.6). If the modified mediational model, in which there is no direct effect of S on B, is valid in the population, collapsing table $SABX$ over the unknown variable S does not change the parameter values which have "B" as superscript. In other words, the parameters τ_{jk}^{BX} and τ_{ijk}^{ABX} do not change after collapsing over S and still unbiasedly estimate the effect of X on B.

However, if the (modified) mediational model is true, the (log-linear) analysis of variance of the difference scores leads to a biased estimate of the effect of X on B. More in particular, when there is no effect of X on B in the mediational model, the "no effect" hypothesis in equation 5.8 $\tau_{ik}^{AX} = \tau_{ik}^{BX}$ will not necessarily be true.

It is more difficult to determine under what kinds of circumstances the unconditional analysis by means of difference scores is most appropriate. Judd and Kenny (1981: 116) defined a "change model" which in a somewhat modified form is depicted in Figure 5.2.

In this change model, it is assumed that the assignment variable S directly influences the pretest and the posttest, be it with the restriction that the effect of S on A is equal to the effect of S on B. It is also very important to notice that, according to the change model, the pretest scores A have no direct influence on the posttest scores B. Judd and Kenny argued that this assignment model is especially applicable when the "experimental" and "control" groups are already existing groups, for example, school classes, or when group assignment has taken place on the basis of stable characteristics, for example, age, education, gender.

Judd and Kenny proved that if the change model is true, standard analysis of variance of the pretest-posttest difference scores gives an unbiased estimate of the effect of X on B; the standard covariance analysis produces a biased estimate in that case. However, for our purposes this result is not very useful.

In the first place, in the change model, any stability of the dependent variable over time is caused only by S and by the correlation between the error terms e_1 and e_2. A direct effect of A on B which does not seem unlikely to occur would destroy Judd and Kenny's formal proof.

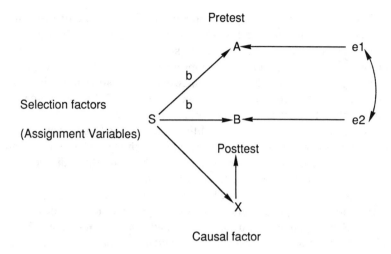

Figure 5.2. The Modified Change Model; Adapted from Judd and Kenny
(1981: 118)

Second, the sources of the correlation between the error terms are most probably unknown latent variables. If these unknown latent variables are correlated with the unknown assignment variable S, again Judd and Kenny's formal proof will be invalid. But how can one ever be certain that these two sets of unknown latent variables are not correlated?

Third, there is a technical problem when translating Judd and Kenny's proof into log-linear terms. It is possible to set up a table $SABXL$ in which L is an extra latent variable and to define for this table a log-linear change model which corresponds to the model in Figure 5.2. In this log-linear model, the direct effect of S on A is equal to the effect of S on B and the latent variable L takes care of the correlation between the error terms. Although this log-linear model meets the requirements of the change model, it will not be generally true that in the absence of an effect of X on B, the parameter τ_{ik}^{AX} in the marginal table AX equals the parameter τ_{ik}^{BX} in the marginal table BX (the "no effects" hypothesis in equation 5.8). Judd and Kenny's formal proof is based on standard path-analytic rules, but these rules are not applicable in the log-linear case (Section 2.7.3).

In sum, Judd and Kenny's change model often will be a very unlikely model and, even if it is applicable, it cannot justify the use of the unconditional log-linear analysis of variance of the pretest-posttest differences.

What kind of underlying assignment model is capable of justifying the use of the unconditional log-linear analysis is unclear. In principle, every model that makes theoretical sense and implies the equality of the pretest and posttest differences is appropriate. But to our knowledge, no such log-linear model has been suggested up to now (nor will it be invented here). This, of course, makes the use of difference scores, intuitively attractive as they may be, rather dubious.

Beside the choice between conditional and unconditional analyses, a second problem that has been discussed extensively in the literature on the nonequivalent control group design is how to adjust the posttest differences for the pretest differences when the latter are unreliable. After all, in terms of the mediational model, if the true degree of anti-Semitism at t_1 is associated with seeing the movie, the true degree of anti-Semitism must also be held constant when determining the effects of seeing the film on the posttest scores, and not the observed, unreliable pretest scores.

In the literature, attention has been focused on unreliable pretest scores, presumably because in experimental designs the causal factor X is manipulated by the experimenter and therefore usually measures what it was intended to measure, and because unreliability of the posttest, the dependent variable, is represented in the error term of the regression or covariance equation. However, in panel studies, the investigator almost never controls the values of the causal factor X and, in log-linear analysis, measurement error in the dependent variable has the same consequences as unreliability of the independent variables.

In principle, unreliability in the measurement of any of the relevant variables can easily be handled by log-linear models with latent variables. The log-linear covariance analyses can be carried out on the latent level by making use of log-linear causal models with latent variables (Section 3.7); in the log-linear difference score analysis, measurement unreliability can be taken into account by making use of latent (quasi-) symmetry models (Section 4.4.3).

Such latent variable analyses are feasible if the underlying variables have been measured by more than one indicator. Otherwise, the parameters of the latent variable models will generally not be identifiable. But even in those cases where only one indicator has been used for each concept, latent variable models can still be used to estimate the consequences of various degrees of unreliability of the variables involved. One might, for example, set up a latent covariance model for the data on degree of anti-Semitism, fix the values of the probabilities of giving a "correct" answer to .90 or .80, and study what the consequences are for the effect of the latent variable X' on the latent variable B'. Some provisional exercises showed that unreliability may indeed affect the conclusions about the effects of X (Hagenaars, 1985: 231-233).

Measurement error can even be disastrous when carrying out difference scores analyses in the sense used in Section 4.2.5. Using Table 5.3, one can construct a "difference variable" which indicates whether an individual's degree of anti-Semitism in May (t_2) was higher than, lower than, or equal to the degree of anti-Semitism in November (t_1). This difference variable can be related to seeing the film or not, the expectation being — which is confirmed by the data — that those who went to see the film have a smaller chance to belong to the difference category "higher" than those who did not see the film.

A comparatively minor problem with this often used approach is that these difference scores are logically related to the starting position, that is, to the pretest and that therefore the effect of X and the pretest may be confounded. One might take this into account by incorporating the pretest as an extra independent variable, if necessary by means of a table with a priori zero cell frequencies. A much more serious problem is that even a very small degree of unreliability in the measurements of the relevant variables may lead to very impressive effects of the causal factor on the "difference variable" even when there is no effect at all. The methods discussed before offer excellent alternatives to this more "traditional" way of analyzing panel data.

5.4 Modified Path Models

The one-group pretest-posttest design and the nonequivalent control group design are used to determine the effect of one particular stable causal factor (X) on one particular changing dependent characteristic (Y_{t1}, Y_{t2}). The analyses of these designs are founded on a causal "mini-model." Very often, investigators will be interested in analyzing panel data by means of more extensive causal models in which a larger number of independent, dependent, and intervening variables occur, much like the causal path models discussed in Section 2.7.3 for cross-sectional data.

Establishing the causal order of the variables in a path model is very often rather problematic. In this respect, the panel design offers more possibilities than the one-shot survey. For example, the causal order among the variables in the causal minimodels underlying the one-group pretest-posttest design and the nonequivalent control group design is unequivocal because of the order in which the measurements have been taken over time. The asymmetry of time is used to determine the asymmetry of the causal order: Events happening at t_2 cannot cause events at t_1. The same is true for larger path models: The causal order of any two variables whose scores have been obtained at different points in time is unambiguous. However, in panel analysis too, the causal order among the variables measured at the same point in time may be unclear.

In principle, analyzing panel data by means of path models is not different from analyzing cross-sectional data using path models (except maybe with regard to the presence of autocorrelation, i.e., with regard to serially correlated errors discussed in Section 5.5.2). Therefore, one simple example should be sufficient. In presenting this example, it will be emphasized that even with panel data some problems with regard to the causal order among the variables remain, even when use is made of the "cross-lagged panel correlation technique" for determining causal order (Hagenaars, 1987).

By way of example the data on political preferences in Table 4.4 will be used (after replacing three observed zero cell frequencies by 0.1).[3] These data pertain to vote intention (1. Christian Democratic; 2. Left Wing; 3. Other) in February (A) and March (B) and preference for a Prime Minister (1. Christian Democratic Candidate; 2. Left Wing Candidate; 3. Other Candidate) in February (C) and March (D). Table 4.4 has been used before to determine whether or not vote preference is a more stable characteristic than candidate preference (Section 4.3.2); whether or not the association between vote preference and candidate preference becomes stronger as the election campaign evolves (Section 4.3.3); whether or not vote preference and candidate preference are indicators of one underlying concept: political preference (Section 4.4.3). In this section, the variables A through D will be treated as four separate variables and the analysis will focus on answering in an exploratory way the question how party and candidate preference influence each other over time.

As depicted in Figure 5.3, the directions of the causal influences between A and B, A and D, C and B, and C and D are fixed by design, as the characteristics in March cannot have affected the characteristics in February. However, the causal order between A and C, and B and D is ambiguous.

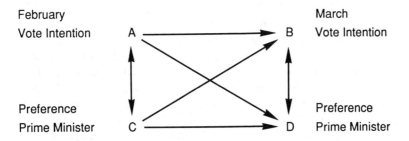

Figure 5.3. Causal Order among Measures of Political Preference in February and March

No matter which causal model is postulated, the unambiguously directed arrows in Figure 5.3 have to be respected. Applying to Figure 3.4 the principles of the modified path analysis set forth in Section 2.7.3, it follows that the causal analysis of the data in Table 4.4 has to start with the analysis of the relationship between A and C in the marginal Table AC.

The most obvious thing to do is to test whether or not vote intention A and preference for Prime Minister C are independent of each other. Model $\{A, C\}$ has to be rejected for the marginal table AC using the observed frequencies f_{i+j+}^{ABCD}: $L^2 = 532.91$, df = 4, $p < .001$ (Pearson-$\chi^2 = 544.10$). Application of the saturated model to table AC indicates that the relation between A and C is very strong. (The parameter estimates are not reported here, only in a very concise way in Figure 5.7; the test statistics for the successive analysis steps are presented in Figure 5.4.) As expected, the preferences for particular parties go hand in hand with the preferences for a Prime Minister allied to that party. Furthermore, the left wing and the Christian Democratic political orientation in particular tend to exclude each other in the sense that the combinations left wing vote intention-preferring a Christian Democratic Prime Minister and a Christian Democratic vote intention-preferring left wing Prime Minister occur far less than expected on the basis of the main effects A and C.

If the relation between A and C is regarded as a causal relation, the strength of this relation is preferably expressed in terms of the parameters of a (polytomous) effect model (Section 2.7). In principle, it has to be decided then whether vote intention (A) or preference for Prime Minister (C) is the dependent variable. However, given the symmetry of the γ parameter in the sense that $\gamma_{11/2}^{A\bar{C}} = \gamma_{11/2}^{C\bar{A}} = (\tau_{11}^{AC})^2$, the practical consequences of this choice are minimal.

This does not mean that different assumptions about the causal order of variables are irrelevant in general. To determine the influences of the "February variables" A and C on the "March variables" B and D the causal order between party and candidate preference has to be established. If the effects of A and C on B are examined, it has to be decided whether or not D has to be held constant. Should the effects of A and C on B be determined by means of table ABC (without holding D constant) or by means of table $ABCD$ (holding D constant)? And, in the same vein, should the effects of A and C on D be investigated by means of table ACD or table $ABCD$?

If it is assumed that in March party preferences cause candidate preferences, that is, $B \rightarrow D$, the "ultimate dependent variable" D must not be held constant when determining the effects of A and C on B and table ABC should be used. But if the opposite causal order $D \rightarrow B$ is assumed, the effects of A and C on the "ultimate dependent variable" B have to be examined while holding D constant by means of table $ABCD$. Analogous remarks apply with regard to the estimation of the effects of A and C on D.

First we will proceed from the assumption that in March party preference causes candidate preference $(B \rightarrow D)$ rather than vice versa. This assumption implies that the effects of A and C on B have to be estimated using table ABC, whereas the effects of A and C on D have to be estimated by means of the full cross-classification $ABCD$.

As explained in Section 2.7, all unsaturated frequency models used to determine the effects of A and C on B must include the term AC. Starting with the saturated model $\{ABC\}$ for the table ABC, it appeared that all τ_{ijk}^{ABC} parameters were rather small and statistically insignificant. Therefore, model $\{AC, AB, BC\}$ has been applied. This model fits the data in table ABC excellently: $L^2 = 2.48$, df = 8, $p = .96$ (Pearson-$\chi^2 = 2.33$). Both the effects of A on B and the effects of C on B are large and statistically significant. There is no need to test other more parsimonious models.

The $\hat{\tau}_{ij}^{AB}$ parameters in model $\{AC, AB, BC\}$ indicate the degree of stability of the party preference between February and March. The estimated relationship between A and B appeared to be strong and in the direction that might have been expected. People have a very strong tendency to remain faithful to their original party preference. Especially

Christian Democrats and left wing adherents tend to avoid each other's parties.

The $\hat{\tau}_{jk}^{BC}$ parameters are smaller than the $\hat{\tau}_{ij}^{AB}$ parameters although still sizeable. The cross-lagged association BC between party and candidate preference follows more or less the same pattern as the cross-sectional relation AC between party and candidate preference described above, albeit that the relation B-C is somewhat weaker than the relation A-C.

Still assuming that B causes D, in the next step the effects of A, B, and C on D are determined. All relevant models now at least contain the term ABC. Given the above results, no three or higher order effects involving D were expected. Model $\{ABC, AD, BD, CD\}$ fits the data in table $ABCD$ rather well (L^2 = 47.56, df = 40, p = .192, Pearson-χ^2 = 51.52), but is needlessly complicated: The effects of A on D appear to be insignificant. The more parsimonious model $\{ABC, BD, CD\}$ also fits the data well (L^2 = 49.38, df = 44, p = .27, Pearson-χ^2 = 53.64) and has to be preferred to the previous model $\{ABC, AD, BD, CD\}$ given the outcome of the conditional test: L^2 = 49.38 − 47.56 = 1.82, df = 44 − 40 = 4, p = .77.

The $\hat{\tau}_{kl}^{CD}$ parameters in model $\{ABC, BD, CD\}$ indicate the degree of stability of candidate preference between February and March. As with party preference, people have a strong tendency to remain faithful to their original candidate. In Section 4.3.2 it was concluded on the basis of a comparison of the two dimensional turnover tables AB and CD that party preference is a more stable characteristic than candidate preference. Comparison of the $\hat{\tau}_{ij}^{AB}$ parameters in model $\{AC, AB, BC\}$ with the $\hat{\tau}_{ij}^{CD}$ parameters in model $\{ABC, BD, CD\}$ leads to the same conclusion, but now within the framework of an explicit causal model.

The association between party and candidate preference in March as measured by $\hat{\tau}_{jl}^{BD}$ in model $\{ABC, BD, CD\}$ is somewhat weaker than the corresponding association AC in February, but a bit stronger than the corresponding cross-lagged association BC. The pattern of association between vote intention and preference Prime Minister is always the same.

The main outcomes of these analyses, departing from the assumption that party preference causes candidate preference, are presented in Figure 5.4a. The validity of the path model in Figure 5.4a as a whole is tested by means of L^*. There is no reason to reject this modified path model. A conspicuous feature of this path model is the absence of any cross-lagged effect of vote intention A on preference for Prime Minister D. A affects D only via B and C.

a).Vote Intention causing Preference for Prime-Minister.

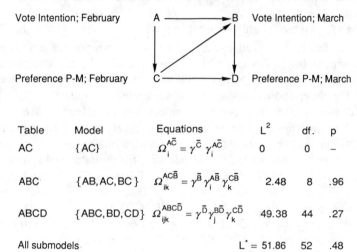

Vote Intention; February   A ────────▶ B   Vote Intention; March

Preference P-M; February   C ────────▶ D   Preference P-M; March

Table	Model	Equations	L^2	df.	p
AC	{AC}	$\Omega_i^{A\bar{C}} = \gamma^{\bar{C}}\,\gamma_i^{A\bar{C}}$	0	0	–
ABC	{AB, AC, BC}	$\Omega_{ik}^{AC\bar{B}} = \gamma^{\bar{B}}\,\gamma_i^{A\bar{B}}\,\gamma_k^{C\bar{B}}$	2.48	8	.96
ABCD	{ABC, BD, CD}	$\Omega_{ijk}^{ABC\bar{D}} = \gamma^{\bar{D}}\gamma_j^{B\bar{D}}\gamma_k^{C\bar{D}}$	49.38	44	.27
All submodels			$L^* = 51.86$	52	.48

b.) Preference Prime Minister causing Vote Intention.

Vote Intention; February   A ────────▶ B   Vote Intention; March

Preference P-M; February   C ────────▶ D   Preference P-M; March

Table	Model	Equations	L^2	df.	p
AC	{AC}	$\Omega_k^{C\bar{A}} = \gamma^{\bar{A}}\,\gamma_k^{C\bar{A}}$	0	0	–
ACD	{AC, AD, CD}	$\Omega_{ik}^{AC\bar{D}} = \gamma^{\bar{D}}\gamma_i^{A\bar{D}}\gamma_k^{C\bar{D}}$	8.35	8	.40
ABCD	{ACD, AB, BD}	$\Omega_{ik\,l}^{ACD\bar{B}} = \gamma^{\bar{B}}\gamma_i^{A\bar{B}}\gamma_l^{D\bar{B}}$	48.16	44	.31
All submodels			$L^* = 56.51$	52	.31

Figure 5.4. Modified Path Models for the Relations among Vote Intention and Preference for Prime Minister in February and March; Data from Table 4.4

A summary of the test results of the analyses steps that have been taken proceeding from the opposite assumption that in March candidate preference causes party preference $(D \rightarrow B)$ is presented in Figure 5.4b. The first step, analyzing table AC, is essentially the same as before. In the next step, the effects of A and C on D are determined using table ACD. Model $\{AC, AD, CD\}$ turned out to be the best choice.

The final step involves estimating the effects of A, C, and D on B by means of table $ABCD$. A choice had to be made between model $\{ACD, AB, BD\}$ without the cross-lagged effect BC and model $\{ACD, AB, BD, BC\}$ with the cross-lagged effect BC. The test results for the more restricted model are $L^2 = 48.16$, df = 44, $p = .31$ (Pearson-$\chi^2 = 69.59$) and for the more unrestricted model $L^2 = 41.54$, df = 40, $p = .40$ (Pearson-$\chi^2 = 45.48$). Conditionally testing the significance of the cross-lagged effect BC by comparing the two models yields the following results: $L^2 = 6.62$, df = 4, $p = .16$.

Just relying on the conditional test, the more restricted model $\{ACD, AB, BD\}$ has to be chosen. However, the unconditional test results for the more restricted model are ambiguous. Pearson-χ^2 and L^2 have rather different values and on the basis of Pearson-χ^2, the restricted model $\{ACD, AB, BD\}$ has to be rejected. Moreover, several of the crucial parameters $\hat{\tau}_{jk}^{BC}$ in model $\{ACD, AB, BD\}$ appeared to be significant. Particularly the combination "preferring the Christian Democratic candidate in February-voting for the Christian Democrats in March" occurs more often and the combination "preferring the Christian Democratic candidate in February-voting for the left wing parties in March" occurs much less often than expected on the basis of the other effects. Altogether, it is not clear which model to choose, which is symbolized by the dotted line in the path diagram in Figure 5.4b.

The patterns of association found for the two path models in Figure 5.4 are more or less the same, as far as these models are comparable. But, of course, there are very important differences between the two path models. The most striking difference is that, depending on which assumption is made with regard to the direction of the instantaneous causal influence between party preference and candidate preference, different conclusions are drawn about the way vote intention and preference for Prime Minister influence each other over time.

Apparently, for estimating the nature and strength of all direct relations among the four variables A through D, it is absolutely necessary to make assumptions about the causal order of the variables involved. If such an assumption is not made explicitly, it is nevertheless made implicitly because of the choice of the marginal tables that are used to estimate a particular direct relationship and the decision to hold particular variables constant or not. The above discussion has shown that the necessary assumptions about the causal order of the variables are not always empirically testable. Both path diagrams in Figure 5.4 present models that fit the data and there is no way of telling empirically which diagram is the right one.

It is often maintained that panel studies provide the opportunity to establish the causal order between all pairs of variables by applying the *cross-lagged panel correlation technique*. The basis for this technique was laid by Lazarsfeld (1972b) in a manuscript written in 1946.

Using Table 4.4, the basic ideas of this technique can be explained as follows. There are discrepancies in February between people's vote intentions and their preferences for Prime Minister in the sense that some people vote for a particular party, but do not prefer the party's candidate for Prime Minister. Inspection of the turnover in Table 4.4 shows whether or not these discrepancies have been resolved in March and, if so, whether this is done by adjusting the vote intention to the preference for Prime Minister, or by adjusting the candidate preference to the party preference. Several indices have been proposed to measure which variable is more "dependent" on the other by counting the number of adjustments in the one and the other direction.

This basic idea has been adapted by Campbell (1963) and Pelz and Andrews (1964) for variables measured at interval level. If the four variables in Figure 5.4 had been measured at interval level, they would suggest comparing the (partial) cross-lagged correlations AD and AC with each other. If the (partial) correlation AD is greater than BC, it is concluded that the influence of vote intention on preference for Prime Minister is greater than the influence of preference for Prime Minister on vote intention and that accordingly party preference causes candidate preference. If the (partial) correlation BC is greater than AD, the opposite conclusion is drawn.

Many applications of this technique can be found, as well as a large number of extensions and critiques. (To mention a few important ones: Coleman, 1984; Duncan, 1969, 1975b, 1985; Greenberg & Kessler, 1982; Hannan, Robinson & Warren, 1974; Kenny & Harackiewicz, 1979; Kessler, 1977; Mayer & Carroll, 1987, 1988; McCullough, 1978; Rozelle & Campbell, 1969; Yee and Gage, 1968.) From these discussions it emerged that the magnitudes of the (partial) cross-lagged correlations and of the indices developed by Lazarsfeld and his associates depend on

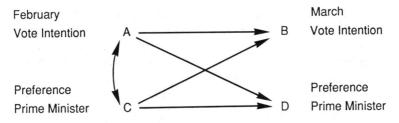

February   March
Vote Intention   A ─────────────────────► B   Vote Intention

Preference   Preference
Prime Minister   C ─────────────────────► D   Prime Minister

Figure 5.5. Causal Model Underlying the Cross-Lagged Panel Correlation
Technique

far more conditions than only the causal order between the relevant
characteristics.

These other conditions are, among other things, the stability over time
of each characteristic and the reliability of the measurements. Moreover,
to be able to infer causal order from differences in cross-lagged correlations, it must be assumed that the system of variables under consideration
is closed and that there are no omitted variables that influence the
relationship between two or more variables of the system. A further
essential assumption is that the "time lag" coincides with the "causal lag,"
that is, the time between the measurements must coincide with the time
needed to make the effects of the one variable on the other visible. If this
is not the case, then the whole basis of the cross-lagged panel correlation
technique falters (Davis, 1978; Heise, 1970, 1975; Pelz & Lew, 1970).

In order to apply this technique it is therefore necessary to set up a
causal model in which all these alternative explanations of the sizes of
the cross-lagged (partial) correlations are taken into account, either by
incorporating these alternative explanations in the model or by explicitly
declaring them nonexistent.

The causal model that most investigators implicitly or explicitly have
in mind when applying the cross-lagged panel correlation technique is
depicted in Figure 5.5.

It is assumed that all variables in the model in Figure 5.5 have been
reliably measured, that there are no instantaneous causal effects between
the variables measured at the same point in time, that the causal process
is adequately observed by means of the time points February-March,
and, finally, that the causal system is closed, that is, that there are no
confounding variables left that are not included in the model. Under
these conditions, the strengths of the cross-lagged direct effects *A-D* and

C-B in Figure 5.5 can be used to determine the causal order between vote intention and preference for Prime Minister.

The validity of model {*AB*, *AC*, *AD*, *BC*, *CD*} in Figure 5.5 has been tested for the data in Table 4.4, with negative results: L^2 = 180.32, df = 52, p < .001 (Pearson-χ^2 = 259.47). The previous analyses of these data mentioned in Figure 5.4 suggest that the failure of model {*AB*, *AC*, *AD*, *BC*, *CD*} is especially caused by neglecting the direct effect between *B* and *D*. The residual frequencies show that the strength of the (marginal) association *B-D* is indeed underestimated by model {*AB*, *AC*, *AD*, *BC*, *CD*}.

The reasons for this outcome may be manifold. It is possible that there are other (omitted) variables involved, that one or more variables have not been reliably measured, or that time lag and causal lag do not coincide. But whatever the reasons are, the fact is that this outcome, underestimation of the strength of the cross-sectional association at t_2, is very often encountered when applying the model in Figure 5.5 (Hagenaars, 1987). With variables measured at interval level, this difficulty is mostly "solved" by correlating the error terms belonging to *B* and *D* with each other. However, this is a mere algebraic, not a substantive solution. In terms of a log-linear path model, this "solution" implies drawing a double-headed arrow between *B* and *D*. But it then becomes unclear, as we saw above, how to determine the effects of *A* and *C* on *B* and *D*.

In sum, in order to determine the cross-lagged direct effects *A-D* and *B-C* by means of which the causal order of party and candidate preference has to be established, it has to be decided a priori whether vote intention (in March) influences preference for Prime Minister (in March) or vice versa.

As mentioned above, unreliability of the measurements might distort the conclusions based on the manifest data. *Path models with latent variables*, as discussed in Section 3.7, can be used to evaluate and correct for the consequences of measurement unreliability. Such models are most appropriate if each underlying concept has been measured by two or more indicators. Below, a rather extreme example will be presented in which each latent variable is connected with one and only one manifest variable, again using the data in Table 4.4. Although this practice is not to be recommended in general, this extreme example has been chosen to facilitate direct comparison of the results of the modified path analysis with and without latent variables.

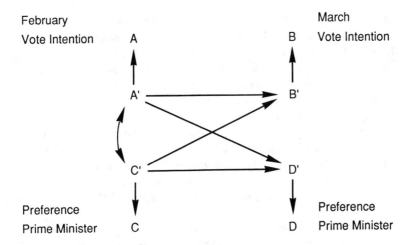

February   March
Vote Intention   A   B   Vote Intention

A' ――――――――――――→ B'

C' ――――――――――――→ D'

Preference   Preference
Prime Minister   C   D   Prime Minister

Figure 5.6. Latent Causal Model Underlying the Cross-Lagged Panel Correlation Technique

Besides the four manifest variables A through D, the path models to be considered below now contain four latent variables denoted as A', B', C', and D'. Vote intention and preference for Prime Minister in February and March are not considered as indicators of just one or two underlying variables as was the case in Section 4.4.3, but each manifest variable is an indicator of a different latent variable.

First it will be determined whether measurement unreliability was the cause of the failure of the standard "cross-lagged panel correlation" model in Figure 5.5. Perhaps it is possible to determine the causal order of vote intention and preference for Prime Minister at the latent level. The relevant model $\{AA', BB', CC', DD', A'B', A'C', A'D', B'C', C'D'\}$ is depicted in Figure 5.6. According to this model, the relations among the latent variables mirror the relations among the manifest variables in the "manifest-only" model in Figure 5.5.

The latent variables A', B', C', and D' are considered to be trichotomies in correspondence with the trichotomous manifest variables. Therefore, there are $3^4 = 81$ latent classes in model $\{AA', BB', CC', DD', A'B', A'C', A'D', B'C', C'D'\}$ and 81 parameters π_t^X to be estimated. In addition, $4 \times 6 = 24$ conditional probabilities ($\pi_{i\,i'}^{\bar{A}A'}$, etc.) have to be independently estimated. The number of knowns, the observed frequencies, is only 81. So the necessary conditions for identifiability are seemingly not met

(Section 3.2.4). However, the 81 parameters π_t^X, can be represented in terms of the much smaller number of log-linear parameters of model $\{A'B', A'C', A'D', B'C', C'D'\}$ that is postulated for the relations among the latent variables. The restrictions on the relations among the latent variables provide 52 extra degrees of freedom, in agreement with the 52 degrees of freedom of the model in Figure 5.5. The model as a whole then has 28 degrees of freedom.

The LCAG program was used to estimate the parameters of the model in Figure 5.6 applied to Table 4.4 and to obtain the test statistics: L^2 = 71.84, df = 28, $p < .001$ (Pearson-χ^2 = 86.80).[4] So the model has to be rejected. Even when correcting for unreliability and analyzing at the latent level, the "cross-lagged panel correlation" technique does not work.

Further analyses can be carried out by imposing different kinds of restrictions on the relations among the latent variables A', B', C', and D' to find a model that does fit the data. An interesting set of restrictions which make it possible to illustrate the possible consequences of unreliability are the restrictions implied by the unsaturated "manifest-only" models in Figure 5.4. Because it is believed that the causal order "party preference → candidate preference" is more plausible than the reverse causal order, the path model in Figure 5.4a has been chosen for comparison.

The appropriate latent variable model has the same structure as the previous latent variable model in Figure 5.6, although a different sub-model has to be specified for the relations among the latent variables in agreement with the "manifest-only" model in Figure 5.4a. All this implies that we have a latent class model in which the score on A depends only on A', the score on B only on B', C only on C', D only on D', and in which a latent modified path model is set up for the relations among the latent variables in such a way that model $\{A'B', A'C', B'C'\}$ is postulated for the latent table $A'B'C'$ and model $\{A'B'C', B'D', C'D\}$ for the complete latent cross-classification $A'B'C'D'$.

The test statistics for this latent variable model applied to Table 4.4 (without replacing three zero cell frequencies by .1 — see note 10) are $L^2 = 38.32$, df = 28, $p = .09$ (Pearson-χ^2 = 39.16). (The number of degrees of freedom has been computed in the usual way, not considering parameter estimates equal to zero as a priori fixed. Treating zero estimates as fixed would increase df to 37 at most.) The conclusion is that the model need not be rejected.

TABLE 5.7 Reliabilities $\hat{\pi}_{i\,i'}^{\bar{A}A'}$, $\hat{\pi}_{j\,j'}^{\bar{B}B'}$, $\hat{\pi}_{k\,k'}^{\bar{C}C'}$, and $\hat{\pi}_{l\,l'}^{\bar{D}D'}$ of Vote Intention and Preference Prime Minister in February and March

(a) $\hat{\pi}_{i\,i'}^{\bar{A}A'}$	Manifest Vote Intention February (A)		
	1. Christ. Democ.	2. Left Wing	3. Other
Latent Vote Intention February (A')			
1. Christian Democratic	.987	.012	.001
2. Left Wing	.026	.886	.088
3. Other	.033	.032	.934

(b) $\hat{\pi}_{j\,j'}^{\bar{B}B'}$	Manifest Vote Intention March (B)		
	1. Christ. Democ.	2. Left Wing	3. Other
Latent Vote Intention March (B')			
1. Christian Democratic	.970	.000	.031
2. Left Wing	.000	1.000	.000
3. Other	.000	.000	1.000

(c) $\hat{\pi}_{k\,k'}^{\bar{C}C'}$	Manifest Preference Prime Minister February (C)		
	1. Christ. Democ.	2. Left Wing	3. Other
Latent Preference Prime Minister February (C')			
1. Christian Democratic	1.000	.000	.000
2. Left Wing	.000	.989	.011
3. Other	.018	.000	.981

(d) $\hat{\pi}_{l\,l'}^{\bar{D}D'}$	Manifest Preference Prime Minister March (D)		
	1. Christ. Democ.	2. Left Wing	3. Other
Latent Preference Prime Minister March (D')			
1. Christian Democratic	.752	.080	.168
2. Left Wing	.003	.977	.020
3. Other	.035	.084	.881

SOURCE: Table 4.4

The estimated conditional response probabilities are presented in Table 5.7. Interpreting the conditional response probabilities as reliability coefficients, it appears from Table 5.7 that B (vote intention in March) and C (preference for Prime Minister in February) were measured almost perfectly reliably. The chances of a misclassification are negligible for these variables.

Vote intention in February (A) was somewhat less reliably measured. The probability of a misclassification in February is largest for those who truly intended to vote for a Left Wing party ($A' = 2$) and amounts to $1 - .886 = .114$. The reliability of preference for Prime Minister in March is the lowest of the four variables and also rather weak in absolute terms. The probability of a misclassification in March for those who truly preferred the Christian Democratic candidate ($D' = 1$) is not less than $1 - .752 = .248$. On the other hand, the true preference for the Left Wing candidate ($D' = 2$) is practically perfectly recorded. For those who truly prefer an Other candidate ($D' = 3$) an intermediate position applies: The chances of a misclassification are $1 - .881 = .119$.

To investigate the consequences of these degrees of unreliability, the two-variable τ parameters indicating the strengths of the relations between the latent variables have been estimated by means of the estimates $\hat{\pi}_{i'j'k'l'}^{A'B'C'D'}$ and compared with the parameter estimates for the "manifest-only" model in Figure 5.4a. (The latter are computed on the original Table 4.4 without replacing three zero cell frequencies by 0.1. See note 3.) The main results are presented in Figure 5.7, using the mean and maximum effects as defined in equations 2.47 and 2.48.

The main thing that emerges from the comparison of the latent and manifest effects in Figure 5.7 is that with one exception the effects at the latent level are larger. The discrepancy between the corresponding latent and manifest effects is largest for the effects involving the variables D and D'. D is the most unreliable indicator and, as expected, the "correction for attenuation" is greatest for the effects involving this variable. The exception concerns the practically perfectly measured variables B and C. The cross-lagged relation between preference for Prime Minister in February (C, C') and vote intention in March (B, B') is smaller at the latent level than at the manifest level. In fact, the latent effect B'-C' was not statistically significant. If the latent variable model employed so far is modified by deleting the direct effect B'-C' (for the latent subtable $A'B'C'$), the test statistics are $L^2 = 40.33$, df $= 32$, $p = .15$, and Pearson-$\chi^2 = 39.16$. A conditional test of the model without the effect B'-C' given the validity of the model with the effect B'-C' yields

Model:

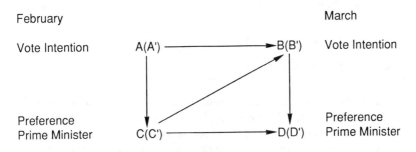

τ – estimates;

Latent				Manifest			
Table	Effect	$\hat{t}^{*}_{max}$	$\hat{t}_{m}$	Table	Effect	$\hat{t}^{*}_{max}$	$\hat{t}_{m}$
A'C'	A'C'	13.61	2.24	AC	AC	12.45	1.99
A'B'C'	A'B'	136.60	4.92	ABC	AB	16.77	2.71
	B'C'	3.11	1.38		BC	5.14	1.52
A'B'C'D'	B'D'	518.03	5.77	ABCD	BD	8.06	1.82
	C'D'	487.35	6.14		CD	7.02	1.95

Figure 5.7. Vote Intention and Preference for Prime Minister in February and March; Comparison of Manifest and Latent Effects; Manifest Variables A, B, C, and D; Latent Variables A', B', C', and D'; Data from Table 4.4

$L^{2} = 2.01$, df = 4, $p = .72$. So there is no reason whatever to maintain the direct relation B'-C'. At the latent level, both cross-lagged relations are absent.

An examination of the separate $\hat{t}$ parameters (not presented here) indicates that the patterns of association at the manifest level are rather similar to the patterns at the latent level. A clear exception is the relation B'-C' compared with B-C. But since the relation B'-C' turned out to be nonsignificant, it does not make much sense to go into this any further. This of course does not imply that unreliability can affect only the sizes of the effects and not the patterns of association. Rather, large discrepancies between the patterns of manifest and latent associations are possible.

5.5 Typical Panel Problems

In the discussion of the analysis of panel data in previous sections, it was taken for granted that the desired panel data are available. However, getting panel data may turn out to be rather difficult. Carrying out a panel study brings along some difficulties that are typical of the panel design and are directly related to the fact that the same persons are being interviewed several times.

There are a lot of *practical problems*. To be able to reinterview the same persons, one must somehow keep track of them over time. To make the results of the various waves comparable, coding procedures have to be carried out in exactly the same ways, requiring detailed and careful documentation. Ongoing research projects pose special problems with regard to staffing. Secondary panel data may be difficult to obtain: Unlike trend studies which may result from a series of cross-sectional studies never intended to be part of a trend study, panel studies have to be explicitly designed as such. Although intimate knowledge of these and other practical problems of planning and organizing panel studies is available at several research institutes, this knowledge has seldom found its way into the literature. Exceptions are the publications by Crider, Willits, and Bealer (1973); Goldstein (1979, Chapters 1 and 2); Kessler and Greenberg (1981, Chapter 12); Kimberly (1976); Rosenberg et al. (1951); and Spencer (1974).

Another major problem is the occurrence of *mismatches* (or nonmatches — Neter, Maynes, & Ramanathan, 1965; Scheuren & Lock, 1975). It is essential for panel studies that cases which carry the same identification number in several waves really are the same person. Mismatches occur when data from different persons are considered to pertain to the same person. Empirical data about the frequency with which mismatches occur in regular panel studies are rare. One gets the impression that the number of mismatches is usually rather small. Detection of mismatches is made easier if a few stable characteristics are independently recorded in each wave. Precise information about multicategory items like years of schooling or date of birth is then more useful than information about the dichotomy gender. In order to prevent mismatches, interviewers should be supplied with precise information about the respondents, for example, with all initials of the names and exact dates of birth, and be instructed not to accept substitutes (See Hagenaars, 1985, Section 4.2.2).

More serious than mismatching are the problems of *panel mortality* (nonresponse) and *repeated interviewing effect* (test-retest effect). Nonresponse also occurs in cross-sectional studies, but in panel studies additional nonresponse has to be expected to occur in each successive wave. As will be shown in Section 5.5.1, nonresponse problems may be very serious in panel studies, although at the same time the panel design offers unique possibilities to correct for the distorting effects of nonresponse.

The mere fact that people have been questioned about certain topics may make them behave differently. The possible consequences of repeated interviewing effects for the analysis results will be dealt with in Section 5.5.2. It will also be made clear that the problem of serially correlated errors is closely related to the problem of test-retest effects.

5.5.1 Nonresponse

In nation-wide social surveys, large percentages of nonresponse are to be expected. Nonresponse percentages of 30% to 40% are no exception. In many panel studies, these figures are even higher. As Ferber (1953), Glock (1952), and Sobol (1959) already noted, in each successive wave there will be additional nonresponse and consequently the percentage of nonresponse based on the original sample size increases with each wave, be it mostly at a decreasing rate. In the end, often 30% or less of the original sample participates in all waves (Hagenaars, Stouthard, & Wauters, 1990).

If nonresponse occurs completely at random, panel mortality only threatens the statistical conclusion validity (Cook & Campbell, 1979). In the case of pure random nonresponse, the people that participate in all waves may be considered as a random sample from the originally intended sample and therefore from the intended population. Pure random nonresponse reduces the sample size and leads to less precise parameter estimates and less powerful statistical tests, but not to a systematic bias of the results.

However, there is usually little reason to expect that nonresponse occurs completely at random. People will have definite reasons and motives not to participate and these reasons and motives may be directly related to the variables under study. Therefore, the fact that respondents

and nonrespondents differ systematically from each other and that conclusions drawn on the basis of the respondent data may not be valid for
the total population must be taken into account.

Research into the differences between respondents and nonrespondents is possible because nonresponse is not necessarily an all-or-nothing
phenomenon. Public records may provide information for all members
of the population including the nonrespondents on background characteristics such as residence, age, and gender. Nonrespondents may be
willing to provide information on just a few important variables in the
context of a special investigation into nonresponse. In panel studies,
much is known about nonrespondents who have participated in at least
one wave, but not in all waves.

In most panel studies in which nonresponse has been reported, rather
small, often insignificant, differences between respondents and nonrespondents have been found. Usually these differences show a systematic pattern: Those who participate in all waves are more interested in
the topics of investigation and know more about them than those who
participate in just one or two waves. Furthermore, in most studies a few
large idiosyncratic differences with respect to particular questions and
topics are reported.[5]

Because even small differences between respondents and nonrespondents can lead to substantial bias when the nonresponse percentage is very
large and because large differences with regard to a few (idiosyncratic)
variables are to be reckoned with, nonresponse may pose a serious threat
to the internal and external validity (Cook & Campbell, 1979). The best
strategy to avoid biased results is to prevent nonresponse. For panel
studies, this means keeping track of the respondents, even over a long
period of time, and finding ways to motivate the respondents to keep
participating in the panel.[6]

Yet sometimes a (high) percentage of nonresponse is inevitable.
Investigators do not always have the time and the money to employ
extensive procedures for preventing nonresponse. Moreover, if use is
made of available data, the secondary analyst has to accept the nonresponse rate as given. What remains in these circumstances is to try to
correct for nonresponse as well as possible at the analysis stage, making
use of whatever information about the nonrespondents is available.

Many methods for correcting for nonresponse have been developed.
Surveys are provided by Anderson, Basilevsky, and Hum (1983), Daniel
(1975), Little and Rubin (1987), and Madow (1983). Every correction
method rests on an assumption about the response mechanism, about how

(non)response came about. An important distinction in this respect is that between *ignorable and nonignorable response mechanisms* (Little, 1982; Rubin, 1974).

A response mechanism is said to be ignorable if the (non)response takes place at random, either within the entire population or within subgroups. In the first case, the missing data are *missing completely at random*; in the second case just *missing at random*. Missing completely at random implies that the multivariate frequency distribution of the variables under study is similar for the (population of) respondents and the (population of) nonrespondents. Missing at random means similarity of the multivariate frequency distribution for respondents and nonrespondents conditional upon the scores on particular (other) variables.

The nonignorable response mechanism can best be described as the reverse of the ignorable response mechanism: The (non)response does not occur at random but is systematically and directly related to the variable whose scores are (partially) missing.

A simple two-variable example can further clarify these distinctions. Say we have an investigation with just two variables Age and Vote Intention. The score on age is known for all cases from public records; the score on vote intention is known only for the respondents, not for the nonrespondents. If the missing data are missing completely at random, the bivariate frequency distribution age-vote intention is expected to be the same for the respondents and nonrespondents. The results from the respondents may be generalized to the whole population without expecting bias to occur.

If the missing data are missing at random within age groups, respondents and nonrespondents are expected to differ with regard to age, but not with regard to vote intention within each age group. The overall distribution of vote intention will be different for respondents and nonrespondents, but once their differences with regard to age have been taken into account and age is held constant, the differences in vote intention disappear. The results from the respondents in a particular age group may be generalized to the whole population of that particular age group without expecting bias to occur.

Finally, in the case of a nonignorable response mechanism, responding or not is directly related to the score on vote intention. The association (non)response-vote intention does not disappear when age is held constant and even within age groups respondents and nonrespondents differ from each other with regard to their vote intention.

An ignorable response mechanism is assumed in most methods of correction, more often than not for mere pragmatic reasons: Otherwise one has to indicate exactly how the (non)response is systematically related to the relevant variables and the selection of a wrong response model may only further enlarge the bias.

Actually, the "correction method" used most often is to simply ignore the nonresponse and base the analyses on just the respondents, using "listwise" or "pairwise" deletion of cases (Lösel & Wüstendörfer, 1974). If nothing at all is known about the nonrespondents, this is the only feasible "correction method," but unless the missing data are missing completely at random, the results will be biased.[7]

Another correction method is assigning weights to the respondents in order to make particular relative (multivariate) frequency distributions in the sample of respondents equal to known population distributions or to known sample distributions from previous waves. In this correction method, the respondents are first divided into subgroups on the basis of those characteristics for which the (multivariate) frequency distribution is known for the total population or for both respondents and nonrespondents. Next, the respondents are weighted with the inverse of the response probability of their subgroup. For example, if the population distribution of age is known, a weight factor will be assigned to the respondents in such a way that the sample distribution of age reflects the population distribution. Or, if the age scores are obtained during the first wave and not all cases have been reinterviewed in the second wave, the respondents who participated in both waves will be weighted so that the age distribution in the first wave is reflected.

This correction procedure does not need to be justified by the strict missing completely at random assumption; the weaker missing at random assumption suffices. In terms of the simple two-variable example used before, respondents and nonrespondents may differ with regard to age and vote intention. Correcting for the age differences between respondents and nonrespondents by a weighting procedure eliminates possible differences in vote intention between respondents and nonrespondents if (non)-response takes place at random within age groups and is directly related only to age but not to vote intention.

Obvious candidates to use as weight factors are background characteristics whose distributions are known from public records. However, it is not guaranteed that nonresponse is directly related only to background variables like age and gender and not directly to variables that are more central to a particular study such as vote intention or political interest. Panel studies have great potentialities in this respect in that it is often possible to use more central variables from previous waves as weight factors.

Imputing scores is another correction method based on the validity of the weaker missing at random assumption. In terms of the simple two-variable example, the data provided by the respondents are used to determine the relationship between age and vote intention. (Remember that the age scores are known for both respondents and nonrespondents, but the scores on vote intention are known only for the respondents.) Next, all nonrespondents are assigned to a particular category of vote intention on the basis of their age score and of the relation between age and vote intention as found among the respondents. For example, a nonrespondent of a particular age is assigned to the (modal) party for which the respondents of the same age intended to vote most frequently. We clearly see the weaker missing at random assumption: Respondents and nonrespondents may differ with regard to age, but, once a respondent and a nonrespondent belong to the same age category, their probability of intending to vote for a particular party is the same.

Lately, there has been a growing interest in a third correction procedure in which one estimates the sufficient statistics for a particular model making use of all information that is available from respondents and nonrespondents (without weighting or imputing scores). This procedure amounts to estimating a covariance matrix in the case of regression or factor analysis and a frequency table in the case of log-linear analysis. Allison (1987); Marini, Olsen, and Rubin (1979); and Muthén, Kaplan, and Hollis (1987) explained this approach for regression and LISREL models. Fay (1986) and Fuchs (1982) explained it for log-linear models with manifest variables; Hagenaars (1986b, 1988c) applied Fuchs's and Fay's approach to log-linear models with latent variables. Because of the great potentials for the log-linear analysis of panel data, the procedures developed by Fuchs and Fay will be elaborated upon below.[8]

Fuchs (1982) started out from the weaker missing at random assumption. Building on the work of Chen (1979); Chen and Fienberg (1974); Hocking and Oxspring (1971, 1974); and Rubin (1974), he uses a variant of the EM-algorithm (see Section 3.2.2) to find the maximum likelihood estimates of the expected frequencies F for a particular hierarchical log-linear model without latent variables, using whatever information is available from respondents and nonrespondents.

For an illustration of his procedure in the case of a three-wave panel study, let us assume that we are interested in the relationships among three variables A, B, and C, where A is a variable from the first wave, B from the second, and C from the third. Moreover, it is assumed that all those who participated in the first wave have a score on A, all those who participated in the second wave have a score on B, and all those who participated in the third wave have a score on C. All relevant cases participated in at least one of the three waves, but not necessarily in all three.

The cases can be divided into mutually exclusive subgroups on the basis of their participation in the three waves, that is, on the basis of the variables on which they scored. For example, the subgroup AC includes all and only those who scored on the variables A and C, but not on B; the subgroup A includes all and only those who scored on A, but not on B and C. In this way seven nonoverlapping subgroups can be defined: A, B, C, AB, AC, BC, ABC (some of which may turn out to be empty).

A special notation is introduced to denote the observed frequencies for each of these subgroups. (Note that this special notation applies only to the observed frequencies f, and not to $\hat{F}$ or $\hat{f}$.) The superscripts attached to the observed frequencies f are used to indicate to which subgroup the observed frequencies refer. For example, f_{ijk}^{ABC} refers to the number of persons who obtained the score (i, j, k) on the joint variable ABC in the subgroup ABC, that is, among those who scored on all three variables and therefore participated in all three waves. The frequency f_{ij+}^{ABC} denotes the number of individuals that scored (i, j) on the joint variable AB, among those who participated in all three waves. The frequency f_{ij}^{AB}, on the other hand, is the number of cases that scored (i, j) on the joint variable AB in the subgroup AB, so among those who only participated in the first and second wave and not in the third. The total number of cases in each subgroup will be denoted by N^A, N^{AB}, etc.

Assume that the hierarchical model we are interested in is the no three-variable interaction model $\{AB, AC, BC\}$. If the analysis is restricted to those who participated in all three waves, that is, to the subgroup ABC, the standard IPF procedure discussed in Section 2.4 can be applied to find the maximum likelihood estimates $\hat{F}_{ijk}^{ABC}$ using the sufficient statistics f_{ij+}^{ABC}, f_{i+k}^{ABC}, and f_{+jk}^{ABC}. However, unless all other subgroups are empty or all subgroups come from the same population (the missing completely at random assumption), the analysis based on the subgroup ABC alone will produce biased results and, even if the missing completely at random assumption is true, the analysis for the subgroup ABC ignores the (partial) information the other subgroups provide about the observed marginal distributions of AB, AC, and AD, resulting in less precise estimates and less powerful statistical tests than possible.

Fuchs proposed the following variant of the EM-algorithm that makes use of all available information, assuming that the postulated hierarchical model is valid in the population and that the missing data are missing at random.

The EM-algorithm starts with appropriate initial estimates $\hat{F}(0)$ for $\hat{f}_{ijk}^{ABC}$, which in this case may all be one as in the standard IPF procedure. The sufficient statistics are estimated in the E-step of the EM-algorithm, using the data of all subgroups. The estimated sufficient statistics for model $\{AB, AC, BC\}$ are the estimated observed marginal frequencies $\hat{f}_{ij+}^{ABC}$, $\hat{f}_{i+k}^{ABC}$, and $\hat{f}_{+jk}^{ABC}$. The estimated observed frequencies, $\hat{f}_{ijk}^{ABC}$ are calculated as follows, making use of the observed frequencies in all subgroups and the values $\hat{F}$ obtained so far (see equations 3.4, 3.5):

$$
\hat{f}_{ijk}^{ABC} = f_{ijk}^{ABC} + \frac{\hat{F}_{ijk}^{ABC}}{\hat{F}_{ij+}^{ABC}} f_{ij}^{AB} + \frac{\hat{F}_{ijk}^{ABC}}{\hat{F}_{i+k}^{ABC}} f_{ik}^{AC} + \frac{\hat{F}_{ijk}^{ABC}}{\hat{F}_{+jk}^{ABC}} f_{jk}^{BC}
$$

$$
+ \frac{\hat{F}_{ijk}^{ABC}}{\hat{F}_{i++}^{ABC}} f_{i}^{A} + \frac{\hat{F}_{ijk}^{ABC}}{\hat{F}_{+j+}^{ABC}} f_{j}^{B} + \frac{\hat{F}_{ijk}^{ABC}}{\hat{F}_{++k}^{ABC}} f_{k}^{C} \qquad (5.9)
$$

$$
\hat{f}_{ijk}^{ABC} = f_{ijk}^{ABC} + \hat{\pi}_{ijk}^{A\overline{BC}} f_{ij}^{AB} + \hat{\pi}_{ijk}^{A\overline{BC}} f_{ik}^{AC} + \hat{\pi}_{ijk}^{\overline{ABC}} f_{jk}^{BC}
$$

$$
+ \hat{\pi}_{ijk}^{A\overline{BC}} f_{i}^{A} + \hat{\pi}_{ijk}^{A\overline{BC}} f_{j}^{B} + \hat{\pi}_{ijk}^{\overline{AB}C} f_{k}^{C}
$$

The desired estimated observed marginal frequencies $\hat{f}_{ij+}^{ABC}$, $\hat{f}_{i+k}^{ABC}$, and $\hat{f}_{+jk}^{ABC}$ are obtained from equations 5.9 by summing $\hat{f}_{ijk}^{ABC}$ over the appropriate subscripts. In computing the estimates $\hat{f}$ all available information from all subgroups is used. The assumption that the missing data are missing at random is essential. For example, it appears from equations 5.9 that two individuals with the same observed scores on A and C are supposed to have the same estimated probability of obtaining the score $B = j$, regardless of the fact that one of them may have participated in the second wave B and the other did not. In general, not having a score on a particular variable may depend on all variables whose scores are actually observed, but not directly on the value of the particular variable whose scores are missing.

The estimated observed frequencies $\hat{f}$ are used in the M-step of the EM-algorithm to improve the estimates $\hat{F}$ obtained so far by adjusting the estimates $\hat{F}$ to the appropriate marginals $\hat{f}$. The newly obtained estimates $\hat{F}$ are then used in equations 5.9 to obtain improved estimates $\hat{f}$, etc., until the results converge. This algorithm provides the maximum likelihood estimates for the parameters of any hierarchical log-linear model taking (partial) nonresponse into account in a simple and intuitively appealing manner, provided the weak missing at random assumption is true.

Problems may arise in the application of this algorithm when in the subgroup ABC empty cells occur in the marginals to be reproduced, even when contributions from other subgroups are made to these cells (Fuchs, 1982: 272-273). In our example, empty cells in the marginals f_{ij+}^{ABC}, f_{i+k}^{ABC}, and f_{+jk}^{ABC} are problematic. In that case, the estimated expected frequencies $\hat{F}$ corresponding with these empty cells will remain zero or get values which are dependent on the initial starting values $\hat{F}(0)$. Fuchs recommended substituting .5 for the observed zero frequencies, but this is a very arbitrary solution. Perhaps the best advice is to try several substitutes and/or initial estimates and study the consequences (see Section 2.9).

It often occurs in panel studies that individuals who do not participate in a particular wave are not contacted in the next waves. Because of this, the nonresponse may show a nested pattern in the sense that the variables can be ordered in such a way that a missing score on a variable implies missing scores on all successive variables. The EM-algorithm described above can be used for nested and nonnested patterns of nonresponse. But in the case of a nested pattern, closed form expressions for the estimates $\hat{f}$ and $\hat{F}$ can also be given if a saturated log-linear model is postulated (Fuchs, 1982: 273). Because of this restriction to saturated models and because of the general applicability of the algorithm given above, the direct estimates will not be discussed any further.

Fuchs's procedure can be extended to log-linear models with latent variables (Hagenaars, 1986b, 1988c). In the case of latent variables there is a double missing data problem: The scores on the latent variables are missing for all cases, and some cases do not have scores on some of the manifest variables.

The E-step of the EM-algorithm in which the estimated observed frequencies $\hat{f}$ are obtained now looks as follows, denoting the manifest variables by A, B, and C and the latent variable by X and combining equations 3.4 and 5.9:

$$\hat{f}_{ijkt}^{ABCX} = \frac{\hat{F}_{ijkt}^{ABCX}}{\hat{F}_{ijk+}^{ABCX}} f_{ijk}^{ABC} + \frac{\hat{F}_{ijkt}^{ABCX}}{\hat{F}_{ij++}^{ABCX}} f_{ij}^{AB} + \frac{\hat{F}_{ijkt}^{ABCX}}{\hat{F}_{i+k+}^{ABCX}} f_{ik}^{AC} + \frac{\hat{F}_{ijkt}^{ABCX}}{\hat{F}_{+jk+}^{ABCX}} f_{jk}^{BC}$$

$$+ \frac{\hat{F}_{ijkt}^{ABCX}}{\hat{F}_{i+++}^{ABCX}} f_i^A + \frac{\hat{F}_{ijkt}^{ABCX}}{\hat{F}_{+j++}^{ABCX}} f_j^B + \frac{\hat{F}_{ijkt}^{ABCX}}{\hat{F}_{++kt}^{ABCX}} f_k^C \qquad (5.10)$$

$$= \hat{\pi}_{ijkt}^{ABC\overline{X}} f_{ijk}^{ABC} + \hat{\pi}_{ijkt}^{AB\overline{CX}} f_{ij}^{AB} + \hat{\pi}_{ijkt}^{A\overline{BCX}} f_{ik}^{AC} + \hat{\pi}_{ijkt}^{\overline{ABCX}} f_{jk}^{BC}$$

$$+ \hat{\pi}_{ijkt}^{A\overline{BCX}} f_i^A + \hat{\pi}_{ijkt}^{\overline{ABCX}} f_j^B + \hat{\pi}_{ijkt}^{\overline{ABCX}} f_k^C$$

The estimates $\hat{f}$ are used in the M-step in the customary way to improve the estimates $\hat{F}$, which in their turn are used to improve $\hat{f}$, etc. until the results converge. This algorithm has been implemented in the program LCAG (Hagenaars & Luijkx, 1987).

So far the discussion has dealt only with the estimation of the expected frequencies $\hat{F}$ assuming the validity of the missing at random assumption and the validity of the postulated log-linear model. But investigators certainly want to test such assumptions. Fuchs, following Chen (1979) and Chen and Fienberg (1974), recommended the following testing procedure, which can also be used in the case of models with latent variables.

First the estimated probabilities $\hat{\pi}_{ijk}^{ABC} = \hat{F}_{ijk}^{ABC}/N$ are calculated, where N is the total number of cases obtained by summing all cases in all mutually exclusive subgroups. By means of $\hat{\pi}_{ijk}^{ABC}$ and the sizes of the subgroups (N^A, N^{AB}, etc.) one then obtains the estimated expected frequencies for each subgroup as follows, for subgroups A, AB and ABC:

Subgroup A: $\qquad \hat{F}_i^A = N^A \sum_j \sum_k \hat{\pi}_{ijk}^{ABC}$

Subgroup AB: $\qquad \hat{F}_{ij}^{AB} = N^{AB} \sum_k \hat{\pi}_{ijk}^{ABC} \qquad (5.11)$

Subgroup ABC: $\qquad \hat{F}_{ijk}^{ABC} = N^{ABC} \hat{\pi}_{ijk}^{ABC}$

Next, the values of the test statistics L^2 and Pearson-χ^2 for each subgroup are computed in the ordinary way by means of the subgroup's estimated expected frequencies and the subgroup's observed frequencies. Finally, the subgroup test statistics are summed to obtain the overall test statistic for all N cases. The number of degrees of freedom of this overall test is equal to the number of cells in all nonempty subgroups minus the number of nonempty subgroups and minus the number of log-linear parameters to be independently estimated.

This overall test is used to test the composite hypothesis that all subgroups originate from the same population and that the postulated log-linear model is true. The first part of this hypothesis corresponds to the "missing completely at random" assumption. This part alone can be tested by postulating the saturated model $\{ABC\}$ instead of an unsaturated model like $\{AB, AC, BC\}$. Contrary to the saturated model for data without nonresponse, there are now degrees of freedom and L^2 or Pearson-χ^2 is not necessarily equal to zero.

If the missing completely at random assumption need not be rejected, conditional tests statistics $L^2_{r/u}$ can be used to test the second part of the composite hypothesis, namely, the validity of the postulated unsaturated model. The subscript r in $L^2_{r/u}$ refers to the restricted unsaturated model to be tested, and u to the unrestricted saturated model. This conditional test statistic has the same number of degrees of freedom as the corresponding nonconditional L^2_r based only on the N^{ABC} observations of subgroup ABC. However, $L^2_{r/u}$ is based on all N observations and is expected to have more power than the corresponding L^2_r for subgroup ABC.

If, given the saturated model, the missing completely at random assumption has to be rejected, it seems no longer possible to test more parsimonious log-linear models. However, as Little and Rubin (1987: 192) noted, the conditional test statistic $L^2_{r/u}$ is still valid provided the weaker missing at random assumption is true. The reason for this is that given the validity of the missing at random assumption, the component of the loglikelihood for the (non)response mechanism cancels out when L^2_u is subtracted from L^2_r to obtain $L^2_{r/u}$. Whatever *ignorable* response mechanism is actually at work, the conditional test statistic $L^2_{r/u}$ (and the log-linear parameter estimates) will not be influenced by it.

In sum, Fuchs has developed a very useful and versatile procedure for testing the validity of unsaturated hierarchical log-linear models and estimating the parameters of these models in the case of (partial) non-response, assuming only the weaker missing at random mechanism. However, his procedure has some disadvantages. In some cases, we may conclude that the missing data are missing completely at random, but mostly it is just assumed that the response mechanism is such that the missing data are missing at random without being able to specify further the nature of the response mechanism. Yet in order to prevent non-response in future investigations it may be important to know exactly which variables are related to nonresponse in what way and with what strength.

Actually, Chen and Fienberg (1974) have already shown how the missing at random response mechanism can be further specified by introducing extra parameters which indicate the probability with which a person belongs to (a particular cell of) a particular response subgroup. More details about the Chen and Fienberg procedure will not be given because below a (log-linear) causal modeling approach suggested by Fay (1986) will be discussed that can be regarded as a reparametrization of the procedures proposed by Fuchs and Chen and Fienberg as far as the causal models imply ignorable response mechanisms, but is more general and versatile than Chen and Fienberg's approach. The procedure proposed by Fay also deals with nonignorable response mechanisms. Another approach to dealing with ignorable and nonignorable response mechanisms making use of Haberman's program NEWTON (Haberman, 1988b) has been presented by Winship and Mare (1989).

Ignorable response mechanisms have received much more attention than nonignorable. As we saw above, a nice feature of the ignorable response mechanism is that even if the wrong model is chosen from the class of ignorable response models, the parameter estimates and the conditional test statistics remain the same. (That is actually the reason why they are called "ignorable.") Furthermore, when the response mechanism is ignorable and a particular log-linear model with or without latent variables is identifiable when applied to the "complete" response group ABC, it is also identifiable when applied to all response groups together.

Nonignorable response models do not have these features. Non-ignorability of the response mechanism may render a log-linear model unidentifiable, and selecting the wrong nonignorable response model can seriously bias the parameter estimates. However, if a particular nonignorable response mechanism has produced the data, postulating an ignorable response mechanism may also result in biased estimates. Therefore, Fay's approach deserves full attention.

The essential idea of Fay (1986) is to add to the set of ordinary variables (A, B, and C) indicator variables each of which designates whether or not the score on a particular ordinary variable is missing, and to set up modified path models (Section 2.7.3) for the relations among all variables, including the relations among the ordinary variables and the indicator variables. The indicator variables do not influence the ordinary variables, but are dependent on the ordinary variables or the other response variables. Three "ordinary" manifest variables A, B, and C are the point of departure. The scores on A or C or both are missing for some cases, but there are no missing scores on B. So we have four mutually exclusive response subgroups, ABC, AB, BC, and B with observed frequencies f_{ijk}^{ABC}, f_{ij}^{AB}, etc. Two indicator variables R and S are needed to designate the nonresponse, where $R = 1$ if the score on A is observed, $R = 2$ if the score on A is missing, $S = 1$ if the score on C is observed, and $S = 2$ if the score on C is missing. Thus, $RS = 11$ for the members of subgroup ABC, $RS = 12$ for subgroup AB, $RS = 21$ for subgroup BC, and $RS = 22$ for subgroup B.

The EM-algorithm is used to obtain the estimated expected frequencies $\hat{F}_{ijkmn}^{ABCRS}$ for the postulated log-linear (path) model. The sufficient statistics, that is, the estimated observed frequencies $\hat{f}_{ijkmn}^{ABCRS}$ are estimated in the E-step by means of equations 5.12, which are very much similar to equations 5.9 used by Fuchs.

$$\hat{f}_{ijk11}^{ABCRS} = f_{ijk}^{ABC}$$

$$\hat{f}_{ijk21}^{ABCRS} = f_{jk}^{BC} \left(\hat{F}_{ijk21}^{ABCRS} / \hat{F}_{+jk21}^{ABCRS} \right)$$

$$\hat{f}_{ijk12}^{ABCRS} = f_{ij}^{AB} \left(\hat{F}_{ijk12}^{ABCRS} / \hat{F}_{ij+12}^{ABCRS} \right)$$

$$\hat{f}_{ijk22}^{ABCRS} = f_{j}^{B} \left(\hat{F}_{ijk22}^{ABCRS} / \hat{F}_{+j+22}^{ABCRS} \right)$$

(5.12)

As usual, the estimates of the expected frequencies F_{ijkmn}^{ABCRS} obtained so far are improved in the M-step by means of the last obtained estimates $\hat{f}$ in equations 5.12. The testing procedure proceeds in the manner suggested by Fuchs, but using $\hat{\pi}_{ijk11}^{ABCRS}$, $\hat{\pi}_{ijk12}^{ABCRS}$, etc., instead of $\hat{\pi}_{ijk}^{ABC}$ (see equation 5.11). For identifiable models, the number of degrees of freedom equals the number of all cells in all response subgroups minus the number of subgroups and minus the number of nonredundant parameters to be estimated (including the effects referring to R and S).

Identifiability of the model may be a problem, at least with nonignorable response models (see Little & Rubin, 1987, Section 11.6). Repeating the EM-algorithm with different starting values of $\hat{F}$ may reveal unidentifiability. If different parameter estimates are obtained leading to the same L^2-value, the model is not identifiable.

If the nonresponse shows a nested pattern, certain observed cell frequencies are zero and the corresponding estimated expected cell frequencies have to be set equal to zero. For example, if nonresponse on A implies nonresponse on C, all observed and estimated cell frequencies for which $RS = 21$ are equal to or have to be fixed to zero.

In principle, extension of this approach to log-linear models with latent variables is possible by modifying equations 5.12 in the same manner that equation 5.9 have been modified to obtain equation 5.10 (Hagenaars, 1988c). The program LCAG can also handle this kind of causal modeling of nonresponse. However, the consequences of this extension need to be worked out.

To get an idea of the kinds of response models that can be evaluated by the causal modeling approach, some possible causal models have been presented in Figure 5.8. In the first three models, the relations among the "ordinary" variables A, B, and C are in agreement with the no three-variable interaction model $\{AB, AC, BC\}$. Various models are postulated involving the indicator variables R (scoring on A or not) and S (scoring on C or not).

In Figure 5.8a, the indicator variables R and S are completely independent of A, B, and C and of each other. This model implies that all missing scores on both variables A and C are missing completely at random. The corresponding path model is $\{R, S, AB, AC, BC\}$.

The model in Figure 5.8b represents a missing at random response mechanism: R is not directly related to A and S is not related to C. Having a missing score on variable C depends on the values of A and B (excluding the three-variable interaction ABS), but not directly on the values of C. Likewise, having a missing score on A depends only on B and C (including the three-variable interaction BCR). Variables R and A are associated with each other, as are S and C, but these associations are fully explained by B and A respectively. The corresponding path model is $\{AB, AS, BS, AC, BC, BCR\}$.

Model c provides an example of a nonignorable response mechanisms for missing scores on C: S is directly dependent on C. Missing scores on A depend only on B and C. The pertinent log-linear model is $\{AB, AC, BC, BR, CR, CS\}$.

Models d and e are models with a latent variable X. Model d is an ignorable model for the same reasons as model b. The parameters of this modified path model d have to be estimated stepwise: model $\{XA, XB,$

a) Missing Completely At Random   b) Missing At Random

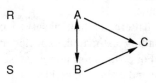

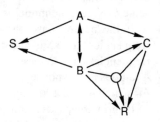

c) Nonignorable Response Mechanism   d) Latent Variable Model;
                                             Ignorable Response
                                             Mechanism

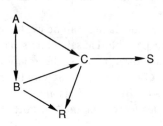

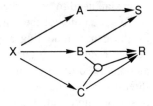

e) Latent Variable Model;
    (Non) Ignorable?

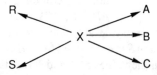

Figure 5.8. Ignorable and Nonignorable Response Models; Manifest Variables
A, B, C; Indicator Variables R (Having Scores on A or Not) and S
(Having Scores on C or Not); Latent Variable X.

XC} for table $XABC$ and model {$XABC, AS, BS, BCR$} for table $XABCRS$.
Model e, model {XA, XB, XC, XR, XS}, seems to be an ignorable model
because the response variables R and S are not directly related to A and
C, but depend directly only on the latent variable X. However, because
X is not directly observed, a model such as e does not have the characteristic features of an ignorable response model.

As these few examples show, the causal modeling approach to non-response allows one to formulate unsaturated log-linear models for the relations among the ordinary variables in combination with a large variety of response models. This approach is probably the most flexible and insightful procedure available for correcting nonresponse. Fay's approach will be especially useful in panel research in which a large amount of nonresponse is common, but in which much is known about the (partial) nonrespondents. The main problem that requires further research is the identifiability of those models which postulate a non-ignorable response mechanism and/or include latent variables.

5.5.2 Test-Retest Effects and Correlated Errors: Local Dependence Latent Class Models

Repeatedly interviewing the same persons can lead to the occurrence of test-retest effects and of serially correlated errors. A great deal has been written about how and why reinterview effects may distort a study's conclusions.[9] Test-retest effects form a threat to the internal, external, and construct validity of an investigation (Cook & Campbell, 1979). Under the influence of previous interviews, respondents may start to think about the matters they have been questioned on, change their opinions and behavior, and react differently — for example, to intervening events — in the next interview than would have been the case without the previous interviews (compare the Socratic effect — Section 4.4.3).

Reinterview effects may also cause differences between pretest and posttests without changing the underlying behavior and attitudes. As applicants for a job know, after having completed a few IQ tests, the next test will show a higher IQ score without a change in true intelligence. Such practice effects have also been noted in marketing research where people recorded their savings, debts, purchases, etc. more accurately the longer they stayed in the panel (Ferber, 1964, 1966). In short term panel investigations, respondents may be inclined to give answers consistent with what they have said in previous waves, a tendency which may be reinforced by being reinterviewed by the same interviewer (Hagenaars & Heinen, 1982: 105; Hyman, 1954: 243-252).

The empirical evidence about the occurrence of test-retest effects is not very clear. Most findings are rather idiosyncratic, typical of a particular investigation. As in the reports on nonresponse, usually no large effects are found, although some exceptions occur. Especially when several months or years lie between the interviews, being interviewed for one or two hours is most probably not such an influential event in the life of a respondent that serious test-retests have to be expected.

However, in the case of continuous participation in a panel and continuously recording particular events like purchases and watching television shows or in the case of five or six interviews being conducted within a short period of time, somewhat larger reinterview effects have to be reckoned with, certainly when the interviews concern topics about which the respondent did not think before or when the respondent knows that the same questions will be asked again. From this perspective, one must perhaps fear most for the distorting influence of previous questions during one and the same interview (see Schuman & Presser, 1981, Chapter 2).

Estimates of the strength of reinterview effects can be obtained by introducing into the panel design control groups which are interviewed only once, although it may be difficult in such a design to separate the effects of panel mortality and reinterviewing. Another possibility is to set up models in which test-retest effects are explicitly taken into account. An example of this latter approach will be given later in this section.

These models with test-retest effects are also useful for attacking the problem of correlated error. Kessler and Greenberg (1981, Chapter 7) identified three sources of serially correlated error in panel studies. The first one is misspecification of the functional form of the relations between variables from different waves. Because log-linear analysis usually does not require relations to have a particular ([curvi]linear) form, this is less of a problem in categorical panel data analysis.

The second source is misspecification of the (causal) model because of an omitted exogenous variable, resulting in correlated prediction errors. If I_1 (initial income at age 30) is predicted from E (education) by means of linear regression, most observations will not lie exactly on the regression line. People will earn more or less money than predicted by their education because factors other than education also determine their income. For example, two people may have attained the same level of education, but one has a higher aspiration level or is better looking than the other with the result that the first one has a higher and the second a lower initial income than predicted from their educational level. Regressing I_2 (final income at age 60) on E and I_1 we may still expect that the person with the higher aspiration level and the better looks has a higher and the other person a lower final income than predicted from E and I_2. Consequently, in a (linear) path model in which I_1 is influenced by E, and I_2 by E and I_1, the error terms connected to I_1 and I_2 will be correlated and the regression coefficients of this path model will be biased because of this correlation between error terms, that is, because of the omitted variables Aspiration Level and Looks.

The total correlation between initial and final income in the above example is caused by education, by the "stability" of income over time, and by the correlation between the error terms, and we would like to know the contribution of each of these components to the total correlation. However, the three-variable path model described above is unidentifiable because there are only three known correlations but four unknowns, namely, three path coefficients and one correlation between error terms. As Kessler and Greenberg point out, the model can be made identifiable by introducing extra assumptions and/or extra (instrumental) variables. Although it has not been worked out and the solutions found for regression equations and path models cannot be directly translated, in principle the same course of action is possible with categorical data and log-linear models, as shown in the example below.

The third source of correlation between error terms mentioned by Kessler and Greenberg is correlated measurement error. If a person is inclined to exaggerate his or her income, this may be expected to occur at both initial and final income, resulting in correlation between the error terms connected with the income variables. In terms of latent structure models, the underlying latent variable does not completely account for the observed correlation between two indicators, but part of the observed correlation is due to the correlation between the measurement error terms connected to the indicators.

Most forms of reinterview effects and correlated error terms have in common that they constitute extra, not directly observed sources of correlation between the observed variables.[10] These sources may be taken into account by using "extra" latent variables or by postulating "extra" direct effects among manifest variables that ordinarily would not have been there. Hagenaars (1988b), using log-linear models with latent variables, provided examples of the latter approach. This method will be briefly outlined here using (again) the data on party and candidate preference in Table 4.4.

A well-fitting four latent variable model for these data has been reported in Section 5.4. The one and two latent variable models presented in Section 4.4.3 and depicted in Figure 4.2 had to be rejected. The reason for this rejection might have been the occurrence of test-retest effects or correlated errors.

A - Vote Intention March   Y, Z - Latent Variables

B - Vote Intention February   A', C'- Quasi-latent Variables

C - Preference Prime Minister March

D - Preference Prime Minister February

a) - Local Independence Model   b)   Local Dependence Model

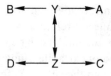   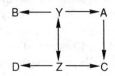

c)   Reformulation of Local Dependence Model

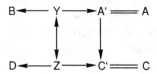

Figure 5.9. Latent Variable Models with Direct Effects between Indicators:
Vote Intention and Preference for Prime Minister; Data from Table 4.4

After all, the basic assumption of latent class analysis is the local
independence assumption. Within a particular latent class, the scores on
the manifest variables are independent of each other. According to the
latent class model, there are no direct relations among the manifest
variables. Yet the test-retest effects and the correlated error terms discussed above may cause associations among the manifest variables that
are not accounted for by the latent variables, thus violating the local
independence assumption (Section 3.5).

The best, but still not acceptable two latent variable model was model
{YZ, YA, YB, ZC, ZD}, in which no true changes in either vote intention
or preference for Prime Minister occurred (Figures 4.2c and 5.9a). The
test statistics reported in Section 4.4.3 were $L^2 = 84.74$, df = 48, $p < .01$.
In principle, inspection of the standardized residual frequencies r_{ijkl}^{ABCD}
($= [f_{ijkl}^{ABCD} - \hat{F}_{ijkl}^{ABCD}]/[\hat{F}_{ijkl}^{ABCD}]^{1/2}$) should provide insight into why the
model failed to fit. Discovering a pattern in the (signs of the) $3^4 = 81$
residuals appeared to be difficult. Therefore, the procedure recom-

mended by Hagenaars (1988b) was followed. Two variable residual tables were formed by summing r^{ABCD}_{ijkl} over the appropriate subscripts. The entries r^{ABCD}_{i+k+} of the residual table AC were definitely the largest, much larger than for the residual table BD. The strength of the observed "positive" association AC was underestimated by model $\{YZ, YA, YB, ZC, ZD\}$.

A possible explanation of this difference between observed and estimated association might be correlated measurement error or, more plausible in this case, a kind of test-retest effect from party preference A to candidate preference C. The voting question was asked earlier in the interview than the question on preference for Prime Minister. It might well be that in February respondents had not yet formed a stable idea about whom to prefer as Prime Minister, but when in doubt adapted their answer on preference for Prime Minister to the answer on party preference given earlier, resulting in the test-retest effect A-C. In March, a month closer to the May Elections, their ideas have become more crystallized and they answer the questions on party and candidate preference "independently" of each other, without generating a test-retest effect B-D. (Given most interpretations of the meaning of correlated error terms one would expect them to operate in both waves. So an interpretation in terms of test-retest effects is preferred.)

The extra test-retest effect A-C is incorporated in the model in Figure 5.9b. However, as it stands, this is a local dependence model (Section 3.5) and its parameters cannot be estimated in the way latent class parameters are usually estimated. A reformulation of the model as depicted in Figure 5.9c provides the solution. Two quasi-latent variables A' and C' are introduced, which are perfectly connected to A and C respectively (Section 3.3). In this four latent variable model in Figure 5.9c, A and B are locally independent of each other, and the test-retest effect is represented by the direct effect between A' and B'.

The parameters of this modified path model with latent variables can be estimated using the program LCAG. There are 81 latent classes, categories of the joint latent variable $YZA'C'$. The conditional response probabilities have to be restricted in such a way that A' and C' are perfectly related to A and C respectively and that the score on B depends directly only on Y and the score on D only on Z. Moreover, the relations among the latent variables have to agree with the model in Figure 5.9c. This means that model $\{YZ, YA\}$ is imposed on the probabilities in table YZA' and model $\{YZA', A'C', ZC'\}$ on table $YZA'C'$, according to the principles of Goodman's modified path analysis approach.

The model in Figure 5.9c has four degrees of freedom less than the original model in Figure 5.9a and it appears to fit the data excellently: $L^2 = 45.97$, df = 44, $p = .39$ (Pearson-$\chi^2 = 44.04$). The final conclusion may be that in contrast to what has been said before the no-latent-change model can be accepted, albeit after introduction of a test-retest effect A-C.

All in all, this method is very attractive. One sets up a particular model, inspects the residual frequencies, introduces extra effects among manifest variables where this seems to be necessary, and in many cases the result will be a well-fitting model (Hagenaars, 1988b). However, this route should not be taken too lightly. This is seen most clearly if it is realized that this approach is equivalent to introducing extra latent variables into the model that have no direct empirical referents, but explain the otherwise unexplained part of association among the manifest variables. These latent variables just stand for what the investigator makes them stand for. Only if there are clear theoretical reasons for expecting test-retest effects and correlated error terms should this strategy be used.

A fine example of interpreting direct effects between indicators is given by Duncan, who in a somewhat different manner takes "response consistency effects" into account (Duncan, 1979b; Duncan, Sloane & Brody, 1982; Hagenaars, 1988b). In his analysis of sets of four dichotomous items, it appeared that the observed number of persons who agree four times (the ++++ response) and disagree four times (– – – –) is clearly underestimated in the models he applied. He concluded that there was a response set toward (dis)agreement, toward "yeasaying" and "naysaying." He then introduced extra effects which refer (exclusively) to these two cells with the extreme response patterns ++++ and – – – –. In terms of latent class analysis, he added two extra latent classes, one that gives the ++++ answer with a probability of 1, and the other with absolute certainty scores – – – –.

Duncan then goes on to make a distinction among those who give the same answer four times (++++ or – – – –) between "ideological respondents" (those who are in the two extra latent classes) and "situational respondents" who give these answers for other than ideological reasons. Starting out from the methodological concept "response set," he tries to explain this response set in terms of substantive theory.

Here, and elsewhere in this and the previous chapter, it appears that log-linear models with or without latent variables applied to panel data can be used to answer many theoretically relevant questions involving (short term) social change. Davis's (1975b) hard judgment on the "panel study fad, which exhausted itself in producing fancy methodology and never got around to analyzing and reporting much in the way of results" (p. 76), was certainly not groundless. The above expositions were intended to show that this is not necessarily true.

Notes

1. To mention just a few: Bennett and Lumsdaine (1975); Bohrnstedt (1969); Campbell and Boruch (1975); Campbell and Erlebacher (1975); Cook and Campbell (1979); Judd and Kenny (1981); Mohr (1982); Reichardt (1979); Struening and Guttentag (1975).

2. Instead of using Table 5.5, all test statistics in Table 5.6 can be obtained by means of the original ABX Table 5.3 employing the methods presented in Section 4.3.1. The symmetry model which postulates symmetry for both conditional turnover tables AB implies the following restrictions on the parameters of model $\{ABX\}$: $\tau_i^A = \tau_i^B$, $\tau_{ik}^{AX} = \tau_{ik}^{BX}$, $\tau_{ij}^{AB} = \tau_{ji}^{AB}$, and $\tau_{ijk}^{ABX} = \tau_{jik}^{ABX}$. In terms of the extended table $ZABX$, where $Z = 1$ refers to the original table ABX and $Z = 2$ to the two transposed conditional turnover tables, the estimated expected frequencies for the symmetry model are obtained by means of model $\{ABX\}$.

The estimates for the quasi-symmetry model postulating that there is quasi-symmetry in both turnover tables can be obtained by applying model $\{ABX, ZAX, ZBX\}$ to the extended table. This quasi-symmetric model implies the following restrictions on the parameters of model $\{ABX\}$ for the original table ABX: $\tau_{ij}^{AB} = \tau_{ji}^{AB}$ and $\tau_{ijk}^{ABX} = \tau_{jik}^{ABX}$.

The estimates for the partial symmetry model in which it is assumed that there is symmetry in both turnover tables as far as the overall differences between the marginal distributions of A and B permit can be found by means of model $\{ABX, ZA, ZB\}$ for the extended table. For the original table ABX, this implies the following restrictions to be imposed in model $\{ABX\}$: $\tau_{ik}^{AX} = \tau_{ik}^{BX}$, $\tau_{ij}^{AB} = \tau_{ji}^{AB}$, and $\tau_{ijk}^{ABX} = \tau_{jik}^{ABX}$. The partial symmetry model differs from the symmetry model in that the parameters τ_i^A and τ_i^B are left free in the former. The partial symmetry model differs from the quasi-symmetry model in that it restricts the parameters τ_{ik}^{AX} and τ_{ik}^{BX} to be equal to each other.

3. One of the analysis steps reported later involves estimating the effects of A, C, and D on B. The relevant logit model reproduces exactly the observed frequency table ACD, which, however, contains a zero cell, namely, cell $ACD = (2, 1, 3)$. This makes it impossible to estimate the effect parameters in the standard way and it was decided to replace the cell frequency zero by a small constant. To make all (sub)analyses mentioned in Figure 5.4 comparable with each other, it was decided to increase three cell frequencies of Table 4.4 by 0.1, namely, $ABCD = (2, 1, 1, 3)$, $(2, 2, 1, 3)$, and $(2, 3, 1, 3)$. All analyses have been performed as far as possible both on the original Table 4.4 and on the "increased" table. The differences were negligible. Only the results for the "increased" table are presented.

4. Because we are dealing with a rather extreme model, the possibility of uni-dentifiability has to be seriously considered. Therefore, several different sets of initial

estimates were used, both "randomly" chosen and according to the best of our knowledge. In all cases the same final solution was obtained. This solution always contained a few estimates for π_t^X equal to zero; also some of the conditional response probabilities were zero. The final estimates may be considered as the maximum likelihood estimates under the condition that these zero estimates are indeed zero in the population. Taking the fixed character of these zero estimates into account, the number of degrees of freedom at the very most equals 43. This does not alter the conclusion that the model has to be rejected.

5. Numerous articles and research reports on the nature and the consequences of nonresponse are available. Important sources are the *Journal of Marketing Research*, *Public Opinion Quarterly*, and the *Proceedings of the American Statistical Association: Social statistics section*. For reports on typical panel mortality see, among others, Arminger (1976); Butler and Stokes (1969, Appendix); Ferber (1953, 1966); Glock (1952); Goudy (1976); Hagenaars et al. (1990, Appendix B); Sewell and Hauser (1975, Chapter 2); Sobol (1959); Williams (1969); Williams and Mallows (1970). Furthermore, the articles by Fitzgerald and Fuller (1982); Smith (1983); and Steeh (1981) are relevant.

6. Although success stories are more readily reported than failures, the literature makes clear that systematically applying a variety of procedures to locate respondents and to induce them to participate can reduce the percentage of nonresponse considerably, even in panel studies with ten or more years between waves (Allison, Zwick, & Brinser, 1958; Clarridge, Sheehy, & Hansen, 1977; Crider, Willits, & Bealer, 1971; Crider & Willits, 1973; Droege, 1971; Eckland, 1968; Freedman, Thornton, & Camburn, 1980; McAllister, Butler, & Goe, 1973; McAllister, Goe & Butler, 1973; Thornton, Freedman, & Camburn, 1982).

7. The extent of the possible bias can be detected by assigning extreme scores to nonrespondents, where "extreme" means assigning to all nonrespondents both a score which is as unfavorable as possible for the null hypothesis and one which is as favorable as possible. Suppose a positive association between the variables A and B is expected. First, table AB is set up for the respondents. Next the nonrespondents are added to this table in such a way that the relationship A-B is as positive as possible and then in such a way that the relationship is as negative as possible. Comparison of the two extreme tables shows the limits within which the actually observed data can be biased by nonresponse.

8. Fuchs's procedure can be regarded as a more sophisticated form of the nonresponse correction that is known as "raking" (Bailar, Bailey, & Corby, 1978; Chapman, 1976). Raking in its turn is a classic application of the iterative proportional fitting procedure in which the frequencies in a particular table are adjusted to one or more marginals of another table (Bishop et al., 1975, Section 3.6; Deming and Stephan, 1940; Haberman, 1979, Chapter 9).

9. A few references are Bailar (1975); Bridge et al. (1977); Buck, Fairclough, Jephcott, and Ringer (1977); Campbell and Stanley (1966); Chaffee and McLeod (1968); Ehrenberg (1960); Glock (1952); Lana (1969); Neter and Waksberg (1964); Nosanchuk (1970); Sobol (1959); Turner and Martin (1984, Vol.1, Section 9.2).

10. As far as test-retest effects result from the fact that respondents try to answer consistently over time, these effects cause "extra" manifest correlation. But as far as having been interviewed leads to changes in underlying attitudes and behavior, test-retest effects do not necessarily lead to extra manifest correlation. These forms of reinterview effects might be modeled by postulating direct effects between the t_1 manifest variables and the t_2 latent variables.

6. Trend Analysis

6.1 Introduction

In trend studies the data of the successive waves refer to different persons who, however, all belong to the same population. Trend studies form the main practical source of information about processes of social change. This is not a consequence of the superiority of the trend design as such. On the contrary, all research questions which can be answered by means of trend data can in principle also be answered using panel or cohort data, but the reverse is not true. For example, if the research questions focus on gross change, only panel data can provide the answers; if net changes are the object of investigation, panel and trend data can be used. However, trend studies, if relevant, have some advantages over panel and cohort studies in that trend data are more readily available and can be analyzed in a simpler way than cohort and panel data.

The investigation of long term social change in particular has to rely on trend rather than panel data. In the first place, long term panel data are scarce. Second, performing categorical panel analyses for more than three waves is in principle possible, but, unless the sample size is very large, multiwave panels give rise to serious problems due to the many (nearly) empty cells in the multivariate turnover tables to be analyzed. In trend analysis, a large number of waves poses no such problems. First there are more cross-sectional surveys available that can be arranged into a long term trend design. Second, in trend designs the total sample size increases with each successive wave, avoiding the empty cells problem.

In general, log-linear trend analyses are less complicated than log-linear panel analyses. The techniques presented in Chapters 2 and 3 within the framework of the one-shot survey can be applied directly to trend data; the only complication is the introduction of the independent variable T(time) (Davis, 1975b, Landis & Koch, 1979). This may be clear from the previous discussions on the pretest-posttest design (Section 5.2) and will be further clarified in Section 6.3.1 where the log-linear analysis of the net changes in one characteristic is presented.

In Section 6.3.2, it will be shown how to specify the trends or net changes for different subgroups. It will be explained how to use log-linear modeling for determining the influence of variables like age, income, gender, etc. on the net changes in a particular characteristic. An extension of these analyses to investigate the simultaneous net changes in two or more characteristics is presented in Section 6.4. A particular variant of trend analysis, namely, the interrupted time-series design, is the topic of Section 6.5. In the final section of this chapter, it will be shown how the above-mentioned trend analyses can be carried out at the latent level taking measurement unreliability into account.

Before explaining all these forms of log-linear trend analyses, attention will be paid to the trend design as such, briefly indicating the framework within which the log-linear trend analyses take place.

6.2 The Trend Design

Using Campbell and associates' notation introduced in Sections 5.2 and 5.3, the trend design may be represented as follows (for five waves):

$$
\begin{array}{llllll}
R & O_1 \\
R & & X_a & O_2 \\
R & & & & X_b & O_3 \\
R & & & & & & X_c & O_4 \\
R & & & & & & & & X_d & O_5 \\
\end{array}
$$

As before, the symbol R in this diagram refers to a random assignment of the respondents over the (five) groups. This symbol is justified if the (five) groups are random samples from the same (statistical) population. O_i refers to the distribution of the dependent characteristic, for example, vote intention, at time i. X_a is an event occurring between t_1 and t_2, X_b an event between t_2 and t_3, etc. It is not necessary that a relevant event occurs between each two successive points of observations. Often attention is focused on the effects of just one particular event.

As in the separate sample pretest-posttest design discussed in Section 5.2, the effects of the events X_a, X_b, etc. are found by comparing the pretest measurement(s) O before the event occurred with the posttest measurement(s) O after the event.

In this trend design, *history* may be an important confounding factor (Campbell & Stanley, 1966). If the trend analysis attempts to estimate the effects of particular well-defined events, for example, X_a and X_b, the occurrences of all kinds of other, often unknown, events imply that the effects of X_a and X_b may be confounded with the effects of these other (unknown) events. When trend analysis is used in a more exploratory way to find out which events might explain the observed net changes, it is difficult to determine whether the designated events and not particular other events one did not think of brought about the changes.

The fundamental difficulty is that in trend analysis the factor time is used as a substitute for all sorts of events which take place between or during the periods of observation.[1] The confounding influence of history can be (partially) controlled by direct measurement of the extent to which the respondents have been exposed to the relevant events or by introducing into the trend design a control group that has not been exposed to the relevant events. However, only seldom will this kind of data be available to the secondary analyst.

Mortality is another important possibly confounding factor in the trend design. It encompasses the nonresponse phenomenon discussed in the previous chapter, but also the circumstance that, particularly in long term trend studies, people from the original population die during the period of investigation and new people are born into it. This means that even if all groups are randomly drawn from the "same" population, for example, the Dutch population, there is not one "population" in the statistical sense. Consequently, the symbol R in the above diagram is not strictly justified and the observed net changes may have been caused by changes in population composition rather than by the events X_a, X_b, etc.

The confounding effects of mortality in the sense of population replacement can be overcome by means of cohort analysis, discussed in the next chapter. Corrections for nonresponse are possible using the methods outlined in Section 5.5.1, albeit that in most trend studies less information about the nonrespondents is available than in panel studies.

6.3 Net Changes in One Characteristic

6.3.1 Time as an Independent Variable

As stated above, trend analyses are very much like ordinary cross-sectional analyses. The only extraordinary thing about trend analysis is that one of the variables is Time which has as categories the moments of observation. This can be illustrated by means of Table 6.1.

Table 6.1 shows the net changes in vote intention among the Dutch electorate during the years 1978-1980, where the many political parties among which the Dutch electorate has to choose (see Table 4.1) have been collapsed into just four categories. The data come from the Amsterdam Continuous Survey, a series of roughly semi-annual investigations started in January 1972. In this Continuous Survey, a fresh nation-wide sample is drawn at each moment of observation (Van der Eijk, 1980).

The Columns Total in Table 6.1 forms the two dimensional table *TV* (Time × Vote Intention) from which the nature of the net changes in vote intention can be learned. If the relative distribution of vote intention is the same at each moment of observation, it is concluded that no net changes in vote intention occurred during the years 1979-1980. In log-linear terms the no-net-change hypothesis implies the validity of the independence model {*T, V*} for the two dimensional table *TV*.

The test results for the independence model {*T, V*} are presented in Table 6.2, model 1. Given these outcomes, a strict employment of the .05 significance level leads to acceptance of the no-net-change hypothesis.

But for several reasons this conclusion may be somewhat premature. The test result is not completely conclusive as the *p* value lies practically on the borderline of accepting or rejecting the null hypothesis. This is the more important as there is the substantive argument that it is not very likely that in a period of three years not a single relevant shift in vote intention took place. During the years 1978-1980 many events occurred in the Dutch political arena — some of which will be mentioned below — that most likely did influence party preferences. Even with the coarse categorization used, vote intention should show some net changes.

Given the nature of the Dutch party system, it is to be expected that these net changes will be rather small. However, even very small shifts can have important political consequences. For a political party, obtaining 4% more votes than during the previous election means a gain of six seats in the Dutch Parliament (which has 150 seats). In the Dutch political setup these are true landslides, which do not occur often. Even

TABLE 6.1 Time (T), Religious Denomination (R) and Vote Intention (V)[a]

T. Time	V. Vote Intention	R. Religious Denomination			T. Time	V. Vote Intention	R. Religious Denomination		
		1. No	2. Yes	Total			1. No	2. Yes	Total
1. Ja-78	1. Left	112	70	182	5. Oc-79	1. Left	143	100	243
1. Ja-78	2. Christ.	4	170	174	5. Oc-79	2. Christ.	8	133	141
1. Ja-78	3. Right	33	33	66	5. Oc-79	3. Right	28	47	75
1. Ja-78	4. Nonv.	23	24	47	5. Oc-79	4. Nonv.	25	19	44
	Total	172	297	469		Total	204	299	503
2. Ju-78	1. Left	123	64	187	6. Ap-80	1. Left	135	78	213
2. Ju-78	2. Christ.	6	143	149	6. Ap-80	2. Christ.	8	135	143
2. Ju-78	3. Right	24	32	56	6. Ap-80	3. Right	26	30	56
2. Ju-78	4. Nonv.	20	31	51	6. Ap-80	4. Nonv.	23	14	37
	Total	173	270	443		Total	192	257	449
3. No-78	1. Left	122	83	205	7. Oc-80	1. Left	133	73	206
3. No-78	2. Christ.	10	114	124	7. Oc-80	2. Christ.	9	119	128
3. No-78	3. Right	30	41	71	7. Oc-80	3. Right	31	29	60
3. No-78	4. Nonv.	18	27	45	7. Oc-80	4. Nonv.	20	29	49
	Total	180	265	445		Total	193	250	443

(continued)

TABLE 6.1 Continued

T. Time	V. Vote Intention	R. Religious Denomination		Total
		1. No	2. Yes	
4. Ma-79	1. Left	137	59	196
4. Ma-79	2. Christ.	3	115	118
4. Ma-79	3. Right	36	39	75
4. Ma-79	4. Nonv.	21	31	52
	Total	197	244	441

T. Time	V. Vote Intention	R. Religious Denomination		Total
		1. No	2. Yes	
8. De-80	1. Left	141	57	198
8. De-80	2. Christ.	11	119	130
8. De-80	3. Right	30	43	73
8. De-80	4. Nonv.	25	18	43
	Total	207	237	444
			Total	3637

a. T. Time   1. January 1978; 2. June 1978; 3. November 1978; 4. May 1979; 5. October 1979; 6. April 1980;
7. October 1980; 8. December 1980

R. Religious Denomination   1. No = does not belong to a religious denomination; 2. Yes = belongs to a religious denomination
V. Vote Intention   1. Left Wing Party; 2. Christian (Democratic) Party; 3. Right Wing Party; 4. Nonvoter
SOURCE: Continuous Survey, University of Amsterdam / Steinmetz Archives Amsterdam

TABLE 6.2 Log-linear Models for the T(Time) × V(Vote intention) Table

Model	L^2	df	p	Pearson-χ^2
1. $\{T, V\}$	31.24	21	.07	31.40
2. $\{T, V\} + \lambda_{11}^{TV}, \lambda_{12}^{TV}$	19.21	19	.44	19.20
3. $\{T, V\}$ + linear TV	23.59	18	.17	23.47
4. $\{T, V\}$ + lin./quadr. TV	15.17	15	.44	15.21
5. $\{T, V\}$ + lin./quadr. TV	16.18	19	.65	16.21
cat. 2 = − cat. 1				
6. (1) - (2)	12.03	2	.002	
7. (1) - (3)	7.65	3	.05	
8. (1) - (4)	16.07	6	.01	
9. (1) - (5)	15.06	2	.001	
10. (3) - (4)	8.42	3	.04	
11. (5) - (4)	1.01	4	.91	

SOURCE: Columns Total in Table 5.1

smaller gains and losses can lead to totally different coalitions among the political parties in order to form a government supported by the majority of Parliament. The purpose of the analysis of table TV (Table 6.1) should be to discover these small but very relevant systematic differences amid the expected sampling fluctuations. In Converse's words, the problem is "finding the very small" (Converse, 1976: 43).

Because of these considerations, it was decided to continue the analysis and not to simply accept the no-net-change hypothesis. The search for alternative models that could explain the data in table TV was simultaneously guided by substantive ideas about the influence of certain events that were reported in the newspaper summaries for the period 1979-1980 and by a close examination of the data. Given the nature of this book, more space will be devoted to explaining how to perform a thorough data search than to an exposition of the main substantive ideas. However, it will be shown how the final model chosen not only fits the data but also fits in with the substantive ideas. It will then become clear what is meant when it is said that the variable Time is a "substitute variable."

The data search started with an examination of the adjusted residuals of the independence model and of the parameters $\hat{\lambda}_{ij}^{TV}$ of the saturated model $\{TV\}$. Of all adjusted residuals of the independence model (not reported here), only the residuals r_{11}^{TV} and r_{12}^{TV} were significant: $r_{11}^{TV} = -2.805$ and $r_{12}^{TV} = 3.360$. In line with this, the parameters $\hat{\lambda}_{11}^{TV}$ and $\hat{\lambda}_{12}^{TV}$ of the saturated model $\{TV\}$ differed significantly from zero. (Among the remaining parameters $\hat{\lambda}_{ij}^{TV}$ (not reported here) only $\hat{\lambda}_{42}^{TV}$ is significant.) From these significant residuals and $\hat{\lambda}$ parameters, it is concluded that at the first observation moment, in January 1978, the observed support for the political left wing is smaller and the observed support for the Christian parties is larger than expected on the basis of the independence model.

To account for these discrepancies between observed and expected frequencies, one may add to the independence model two two-variable effects which refer only to these two cells. An easy way to accomplish this is to use the parametrization in which the two-variable parameters λ_{ij}^{TV} are added to the independence model, but with the restriction that all λ_{ij}^{TV} parameters are fixed to zero except λ_{11}^{TV} and λ_{12}^{TV}. These latter two parameters are estimated from the data without further restrictions. Because of the introduction of these two effect parameters to the independence model, the frequencies f_{11}^{TV} and f_{12}^{TV} are exactly reproduced. Haberman's program FREQ can be used to estimate the parameters of this extended independence model. The test statistics of this extended independence model are presented in Table 6.2, model 2. Model 2 fits the data rather well and also significantly better than the original independence model 1, as appears from the conditional test outcomes (model 6).

Adding parameters in this way ought to be justified by substantive theoretical ideas about the pertinent phenomenon. In this case it should be made plausible that the beginning of 1978 was an exceptional period with regard to people's political leanings compared with the remainder of the years 1978-1980. Moreover, it should be made plausible that during the period June 1978 - December 1980 the aggregate vote intention remained stable. In the light of what was said before, this latter implication of the extended independence model in particular has to be doubted. Thus, on substantive grounds, model 2 is rejected.

A further examination of the (mostly nonsignificant) adjusted residuals of the original independence model 1 and of the (mostly nonsignificant) two-variable parameters of the saturated model $\{TV\}$ vaguely suggested that support for the left wing and for the Christian parties followed a (curvi)linear pattern over time. However, it is extremely difficult to establish the exact nature of this pattern by just looking at these parameters and residuals, especially in the presence of relatively large sampling fluctuations. One can, however, inspect the data for the presence of systematic patterns of net change by means of more powerful tests than the "interocular test" (Gottman, 1981: 57).

A simple and useful starting point for the detection of systematic patterns of net change is a model in which it is assumed that the log-linear effects of T(time) on vote intention are linear (see Section 2.8, and Goodman, 1981a, 1983). It is then assumed that the log-odds of choosing one party rather than another follow a linearly increasing (decreasing) pattern over time. The odds themselves increase (decrease) over time by a constant factor and consequently lie on a logistic curve. Starting out from the saturated model $\{TV\}$, the following restrictions are imposed on the log-linear and multiplicative two-variable parameters:

$$\lambda_{ij}^{TV} = \lambda_j^{*V} t_i \qquad (6.1)$$

In equation 6.1 λ_j^{*V} is a parameter to be estimated and t_i is the score for category i of variable T.

If the distances between the observation periods are equal, any set of increasing numbers with a constant difference between two consecutive numbers can be used for t_i. If, as in Table 6.1, the measurement moments lie at unequal distances from each other, the scores t_i must reflect these unequal distances. Often the scores t_i will be selected in such a way that they sum to zero in agreement with the restriction that the log-linear effect parameters are zero after summation over any of their subscripts:

$$\sum_i \lambda_{ij}^{TV} = \sum_i \lambda_j^{*V} t_i = \lambda_j^{*V} \sum_i t_i = 0$$

$$\sum_j \lambda_{ij}^{TV} = \sum_j \lambda_j^{*V} t_i = t_i \sum_j \lambda_j^{*V} = 0 \qquad (6.2)$$

The model in which the log-odds of voting for one party rather than another depend linearly on time may then be represented as follows:

$$G_{ij}^{TV} = \theta + \lambda_i^T + \lambda_j^V + \lambda_j^{*V} t_i$$

$$F_{ij}^{TV} = \eta \tau_i^T \tau_j^V (\tau_j^{*V})^{t_i} \qquad (6.3)$$

The parameters in model 6.3 can be estimated by means of the FREQ program (Haberman, 1979), after having set up the correct design matrix.[2] To give an impression of how the linear restrictions in equation 6.1 appear in the design matrix, the following example may be useful (see also Section 2.4). For this example, it is assumed that table TV is a 4×3 table and that the distances t_1-t_2, t_2-t_3, and t_3-t_4 are equal to each other.

In that case, model 6.3 can be rendered as follows using the design matrix **X**:

$$\ln \begin{bmatrix} F_{11} \\ F_{12} \\ F_{13} \\ F_{21} \\ F_{22} \\ F_{23} \\ F_{31} \\ F_{32} \\ F_{33} \\ F_{41} \\ F_{42} \\ F_{43} \end{bmatrix} = \begin{bmatrix} 1 & 1 & 0 & 0 & 1 & 0 & -3 & 0 \\ 1 & 1 & 0 & 0 & 0 & 1 & 0 & -3 \\ 1 & 1 & 0 & 0 & -1 & -1 & 3 & 3 \\ 1 & 0 & 1 & 0 & 1 & 0 & -1 & 0 \\ 1 & 0 & 1 & 0 & 0 & 1 & 0 & -1 \\ 1 & 0 & 1 & 0 & -1 & -1 & 1 & 1 \\ 1 & 0 & 0 & 1 & 1 & 0 & 1 & 0 \\ 1 & 0 & 0 & 1 & 0 & 1 & 0 & 1 \\ 1 & 0 & 0 & 1 & -1 & -1 & -1 & -1 \\ 1 & -1 & -1 & -1 & 1 & 0 & 3 & 0 \\ 1 & -1 & -1 & -1 & 0 & 1 & 0 & 3 \\ 1 & -1 & -1 & -1 & -1 & -1 & -3 & -3 \end{bmatrix} \cdot \begin{bmatrix} \theta \\ \lambda_1^T \\ \lambda_2^T \\ \lambda_3^T \\ \lambda_1^V \\ \lambda_2^V \\ \lambda_1^{*V} \\ \lambda_2^{*V} \end{bmatrix}$$

$$\mathbf{G} \quad = \qquad\qquad\qquad \mathbf{X} \qquad\qquad\qquad\qquad \boldsymbol{\lambda} \qquad\qquad (6.4)$$

The first six columns of the design matrix **X** in equation 6.4 represent the one-variable effects (that also appear in the independence model). With regard to these effects, straightforward effect coding (instead of dummy coding) is used and no linear restrictions are imposed on the parameters λ_i^T and λ_j^V. The last two columns of **X** represent the linear effect of T on V according to restriction 6.1. The scores used for t_i are $-3, -1, 1, 3$. Restrictions 6.2 are satisfied because the scores t_i sum to zero and $\lambda_3^{*V} = -(\lambda_1^{*V} + \lambda_2^{*V})$

If "linear" model 6.3 is applied to table TV (Table 6.1, Columns Total), the test results are rather favorable. As the outcomes for model 3 in Table 6.2 show, model 3 has three degrees of freedom less than the independence model 1, because variable V has four categories and consequently three more parameters λ_j^{*V} have to be estimated independently. The sizes of the test statistics L^2 and Pearson-χ^2 are such that the model need not be rejected. A comparison of the independence model 1 with the "linear" model 3 favors somewhat the latter model (see model 7).

Although the "linear" model is an acceptable model, it is possible that a "curvilinear" model will be an even better choice. Therefore, a test was done to determine whether adding a quadratic term to the linear model would result in a (still) better fit. In this curvilinear model, it is assumed that the effects of T on V and the log-odds of choosing one party above the other do not linearly increase (decrease) over time, but instead follow a curvilinear pattern, or stated more precisely: lie on a parabola. Starting out from the saturated model $\{TV\}$, in this curvilinear model the following restrictions are imposed on the two-variable parameters:

$$\lambda_{ij}^{TV} = \lambda^*{}_j^V t_i + \lambda^{**}{}_j^V t_i' \tag{6.5}$$

Just as in the linear model, the scores t_i are assigned in such a way that the effect of T on V is linear. The scores t_i' are chosen in such a way that the effect is quadratic. One could simply use $t_i' = (t_i)^2$, but it is preferable to make the sums of both t_i and t_i' equal to zero. Unequal distances between the measurement moments should also be reflected in the scores t_i' (see note 2). In the case of equal distances, standard tables with the coefficients of orthogonal polynomials can be used to find appropriate scores for t_i and t_i' (Hays, 1981, appendix D).

This curvilinear model, model 4 in Table 6.2, fits the data excellently and, according to the test statistics for models 8 and 10, significantly better than the independence model 1 or the linear model 3.

Examination of the sizes of the linear and quadratic two-variable effects $\hat{\lambda}^*{}_j^V$ and $\hat{\lambda}^{**}{}_j^V$ (not reported here) made it clear that the effects $\hat{\lambda}^*{}_3^V, \hat{\lambda}^*{}_4^V, \hat{\lambda}^{**}{}_3^V$, and $\hat{\lambda}^{**}{}_4^V$ were very small and also only about half the sizes of their respective standard errors. Apparently, the total support for the right wing parties and the total proportion of nonvoters did not change significantly during the years 1978-1980. The two-variable effects pertaining to the left wing and the Christian parties were significant. The two-variable effects for the left wing parties had about the same size as the corresponding effects for the Christian parties, albeit with the opposite sign. These influences of time on vote intention are well in agreement with the fact that the (socialist) Labour Party and the Christian Democratic Party were the main (opposite) political forces during the period 1979-1980.

This lack of influence of time on nonvoting and support of the right wing parties, and the equality of the opposing effects of time on the left wing and Christian parties, can be formalized by starting out from the saturated model $\{TV\}$ in which the following restrictions are introduced:

$$\lambda_{ij}^{TV} = \begin{cases} 0 & \text{for } j = 3, 4 \\ \lambda^{*V}_1 t_i + \lambda^{**V}_1 t'_i & \text{for } j = 1 \\ -\lambda^{*V}_1 t_i - \lambda^{**V}_1 t'_i & \text{for } j = 2 \end{cases} \qquad (6.6)$$

The test results, obtained by means of FREQ, are reported in Table 6.2, model 5. Model 5 appears to fit the data excellently and, given the outcomes of the conditional tests, it has to preferred above all models considered so far (Table 6.2, models 9, 10). In every respect, model 5 provides a very good and parsimonious description of the data in table *TV*. Moreover, adding cubic or even higher order terms to the curvilinear (parabolic) model did not significantly improve the fit.

The most interesting effects in model 5 are of course the effects $\hat{\lambda}_{ij}^{TV}$ (subjected to restrictions 6.6). These effects are depicted in Figure 6.1 (the solid curve) in the form of the β parameters from the corresponding logit model for the (log-)odds of preferring a left wing rather than a Christian party, where

$$\hat{\beta}_{i\,1/2}^{\overline{TV}} = \hat{\lambda}_{i\,1}^{TV} - \hat{\lambda}_{i\,2}^{TV} = (\hat{\lambda}^{*V}_1 t_i + \hat{\lambda}^{**V}_1 t_i) - (\hat{\lambda}^{*V}_2 t_i + \hat{\lambda}^{**V}_2 t_i)$$

Because of restriction 6.6, these effects lie exactly on a parabolic curve. From January 1978 through October 1979 the (log-)odds left wing/Christian party changed in favor of the left wing parties, although at an ever decreasing rate. At the end of 1979, the left wing received the largest support, but from then on the odds changed in favor of the Christian parties.

The growing popularity of the left wing parties in 1978 was a continuation of an existing trend. The March 1977 elections were clearly won by the (socialist) Labour Party. Nevertheless, in December 1977, a Christian Democratic/Conservative government was formed instead of a socialist/Christian Democratic government. Even many Christian Democratics felt this to be the wrong coalition. Moreover, people were concerned about matters having to do with rising unemployment and with the (international) arms race and the use of nuclear power, matters that concerned the Socialists more than the Christian Democratics. On the other hand, the ongoing economic recession and the growing belief in the necessity of cuts in governmental expenditures eventually increased support for the government and the Christian Democratic Party.

The influence of time can thus be translated into a number of ongoing processes. Whether the processes designated above are indeed responsible for the curvilinear pattern of the net changes in the log-odds left

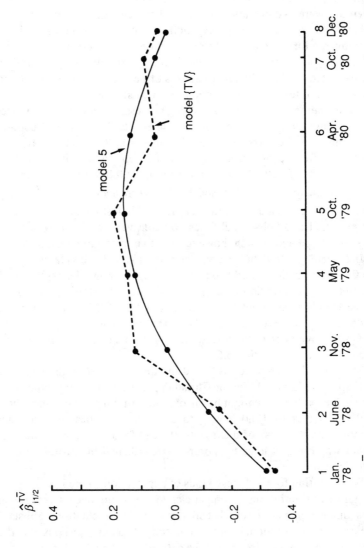

Figure 6.1. Effects of $\beta_{i\,1/2}^{T\bar{V}}$ of Time on Vote Intention; Solid Line: Model 5 in Table 6.2; Dotted Line: Saturated Model $\{TV\}$; Data from Table 6.1

wing/Christian parties depends on the soundness of the substantive
theoretical arguments and on the outcomes of additional empirical tests.
Opinion polls can indicate whether or not the net changes in vote
intention were accompanied by parallel net changes in concern about
nuclear energy, government expenditures, etc., as is implied above.
Moreover, a further elaboration of the theoretical argument may lead to
introducing additional variables and forming particular subgroups which
are expected to show different patterns of net change. How to perform
these kinds of tests will be shown in the next sections.

Beside being a substitute for ongoing processes, Time can also be used
to detect short term influences of discrete events. This can be illustrated
by comparing the expected log-odds left wing/Christian parties for
model 5 with the observed log-odds. The dotted line in Figure 6.1
represents the effects of Time on the log-odds left wing/Christian par-
ties according to the saturated model $\{TV\}$, where again $\hat{\beta}_{i\,1/2}^{\overline{TV}} = \hat{\lambda}_{i\,1}^{TV} - \hat{\lambda}_{i\,2}^{TV}$.
As was to be expected given the very favorable test results for model 5,
the expected and "observed" lines coincide very well. There are two
minor discrepancies which, however, have an interesting explanation.

In November 1978 the observed odds are somewhat less favorable for
the Christian parties than the expected odds. This might be explained by
the fact that the interviews took place right after the startling disclosure
in the Dutch press about the misconduct of the Christian Democratic
leader during the Second World War. The difference between the two
curves in November 1978 indicates the effect that this event had on vote
intention (see also Section 6.5).

The second discrepancy is found in April 1980. Now the observed
odds are more favorable for the Christian parties than the expected odds.
A possible explanation might be the outbursts of violence in Amsterdam
in the first part of 1980 having to do with the widespread squatter
phenomenon. Because of these riots, the call for law and order became
louder which might have temporarily strengthened the position of the
party in power.

When putting forward these kinds of explanations there is always the
danger of attributing the discrepancies to the wrong events and giving
substantive explanations of what are just sampling fluctuations. Substantive arguments, additional tests, and specifying as much as possible the
relevant events before looking at the data are the remedies for "history."

6.3.2 Net Changes in Subgroups: Introduction of Additional Variables

In the previous section, the net changes in vote intention were investigated by themselves. However, very often an investigator wants to introduce additional variables to examine how these net changes are related to other characteristics and how they vary among various subgroups. How to take additional variables into account will be illustrated by means of the complete three-dimensional Table 6.1. The methods proposed here for analyzing this table can be extended in a straightforward manner to the analysis of tables with four or more variables.

Table 6.1 contains the three-variate frequency distribution Time, Vote Intention, and Religion. The analysis will focus on the question whether or not religion is related to vote intention and more specifically to the net changes in vote intention. Variable V (vote intention) is the dependent variable and T (time) and R (religion) are treated as independent variables. Therefore, in all the effect models to be considered, the term $\{TR\}$ must be included (see Section 2.7).

Before looking at the direct effects of T and R on V, it is advisable to inspect the effects in the two-variable marginal tables RV and TR. The relation between time and vote intention in table TV has been extensively discussed in the previous section.

According to the marginal table RV, religion and vote intention are strongly related. The independence model $\{R,V\}$ must definitely be rejected: $L^2 = 1113.64$, df = 3, $p = .00$ (Pearson-$\chi^2 = 947.87$). From the parameters of the saturated model $\{RV\}$ it appears that compared with those who are a member of a religious denomination, nonmembers show a somewhat greater tendency not to vote ($\hat{\lambda}_{14}^{RV} = 0.276$), are somewhat more inclined to vote for a right wing party ($\hat{\lambda}_{13}^{RV} = 0.220$), have a very strong preference for a left wing party ($\hat{\lambda}_{11}^{RV} = 0.617$), but almost totally avoid the Christian parties ($\hat{\lambda}_{12}^{RV} = -1.113$).

Testing the independence model $\{T, R\}$ for the marginal Time × Religion table yields a borderline result: $L^2 = 13.60$, df = 7, $p = .06$ (Pearson-$\chi^2 = 13.57$). The parameters of the saturated model $\{TR\}$ suggest a gradual increase over time of the odds nonmember/member ($\hat{\lambda}_{11}^{TR} = -.107, \ldots, \hat{\lambda}_{81}^{TR} = 0.099$). Accordingly, a linear model in which the odds nonmember/member linearly increase over time fits the data excellently and significantly better than the independence model: $L^2 = 3.84$,

TABLE 6.3 Log-linear Models for the T(Time) × R(Religious Denomination) × V(Vote Intention) Table

Model	L^2	df	p	Pearson-χ^2
1. $\{TR, V\}$	1171.22	45	<.001	998.93
2. $\{TR, TV\}$	1139.98	24	<.001	968.54
3. $\{TR, RV\}$	57.58	42	.06	56.76
4. $\{TR, RV, TV\}$	29.63	21	.10	29.06
5. $\{TR, RV, TV\}$, TV = lin./quadr. cat. 2 = –cat. 1	46.95	40	.21	45.69
6. $\{TRV\}$ TV, TRV = lin./quadr. cat. 2 = –cat. 1	46.00	38	.17	44.87
7. (1) - (2)	31.24	21	.07	
8. (3) - (4)	27.95	21	.14	
9. (3) - (5)	10.63	2	.005	
10. (5) - (4)	17.32	19	.57	
11. (5) - (6)	0.95	2	.62	

SOURCE: Table 6.1

df = 6, p = .70 (Pearson-χ^2 = 3.85). Adding a quadratic term to fit a curvilinear (parabolic) model does not in any way improve the fit. So nonreligiosity appears to increase linearly during the years 1978-1980. Given this linear relationship between time and religion and the strong influence of religion on vote intention, especially on the preferences for Christian and left wing parties, it becomes interesting to see whether the conclusions of the previous section about the net changes in vote intention have to be modified when religion is explicitly taken into account.

A baseline model for investigating the simultaneous influences of time and religion on vote intention is model $\{TR, V\}$, in which it is assumed that neither T nor R is in any way related to V. As the test outcomes in Table 6.3, model 1, show, this model must definitely be rejected.

In model $\{TR, TV\}$ (Table 6.3, model 2) the restriction that T has no influence on V is relaxed. Vote intention is allowed to change over time, but religion still has no direct influence on vote intention. According to model 2, the marginal association between vote intention and religion is spurious and disappears at each particular observation moment, that is, when T is being held constant. Not surprisingly in view of the very strong marginal association R-V and the relatively weak relations T-R and T-V, model 2 must be rejected.

It is interesting to note that the conditional test statistic $L^2_{1/2}$ (Table 6.3, model 7) can be used to test the independence of time and vote intention and that it is identical with the unconditional test statistic L^2 for testing the independence of T and V in the marginal table TV (Table 6.2, model 1). This is no coincidence, but follows from the partitioning calculus of L^2 (Allison, 1980; Goodman, 1971b).

Models $\{TR, RV\}$ and $\{TR, RV, TV\}$ (models 3 and 4 in Table 6.3) are more realistic models in that they both allow for a direct association between religion and vote intention. The difference between models 3 and 4 is that, in the latter, T is supposed to have an influence on V whereas, in the former, no net changes in vote intention are expected. In model 3 it is assumed that the observed marginal net changes in vote intention disappear when religion is held constant; the observed marginal net changes in vote intention are completely caused by the fact that there are net changes in religiosity which consequently lead to net changes in vote intentions: $T \rightarrow R \rightarrow V$. The test statistics for model 3 are much smaller than for the models without the effect R-V, but still the validity of model 3 is doubtful: $p = .06$. Model 4 performs somewhat better: $p = .10$. However, the conditional test $L^2_{3/4}$ points toward preferring the more restricted model 3: $p = .14$ (Table 6.3, model 8).

It was argued in Section 6.3.1 that the test outcome for the hypothesis that T and V are independent from each other in the marginal table TV was not conclusive (Table 6.2, model 1, Table 6.3, model 7). $L^2_{3/4}$ (Table 6.3, model 8) is used to test whether T and V are independent from each other when religion is held constant. The conditional independence hypothesis of T and V seems to receive somewhat stronger empirical support than the marginal independence hypothesis. This suggests that at least some of the net changes in vote intention are due to net changes in religious membership.

But before definitively accepting the hypothesis of conditional independence of T and V, it seems worthwhile in the light of the results in Section 6.3.1 to perform more powerful tests of the existence of a direct association between T and V by postulating the validity of model $\{TR, RV, TV\}$ but with restrictions 6.6 imposed upon the two-variable parameters TV. As we have seen, these restrictions imply that time has a curvilinear (parabolic) influence on the preferences for Christian and left wing parties with the same strength but in opposite directions, but there are no significant net changes in nonvoting and preferring a right wing party. This model appears to fit the data rather well (Table 6.3, model 5) and it has to be preferred above all other models considered so far (see Table 6.3, models 9 and 10). Comparing the partial two-variable effects

TV in model 5 with the corresponding marginal effects (Figure 6.1), it appears that for the partial effects more or less the same (parabolic) curve is found. Discrepancies similar to those in Figure 6.1 are found at $T = 4$ and $T = 6$ between $\hat{\beta}^{TV}_{i\,1/2}$ from model 5 and $\hat{\beta}^{TV}_{i\,1/2}$ from model $\{TR,RV,TV\}$ (model 4 in Table 6.3).

A small difference between the "partial" and the "marginal" parabolic effects of time on the odds left wing/Christian party is that the decline of the curve after October 1979 is somewhat steeper. Under the influence of the various events and processes discussed in Section 6.3.1, the odds left wing/Christian party become continuously more favorable for the Christian parties after October 1979 both for members and for nonmembers of a religious denomination, a trend that is somewhat weakened in the total table TV, because of the fact that the proportion of nonmembers with their relatively strong preference for a left wing party becomes larger over time.

Accepting model 5 implies accepting that the conditional net changes of vote intention are exactly the same among members and nonmembers. Model 5 contains no three-variable interaction term TRV and, given the test outcomes for model 5, it is not necessary to include such a term. However, a more powerful conditional test of the significance of a possible three-variable effect can be obtained as follows. It is assumed that the by now familiar curvilinear (parabolic) restrictions on the relation between time and vote intention (the odds left wing/Christian parties) apply for both members and nonmembers of a religious denomination, but contrary to model 5 it is not assumed that the conditional two-variable effects TV have exactly the same values for both subgroups. Starting out from the saturated model $\{TRV\}$, the two-variable effects TV are subjected to restrictions 6.6 and on the three-variable effects TRV the following restrictions are imposed:

$$\lambda^{TRV}_{ijk} = \begin{cases} 0 & \text{for } k = 3, 4 \\ \lambda^{*RV}_{11}t_i + \lambda^{**RV}_{11}t_i' & \text{for } jk = 11, 22 \\ -\lambda^{*RV}_{11}t_i - \lambda^{**RV}_{11}t_i' & \text{for } jk = 12, 21 \end{cases} \quad (6.7)$$

Not surprisingly, this model fits the data (Table 6.3, model 6). However, it does not significantly improve the fit of the corresponding model without the three-variable interaction term, as follows from the conditional test $L^2_{5/6}$ (Table 6.3, model 11).

From these analyses, it can be concluded that both among members and among nonmembers of a religious denomination the same systematic net changes in vote intention occurred, at least with regard to preferring a left wing rather than a Christian party. In contrast to the marginal net changes in vote intention, the partial net changes are not confounded with the net changes in religiosity. Other variables might be (simultaneously) introduced in an analogous manner to find an explanation of the net changes in vote intention and to provide a still purer estimate of the effects of those events and processes that were designated in the previous section as causes of the effects of time on vote intention.

6.4 Net Changes in More than One Characteristic

Instead of focusing on the net changes in one particular characteristic, attention will be paid in this section to the simultaneous analyses of the net changes in more than one characteristic. The data in Table 6.4 will be used as an illustration. These data come from the Amsterdam Continuous Survey used above and pertain to concern about abuse of social security and concern about (rising) crime during the period November 1977 through June 1981.

In the Amsterdam Survey respondents were asked whether they were concerned about eleven topics, among them abortion, nuclear energy, crime, and the abuse of social security. The latter two items were chosen here with specific hypotheses in mind. It was thought that in reaction to the permissive society of the 1960s and early 1970s and as a consequence of the worsening economic situation, a stronger right wing "conservative" ideology was developing in the late 1970s stressing the importance of an orderly society in which "law and order" prevailed and everybody had to pull their own weight. A growing concern about crime and about the possible misuse of social security provisions were thought to be indicative of this rising conservative ideology. More specifically, the following hypotheses are entertained:

1. During the period November 1977 - June 1981 concern about crime and the abuse of social security increased.

2. As indicators of a more conservative ideology, concern about crime and about the abuse of social security are influenced in the same direction and perhaps even to the same extent by the same kinds of events, namely, those events that affect the "conservative feelings" of the people; concern about crime and about social security will show similar net changes over time.

TABLE 6.4 Time (T), Concern about Abuse of Social Security (S) and Crime (C)^a

	T = 1 No-77	T = 2 Ja-78	T = 3 Ju-78	T = 4 No-78	T = 5 Ma-79	T = 6 Oc-79	T = 7 Ap-80	T = 8 Oc-80	T = 9 De-80	T = 10 Ju-81
(a) Marginal Table Time (T) × Concern about Abuse of Social Security (S)										
S = 1	14	18	15	21	18	14	15	13	16	17
S = 2	29	28	27	30	25	31	33	32	33	40
S = 3	57	54	58	49	57	55	52	56	50	43
100% =	523	587	595	584	591	609	567	602	579	595
(b) Marginal Table Time (T) × Concern about Crime (C)										
C = 1	6	8	7	12	11	8	10	9	14	9
C = 2	24	26	24	29	26	31	30	33	27	40
C = 3	71	67	69	59	64	61	61	58	59	51
100% =	523	587	595	584	591	609	567	602	579	595

(continued)

290

TABLE 6.4 Continued

	T = 1 No-77	T = 2 Ja-78	T = 3 Ju-78	T = 4 No-78	T = 5 Ma-79	T = 6 Oc-79	T = 7 Ap-80	T = 8 Oc-80	T = 9 De-80	T = 10 Ju-81
(c) Time (T), Concern about Abuse of Social Security (S), and Concern about Crime (C)										
S = 1, C = 1	12	26	28	38	38	25	33	27	45	26
S = 1, C = 2	26	37	29	39	35	38	35	37	30	52
S = 1, C = 3	35	44	32	46	35	22	19	12	19	20
S = 2, C = 1	6	9	4	16	11	12	13	18	22	18
S = 2, C = 2	61	74	80	80	62	96	98	101	84	127
S = 2, C = 3	86	82	75	77	72	82	77	71	87	95
S = 3, C = 1	11	11	9	15	15	11	8	9	11	11
S = 3, C = 2	37	39	35	51	55	56	37	62	43	60
S = 3, C = 3	249	265	303	222	268	267	247	265	238	186
Total	523	587	595	584	591	609	567	602	579	595

a.T. Time

S. Concerned About Abuse Social Security
C. Concerned About Crime

1 = no (big) problem
2 = big problem
3 = very big problem

1. November 1977; 2. January 1978; 3. June 1978; 4. November 1978; 5. May 1979; 6. October 1979; 7. April 1980; 8. October 1980; 9. December 1980; 10. June 1981

SOURCE: Continuous Survey, University of Amsterdam / Steinmetz Archives Amsterdam

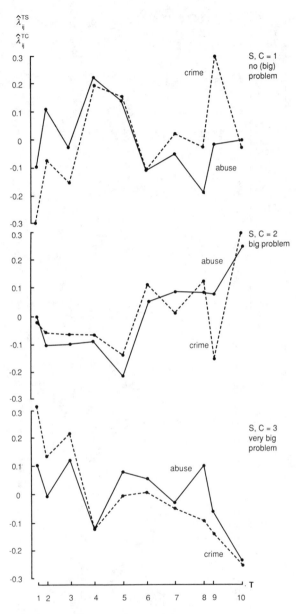

Figure 6.2. Log-Linear Effects of Time (T) on Concern about Abuse of Social Security (S) and on Concern about Crime (C) in the Saturated Model $\{TS\}$ for Table TS (solid line) and in the Saturated Model $\{TC\}$ for Table TC (dotted line); Data from Tables 6.4a and 6.4b

3. As the new conservative ideology becomes more crystallized over time, the correlation between concern about crime and concern about social security abuse becomes stronger (at the individual level).

The first hypothesis about the growing concern about crime (C) and social security abuse (S) over time (T) was tested by means of the marginal tables TS (Table 6.4a) and TC (Table 6.4b). The first question was whether there were any net changes at all. The independence model $\{T, S\}$ had to be rejected for table TS: $L^2 = 73.99$, df = 18, $p < .001$ and the same happened to model $\{T, C\}$ for table TC: $L^2 = 103.78$, df = 18, $p < .001$.

So, concern about crime and social security abuse did show net changes over time. However, inspection of the two-variable parameters of the saturated models $\{TS\}$ and $\{TC\}$ depicted in Figure 6.2 did not suggest an increase in concern about crime and the abuse of social security over time.

Nevertheless, in view of hypothesis 1, models were tested in which time linearly affected concern about crime and the abuse of social security. But these models had to be rejected, as were models in which time was supposed to have curvilinear (quadratic, cubic, up to fifth order) effects. The conclusion must be that there is no indication that concern about crime and the abuse of social security systematically increased during the period November 1977 - June 1981. If in this period a "new" conservative ideology was developing, it was not becoming more popular.

Although the pattern of the parameter estimates in Figure 6.2 undermines the first hypothesis, it supports hypothesis 2 about the concerns for crime and social security abuse being similarly affected by the same kinds of events. The solid ("abuse") and the dotted ("crime") lines in Figure 6.2 are very much identical, at least they show a remarkable parallelism: Almost every increase (decrease) of the curve "abuse" is matched by a simultaneous increase (decrease) of the curve "crime." This strongly suggests that concern about crime and concern about social security abuse are influenced by the same kinds of events in more or less the same manner, supposedly the events that affected "conservative feelings." (The nature of these events will not be discussed in the way it was done in Section 6.3; instead, attention will be restricted to investigating the similarity of the two curves.)

However, before interpreting this similarity of the curves as support for the validity of hypothesis 2, a major alternative explanation of this

TABLE 6.5 (Quasi-)Symmetry Models for Concern About Abuse of Social Security (S) and Concern About Crime (C) for Several Points in Time (T), Using the Extended Table TZSC

Model		L^2	df	p	Pearson-χ^2
1. Symmetry	{SC, T}	481.80	75	<.001	481.29
2. Quasi-sym.	{SC, ZS, ZV, T}	219.63	73	<.001	211.48
3. Symmetry	{TSC}	313.66	30	<.001	300.08
4. Quasi-sym.	{TSC, TZS, TZC}	22.44	10	.01	22.87
5. Part. sym.	{TSC, ZS, ZC}	51.49	28	.004	51.52
6. Marg. hom.	(1) - (2)	262.17	2	<.001	
7. Marg. hom.	(3) - (4)	291.22	20	<.001	
8. Marg. hom.	(5) - (4)	29.05	18	.05	

SOURCE: Table 6.4a

similarity must be refuted. It is possible that the similarity is caused by a variation over time in general concern without any specific reference to the right wing ideology. That something like a general concern might exist came to mind when it was discovered that in the November 1977 Survey ($T = 1$) all (product-moment) correlations (except one) among all eleven "concern items" were positive despite the fact that expressing concern about some topics suggested a totally different political orientation than expressing concern about other topics. Concern about crime and about the abuse of social security may show similar net changes not because they reflect the net changes in conservative feelings but because they reflect the net changes in general concern.

In order to test this alternative hypothesis, the net changes over time for several other topics of concern (abortion, nuclear energy) were depicted in the same manner as in Figure 6.2. These other curves (not shown here) did not vary in any regular way with the curves for abuse and crime and we may adhere to the original interpretation of S and C as indicators of the "new" right wing ideology.

A strict interpretation of hypothesis 2 is that the effects of time on concern about crime are identical to the effects of time on concern about social security abuse: The curves for crime and abuse in Figure 6.2 are not just parallel but are exactly the same in the population. This strict hypothesis means that the marginal tables TS (Table 6.4a) and TC (Table 6.4b) are similar in certain respects, namely, the two-variable effects TS in the saturated model {TS} for Table TS are identical to the corresponding two-variable effects TC in the saturated model $\{TC\}$ for table TC:

$$\lambda_{ij}^{TS} = \lambda_{ik}^{TC} \qquad \text{for all } j = k \qquad (6.8)$$

Because Tables 6.4a and 6.4b contain the same people, the equality hypothesis in equation 6.8 cannot be formally tested in a simple direct way, but marginal homogeneity and (quasi-)symmetry models have to be applied to table TSC, that is, to Table 6.4c which gives the joint distribution of abuse and crime for each point of time. As shown in Sections 4.2.3 and 4.3, an easy and simple way to estimate and denote (quasi-)symmetry models is to make use of an extended table, in this case table $TZSC$, where $Z = 1$ refers to the original conditional table SC at each point of time and $Z = 2$ to the transposed conditional table SC.

The test outcomes for various (quasi-)symmetry models are presented in Table 6.5. Model 1 in Table 6.5 implies that at each point in time the conditional table SC is symmetric, but, moreover, that all conditional tables SC are the same except for the differences caused by the differences in sample size. Model 2 postulates similar quasi-symmetric conditional tables SC. Models 1 and 2 can be used to obtain marginal homogeneity model 6, which implies that the relative distribution of S in the marginal table TS (Table 6.4a) at each particular moment of observation is identical to the relative distribution of C in the marginal table TC (Table 6.4b) and that, moreover, there are no net changes over time, neither in table TS nor in table TC.

Acceptance of model 6 implies accepting the equality of the effects of time on abuse and on crime (there are no effects of time in either case), but rejecting model 6 does not necessarily imply that equality 6.8 is not true. It is clear from the test outcomes in Table 6.5 that model 6 has to be rejected.

Models 3, 4, and 7 can be used to test the weaker hypothesis that there are net changes in both abuse S and crime C but in such a way that at each particular point of time the relative distribution of S in the marginal table TS is exactly the same as the relative distribution of C in the marginal table TC. Models 3 and 4 imply that the conditional table SC is (quasi-)symmetrical at each point of time without imposing the restriction that the conditional tables are the same for all periods. Marginal homogeneity model 7 can then be used to test the weaker hypothesis mentioned above.[3] As is clear from Table 6.5, this simpler hypothesis also has to be rejected.

However, although one must conclude that the marginal tables TS and TC differ from each other, this still does not necessarily mean that time has different effects on abuse and crime and that equation 6.8 is not true. It may be that at each point of time there is a systematic difference between the relative distributions of S and C, for example, people are less concerned about the abuse of social security than about crime, but that this difference is the same for all points of time and that there are otherwise no differences between the relative distributions of S and C. In terms of the saturated models $\{TS\}$ and $\{TC\}$ for tables TS and TC, λ_i^S differs from λ_i^C, but λ_{ij}^{TS} equals λ_{ij}^{TC}. This possibility can be investigated by means of models 4, 5, and 8.

In quasi-symmetry model 4 it is assumed that all conditional tables SC are symmetrical as far as the differences between the distributions of S and C at each point of time permit. Partial symmetry model 5 postulates that all conditional tables SC are symmetric as far as the differences between the distributions of S and C permit, assuming that these differences are the same for all moments of observation. Marginal homogeneity model 8 is used to test whether the marginal tables TS and TC are similar as far as the differences in sample size and a uniform difference over time between the relative distributions of S and C permit. Accepting model 8 means accepting equality 6.8; rejection of model 8 implies rejecting the hypothesis that the effects of time on abuse are exactly the same as the effects of time on crime.

Although the outcomes in Table 6.5 show that model 8 performs much better than the other marginal homogeneity models, it is not clear whether or not to reject model 8: The p-value lies on the borderline of accepting or rejecting the model; moreover, the use of the conditional test (model 8) may be problematic because the unrestricted model (model 4) has to be rejected.

If model 8 is accepted (because the differences between the two curves in Figure 6.2 are rather small — see below), an estimate of the similar effects of Time on abuse and crime must be obtained. However, the conditional testing procedures used do not provide estimated expected frequencies. As proposed in Section 4.3.2, a solution might be obtained by considering the two marginal tables TS (Table 6.4a) and TC (Table 6.4b) as one three-dimensional table TDH, where T as usual refers to the ten points of time, D denotes the three degrees of concern: 1. no (big) problem, 2. big problem, 3. very big problem; and H is an auxiliary variable: $H = 1$ refers to the frequencies for S (Table 6.4a) and $H = 2$ to the frequencies for C (Table 6.4b).

To estimate the similar effects of time on concern, model $\{TD, TH, DH\}$ is applied to table TDH. The term TH in model $\{TD, TH, DH\}$ merely serves to reproduce exactly the sample size at each point of time for both "subtables" (Tables 6.4a and 6.4b). The two-variable effects $\hat{\lambda}_{jk}^{DH}$ indicate the uniform differences over time between the relative distributions of abuse and crime (which are such that on average people are more concerned about crime than about the abuse of social security). The two-variable effects $\hat{\lambda}_{ij}^{TD}$ are estimates of the identical effects of time on concern about the abuse of social security and concern about crime. If the effects of time on degree of concern had been drawn in Figure 6.2, the resulting curves would lie between the solid and the dotted curves.

If model 8 is rejected, separate analyses for Tables 6.4a and 6.4b have to be carried out to determine the different effects of time on abuse and crime, the results of which were depicted in Figure 6.2. Alternatively, one can apply model $\{TDH\}$ to table TDH described above. The three-variable effects TDH provide estimates of the size and direction of the differences between the effects of time on abuse and the effects of time on crime. Actually, these three-variable effects TDH are nothing but the (halved) differences between the solid and dotted curves in Figure 6.2. It turns out that two thirds of the absolute values of the $\hat{\lambda}_{ijk}^{TDH}$ parameters are smaller than .05, a few are between .05 and .10, and only two are larger than .10: $\hat{\lambda}_{911}^{TDH} = -0.162$ and $\hat{\lambda}_{921}^{TDH} = 0.118$. As can be seen in Figure 6.2, these small three-variable interaction effects are not enough to destroy the parallelism of the curves for abuse and crime. It seems safe to accept model 8 and ignore these small effects, with perhaps two exceptions.

The largest differences between the two curves occurred at $T = 9$. the events between time points 8 and 9 may have had a somewhat different influence on abuse than on crime and it may be worthwhile to look for an explanation. It may also be useful to find an explanation of a very small but systematic difference between the two curves. The curve indicating the effects of time on finding crime "no (big) problem" lies below the corresponding curve for abuse during the first four periods, after which it consistently lies above. For the curves "very big problem" the situation is reversed. Relatively speaking, at first the events cause a somewhat larger concern about crime than about the abuse of social security, later on the opposite is true.

TABLE 6.6 Log-linear Models for Table *TSC*

Model	L^2	df	p	Pearson-χ^2
1. {*TS, TC*}	1391.76	40	<.001	1529.21
2. {*SC, T*}	205.77	72	<.001	202.57
3. {*SC, TC*}	101.99	54	<.001	102.57
4. {*SC, TS, TC*}	50.25	36	.06	49.90
5. {*TSC*} *TSC*—lin.	31.43	32	.50	31.55
6. (4) - (5)	18.82	4	<.001	

SOURCE: Table 6.4

In sum, hypothesis 2 is supported by the data. Concern about crime and about the abuse of social security are affected by the intermediate events in the same direction and to a large extent with the same strength. Log-linear models proved to be useful for investigating in detail the parallelism of the effects of time on abuse and crime.[4]

Whereas hypotheses 1 and 2 concerned the marginal net changes in abuse and crime, hypothesis 3 pertains to changes in the bivariate distribution "abuse(S) × crime(C)." More specifically, hypothesis 3 states that the positive association between S and C increases over time. To test this hypothesis the three dimensional table TSC (Table 6.4c) has to be used, not as above to test hypotheses about the marginal tables TS and TC, but to test an hypothesis about over-time differences between the conditional tables SC.

Several models for table TSC, Table 6.4a, will be discussed below, explaining their relevance for hypothesis 3 and, if applicable, comparing them with the previous "marginal" analyses. The test results are shown in Table 6.6.

According to hypothesis 3, concern about the abuse of social security and concern about crime are associated with each other. The opposite hypothesis that at each point of time abuse and crime are independent of each other is tested by means of model 1 in Table 6.6. Model 1 does not fit the data at all; abuse and crime are associated with each other, at least at some points of time.

Hypothesis 3 explicitly states that the association S-C increases over time. Model 2 represents an opposite hypothesis: The conditional tables SC are exactly the same at all moments of observations after taking differences in sample size into account. Although the test results for model 2 are better than for model 1, model 2 has to be rejected.

In model 3 the restrictions imposed by model 2 are weakened in the sense that model 3 allows for an effect of time on crime. Model 3 also has to be rejected. According to model 3, all conditional tables SC are the same as far as differences in sample size and net changes in the marginal distribution of C permit. Comparing models 2 and 3 by means of the conditional test statistic $L^2_{2/3}$ thus provides a test of the significance of the net changes in C; $L^2_{2/3}$ is identical to the unconditional test statistic L^2 for model $\{T, C\}$ applied to the marginal table TC (see above). Model 3 does not provide a test of the significance of the marginal net changes in S. According to model 3, T and S are independent of each other in table TSC holding S constant, but not in the marginal table TS.

Model $\{SC, TS\}$ in which only the marginal distribution of S shows net changes must also be rejected (test statistics not reported in Table 6.6). In model 4 both marginal distributions S and C are allowed to change over time but not the association between S and C. The test outcomes for model 4 are inconclusive: $p = .06$. This is annoying because if model 4 is accepted, hypothesis 3 is rejected.

A more powerful, and hopefully more conclusive, test of hypothesis 3 can be obtained by directly testing whether or not the association between S and C increases linearly over time. The point of departure is the saturated model $\{TSC\}$ in which the following restrictions are imposed upon the three variable effects λ^{TSC}_{ijk}:

$$\lambda^{TSC}_{ijk} = \lambda^{*SC}_{jk} t_i \qquad (6.9)$$

This "linear" model, model 5 in Table 6.6, has to be preferred in all respects. It fits the data excellently. It also fits significantly better than model 4 in which no net change in the association S-C is postulated (see model 6, Table 6.6). The introduction of quadratic or even higher order terms does not significantly improve the fit of the linear model.

If we look at the estimated parameters of this linear model it becomes clear that the nature of the association between concern about the abuse of social security and concern about crime and the nature of the effects of time on this association completely agree with hypothesis 3. By means of the parameter estimates $\hat{\lambda}^{SC}_{jk}$ and $\hat{\lambda}^{TSC}_{ijk}$, the conditional effects $\hat{\lambda}^{SC/T}_{jk\,i}$ can be computed. At each point of time, the relationship between S and C is positive — perhaps an even simpler model than model 5 could be found by imposing linear restrictions on the relation S-C — and the size of the positive association clearly increases (linearly) over time. To illustrate this for the first and the last moment of observation:

$\lambda^{SC/T}_{jk\,1}$				$\lambda^{SC/T}_{jk\,10}$		
	C				**C**	
	1	2	3		1	2	3
S				**S**			
1	0.80	−0.17	−0.63	1	1.02	−0.00	−1.02
2	−0.45	0.56	−0.11	2	−0.20	0.26	−0.06
3	−0.35	−0.39	0.74	3	−0.82	−0.26	1.08

From these analyses it appears that concern about abuse and concern about crime have become more strongly related to each other. It might be interesting to see whether this trend continued into the 1980s. From the methodological point of view, it became apparent that log-linear models provide flexible tools for investigating interesting questions about the simultaneous changes in two or more characteristics. Although only the case of two characteristics has been presented, extensions of quasi-symmetry models by Haberman and Bishop et al., mentioned in the previous chapter, make it possible to investigate simultaneously the net changes in more than two characteristics.

6.5 Interrupted Time-Series Design

The interrupted time-series design is a direct extension of the separate-sample pretest-posttest design briefly discussed in Section 5.2. As explained there, the separate-sample pretest-posttest design is a trend design that is used to determine the effects of one particular event X_e. For example, an investigator wants to know in what way people's opinions about the use of nuclear energy were influenced by the catastrophe at the nuclear plant in Chernobyl in the Soviet Union on April 26, 1986. Assume that there was an opinion poll questioning people about the use of nuclear energy one month before and a poll one month after the disastrous accident. According to the first poll, 30% of the people were opposed to the use of nuclear energy and according to the second poll, 55%. The effect of the accident is then measured by the difference between the pretest and the posttest, in log-linear terms by $\hat{\gamma}^{TU}_2 = 1.689$ or $\hat{\beta}^{TU}_2 = 0.524$, where T as usual refers to Time and U to Opposing the Use of Nuclear Energy (1. yes; 2. no).

As explained in Section 5.2, the difference between the pretest and the posttest measurement does not always provide an unbiased estimate of the effect of X_e. Other relevant events might have occurred between t_1 and t_2 whose effects are confounded with the effects of X_e (although in this example few or none can be expected to have the same impact as the disaster in Chernobyl). It is also possible that there existed a long term trend of increasing concern about the use of nuclear power and that at least part of the difference $O_2 - O_1$ has to be attributed to this trend rather than to the accident. Finally, most time series show random peaks and valleys around a trend. If O_1 happens to be a low point and O_2 a peak, random fluctuations will be confounded with the effects of X_e. Campbell and associates elaborated upon these points and showed that extending the number of pretests and posttests might (partially) overcome the problems of the separate-sample pretest-posttest design (Campbell, 1963, 1969; Campbell & Stanley, 1966; Cook & Campbell, 1979).

In terms of our example, we have to set up a time series in which opinions about the use of nuclear energy are established a number of months before and a number of months after the disaster in Chernobyl (where at each moment of observation a new sample has been drawn from the same population). The logic underlying the interrupted time-series approach is that the variation in the before measurements is determined by two components — a systematic trend and random fluctuations around this trend — whereas the after measurements are determined in the same manner by the same two components plus the effect of X_e.[5] It is assumed that without X_e the before and after measurements would be similar. The event X_e somehow interrupts the ongoing process and the extent and the nature of the interruption is used as an estimate of the effect of X_e.

The interrupted time-series design overcomes many of the problems of the separate-sample pretest-posttest design. Trend and random fluctuations are taken into account and their effects are no longer confounded with the effects of X_e. The influences of other events are also reckoned with as far as these influences behave like random fluctuations. However, when, beside X_e, a few unknown events systematically influence the variation in the time series (before or after X_e), even time-series analysis will generally provide a biased estimate of the effect of X_e.

In practice, it will be difficult to exploit fully the potentialities of the time-series design with secondary survey data. In order to establish reliably the nature of the trend and the size of the error variance, perhaps as many as 30 before and 30 after measurements are required (Gottman,

TABLE 6.7 Time (*T*) and Concern about the Use of Nuclear Energy (*U*)

The Use of Nuclear Energy (U)	T = 1 Nov 77	T = 2 Jan 78	T = 3 Jun 78	T = 4 Nov 78	T = 5 May 79	T = 6 Oct 79	T = 7 Apr 80	T = 8 Oct 80	T = 9 Dec 80
1.Very big problem	35	48	43	38	53	49	46	43	45
2.Not a very big problem	65	52	57	62	47	51	54	57	55
100% =	528	580	594	581	587	611	568	607	582

SOURCE: Continuous Survey, University of Amsterdam / Steinmetz Archives Amsterdam

1981: 361). Aside from a few exceptions like vote intention, secondary data will not enable us to form such long time-series for the characteristics we are interested in: The necessary data are just not available. Moreover, the logic of the interrupted time series requires that the time series is decisively influenced only by the one particular event X_e we are interested in and not by other (unknown) events. Again, real-world data does not often fulfill these requirements.

More exact details on how to analyze time-series data using log-linear models will be discussed below using an example which will show the improvement of the time-series design over the separate-sample pretest-posttest design, but at the same time will make clear the practical difficulties in carrying out time-series analyses.

Several waves of the Amsterdam Continuous Survey contained a question about the extent to which people are concerned about the use of nuclear power (*U*). The data are presented in Table 6.7.

About halfway into the period November 1977 (*T* = 1) to December 1980 (*T* = 9) on March 28, 1979, there was an accident at the nuclear plant in Harrisburg in the United States that might have evolved into a major catastrophe. The accident got a lot of attention in the press and one might wonder what influence this accident had on people's opinions about using nuclear energy.

In November 1978, the last measurement before the accident, 38% of the respondents were very concerned about the use of nuclear power (*U* = 1), whereas in May 1979, the first measurement after the accident, 53% found nuclear power a very big problem. Restricting the analysis to these two points in time, the effect of the accident in Harrisburg is estimated as $\hat{\gamma}_2^{\overline{TU}} = 1.369$ or $\hat{\beta}_2^{\overline{TU}} = 0.314$.

Looking at the percentages in Table 6.7 for all time points, it becomes clear that the rise in concern between November 1978 and May 1979 is not the only source of variation of the time-series U. Somehow, the difference between the measurements of November 1978 and May 1979 has to be evaluated against these other sources of variation. A good starting point for the analysis of Table 6.7 is the independence model $\{T, U\}$. If this model is accepted, it is concluded that the variation of the time series is caused only by (random) sampling fluctuations, that there are no net changes in concern about nuclear power in the population and that consequently "Harrisburg" did not have any influence on U. But model $\{T, U\}$ has to be rejected: $L^2 = 59.92$, df = 8, $p < .001$. The nature of the (significant) net changes can be found by estimating the two-variable effects TU in the saturated model $\{TU\}$. These effects are depicted in Figure 6.3.

The essence of the log-linear interrupted time-series approach is to decompose the systematic effects of time $\hat{\beta}_i^{\overline{TU}}$ into two components: a trend component and an X_e effect component (for a related approach, see Hibbs, 1977). A simple example of this decomposition is

$$\lambda_{ij}^{TU} = \lambda_j^{*U} t_i + \lambda_j^{**U} x_i \qquad (6.10)$$

The first part at the right hand side of equation 6.10 is a (simple) trend component. If t_i is set to zero for all values of i, it is assumed that there are no systematic net changes in concern about the use of nuclear energy except for changes caused by X_e. If an increasing set of numbers is assigned to t_i reflecting the intervals between the moments of observation, it is assumed that the log-odds of finding nuclear energy a very big problem rather than finding it not so problematic increase linearly over time (disregarding the effects of X_e). As shown in Section 6.3.1, the simple linear trend component $\lambda_j^{*U} t_i$ can easily be extended to account for parabolic trends, cubic trends, etc. (see also Clogg, 1979a, Chapter 7; Simonton, 1977; Swaminathan & Algina, 1977).

The second part at the right hand side of equation 6.10 represents the effect of the event X_e. If no effects are expected from the Harrisburg event, x_i is set to zero for all time points i. In that case, the increase in concern between November 1978 and May 1979 is completely accounted for by the trend component. If it is assumed that "Harrisburg" did cause net changes in concern about nuclear power over and above the trend, scores have to be assigned to x_i in correspondence with the expected nature of the influencing process. A few examples will be given below (see also Gottman, 1981: 50; Judd & Kenny, 1981: 157; McCleary & Hay, 1980, Chapter 3).

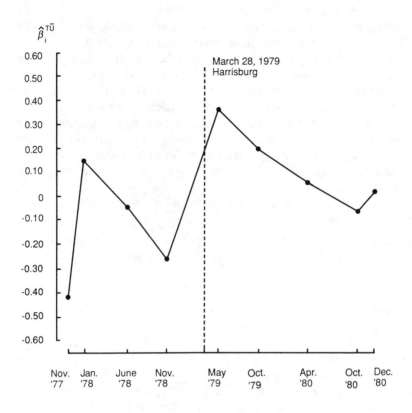

Figure 6.3. Log-Linear Effects of Time (T) and Concern about the Use of
Nuclear Energy (U); Data from Table 6.7

If it is expected that "Harrisburg" brought about a permanent stable
increase in concern about nuclear energy, x_i receives the value 0 for all
time points before X_e, and 1 for all measurements after "Harrisburg."[6] A
one-shot effect immediately after the event X_e that disappears abruptly
can be modeled by setting $x_i = 1$ for i immediately after the event and
$x_i = 0$ for all other values of i. If a linear trend is present in the data and
the event X_e not only brought about a permanent stable increase in
concern but also speeded up the linear increase — the linear trend is
continued after the event X_e but at a higher level and with a steeper slope
— the terms at the right-hand side of equation 6.10 have to be defined

in such a way that the linear trend and the permanent increase are realized and a third (interaction) term $\lambda^{***U}_j (t_i)(x_i)$ is added to represent the increase of the slope. Also interesting is the possibility to postulate decaying effects of X_e. A linear decay is modeled by assigning the value zero to all x_i referring to the pretests and by increasing numbers in correspondence with the spacing between the observations over time to all x_i pertaining to the posttests. Exponential decay is possible by setting the "posttest" x_i equal to e^{-c_i}, where c_i reflects the distances between the moments of observation; all "pretest" x_i are set to zero.

Combining the several ways of defining the trend component with the many ways in which the X_e effect component can be specified yields a large number of models. If the following conjectures about the trend component: no trend, linear trend, parabolic trend, . . . up to a fifth order polynomial are combined with the following hypotheses about the X_e effect component: no effect, permanent, stable effect, one-shot effect, exponentially decaying effect, 20 models result. Most of these models are not chosen for substantive reasons but to explore to the full extent the possibilities of the interrupted time-series design for the data in Table 6.7. Setting up the appropriate design matrices and using the FREQ program, the parameters of these models have been estimated for Table 6.7. However, none of the 20 models yielded a satisfactory fit: $p < .002$ for all models. The adjusted residuals r^{TU}_{1j} and r^{TU}_{2j} in particular were almost always significant.

An explanation of these significant residuals may be the rather heated political debates in the Netherlands in January 1978 about the delivery of enriched uranium to Brazil. As a consequence people might have become significantly more concerned about the use of nuclear energy between November 1977 and January 1978. If an extra two-variable effect λ^{TN}_{2j} is added to the models, for example, to the model in which the trend is linear and the X_e effect decays exponentially, the result is a well-fitting model.

However, this kind of ad hoc and ex post facto fitting of curves violates the logic underlying the interrupted time-series approach as outlined in the beginning of this section. From a causal point of view, the estimates of the effect of X_e obtained in this way are not much better than those obtained by means of the separate-sample pretest-posttest design (or the trend analysis discussed in Section 6.3.1). As has been argued above, this may often be the case when use has to be made of secondary survey data.

6.6 Models With Latent Variables

In principle, all analyses carried out so far in this chapter can be "translated" into the latent level. As explained in Section 3.8, it is possible to impose linear restrictions on the relationship between a latent and a manifest variable. In this way, it can be investigated whether the manifest variable Time is (curvi)linearly related to a particular latent characteristic. Latent (quasi-)symmetry models (Section 4.4.3) can be used to investigate whether or not the net changes in a particular latent characteristic are similar to the net changes in another latent characteristic. The advantages of this translation are evident: The effects of time can be determined after correction for unreliability in the measurements of the characteristics whose net changes we want to study. There are, however, several practical difficulties in carrying out this translation caused by the complexity of the data and the lack of appropriate standard computer programs.

These points will be illustrated by means of Table 6.8, again coming from the Amsterdam Continuous Survey. The data in Table 6.8 pertain to four indicators of the basic political orientation of people: the self-score on a Left/Right scale (L) and agreement with the political ideas of the three main political leaders of the period 1977-1981 in the Netherlands: Hans Wiegel, the leader of the Conservative Party (C); Joop den Uyl, the leader of the Socialist Party (S); and Dries van Agt, the leader of the Christian Democratic Party (D).

The latent class analyses that will be carried out for Table 6.8 have much in common with the "Simultaneous analyses in several groups" (Section 3.6). The several moments of observation, the variable Time, replace the division in several (sub)groups, the stratification factor S in Figure 3.7.

First it must be determined whether the four indicators L, C, S, and C can indeed be regarded as indicators of one underlying latent variable Political Orientation (X). A three latent class model $\{XL, XC, XS, XD\}$, where the three latent classes refer to a basically left wing political orientation ($X = 1$), to a middle, Christian Democratic orientation ($X = 2$), and to a right wing, conservative orientation ($X = 3$), should be valid at each point of time. If no further restrictions are imposed, it is assumed that model $\{TXL, TXC, TXS, TXD\}$ (model 1a in Figure 3.7) holds for the total Table 6.8. The test results show that model $\{TXL, TXC, TXS, TXD\}$ need not be rejected: $L^2 = 55.73$, df = 48, $p = .21$ (Pearson-$\chi^2 = 59.63$).

TABLE 6.8 Time (*T*), Left/Right-score (*L*), Agreement with the Political Ideas of the Conservative Leader (*C*), the Socialist Leader (*S*) and the Christian Democratic Leader (*D*)

L	C	S	D	T = 1 No-77	T = 2 Ja-78	T = 3 Ju-78	T = 4 No-78	T = 5 Ma-79	T = 6 Oc-79	T = 7 Ap-80	T = 8 Oc-80	Total
1	1	1	1	6	13	6	8	3	12	9	7	64
1	1	1	2	8	5	17	7	10	4	10	5	66
1	1	2	1	3	1	2	0	2	3	2	1	14
1	1	2	2	1	1	3	2	0	0	1	0	8
1	2	1	1	16	9	21	16	10	17	19	12	120
1	2	1	2	76	68	67	83	73	91	67	77	602
1	2	2	1	0	1	0	0	1	1	1	2	6
1	2	2	2	2	1	1	2	1	4	3	3	17
2	1	1	1	72	69	64	65	65	63	76	76	550
2	1	1	2	20	19	19	11	13	11	15	9	117
2	1	2	1	24	21	34	29	26	43	37	47	261
2	1	2	2	5	2	4	1	4	2	3	3	24
2	2	1	1	44	33	49	47	42	42	47	45	349
2	2	1	2	26	41	41	47	38	53	31	53	330
2	2	2	1	7	6	9	10	12	17	27	26	114
2	2	2	2	3	3	2	4	3	3	4	6	28
3	1	1	1	49	79	50	46	52	46	41	44	407
3	1	1	2	5	4	2	3	3	4	1	2	24
3	1	2	1	51	81	76	86	86	89	74	60	603
3	1	2	2	1	3	1	2	1	3	1	1	13
3	2	1	1	16	13	14	21	19	7	8	9	107
3	2	1	2	3	5	5	9	8	5	4	6	45
3	2	2	1	16	9	9	13	15	20	16	16	114
3	2	2	2	1	0	0	0	3	0	1	4	9
Total				455	487	496	512	490	540	498	514	3992

a. *T*. Time   1. November 1977; 2. January 1977; 3. June 1978; 4. November 1978; 5. May 1979; 6. October 1979; 7. April 1980; 8. October 1980.

L. Left/Right-score   1. Left (scores 1,2,3) 2. Middle (scores 4,5,6)
    10-point scale   3. Right (scores 7,8,9,10)
Agreement with political   1 = mostly or sometimes agrees
    Ideas of leader (C, S, D)   2 = seldom or never agrees
C. Leader Conservative Party (VVD — Hans Wiegel)
S. Leader Socialist Labour Party (PVDA — Joop Den Uyl)
D. Leader Christian Democratic Party (CDA — Dries Van Agt)
SOURCE: Continuous Survey, University of Amsterdam / Steinmetz Archives Amsterdam

It would take too much space to represent the parameters of the latent class models for all time points. Instead the three latent class model is applied to the Column Total in Table 6.8 ($L^2 = 10.11$, df = 6, $p = .12$, Pearson-$\chi^2 = 10.39$) and its parameter estimates presented in Table 6.9.

TABLE 6.9 Standard Three Latent Class Model

Latent Class X	$\hat{\pi}_t^X$	L. Left/Right score $\hat{\pi}_{it}^{\bar{L}X}$			C. Conservative Leader $\hat{\pi}_{jt}^{\bar{C}X}$		S. Socialist Leader $\hat{\pi}_{kt}^{\bar{S}X}$		D. Christ. Dem. Leader $\hat{\pi}_{lt}^{\bar{D}X}$	
		1. L	2. M	3. R	1. Agree	2. Dis.	1. Agree	2. Dis.	1. Agree	2. Dis.
1	.291	.660	.302	.039	.088	.912	.972	.028	.125	.875
2	.375	.088	.785	.128	.609	.391	.783	.218	.837	.163
3	.334	.000	.186	.814	.852	.148	.361	.639	.983	.017

$L^2 = 10.11$, df = 6, $p = .12$, Pearson-$\chi^2 = 10.39$
SOURCE: Table 6.8, column Total

Although the outcomes of the eight separate latent class analyses vary somewhat between the time points, they all show the pattern present in Table 6.9.

Summed over the eight periods, each latent class has about the same size. Latent class $X = 1$ in Table 6.9 represents the left wing political orientation. Members of class 1 frequently characterize themselves as left (L =1), they endorse the ideas of the socialist party leader, but not those of the conservative or the Christian Democratic leader. The latent class 3 members belong to the political right wing: they characterize themselves as such ($L = 3$), they reject the ideas of the socialist leader but agree with the ideas of the conservative and the Christian Democratic leaders (during this period there was a conservative/Christian Democratic government: the Van Agt/Wiegel Cabinet). The remaining class ($X = 2$) consists of people from the political middle: they identify themselves as such ($L = 2$) and they do not clearly reject the ideas of any of the political leaders.

Given that at each point of time the latent variable X can be interpreted as the basic political orientation, the parameter estimates $\hat{\pi}_{i\,t}^{TX}$ from the complete model $\{TXL, TXC, TXS, TXD\}$ form the entries of table TX by means of which the net changes in latent political orientation can be investigated. However, before attaching too much meaning to the net changes in this table TX, we would like to know whether the estimated net changes are significant, whether or not T and X are independent of each other.

Testing the independence of T and X requires a modified path analysis approach. Model $\{TXL, TXC, TXS, TXD\}$ is applied to the data in Table 6.8 but with the additional restriction that model $\{T, X\}$ holds for the marginal table TX (see Figure 3.7,1b). In Chapter 3 it is shown how to obtain the estimates of such a modified path model by means of the LCAG program.[7] The test results are $L^2 = 70.60$, df = 62, $p = .21$ (Pearson-$\chi^2 = 75.98$). The conditional test of this model compared with the previous one without the independence restriction between T and X yields $L^2 = 70.60 - 55.73 = 14.87$, df = 62 - 48 = 14, $p = .39$. There is no reason to reject the hypothesis that T and X are independent of each other. At the latent level, no net changes in the basic political orientations of people have occurred during the period November 1977 - October 1980. The absence of any net change in the manifest vote intention during a long period was thought to be highly unlikely (Section 6.3.1). Much more likely and in agreement with our ideas about what constitutes a latent variable is the stability of X — the underlying political orientation.

Nevertheless, even the conclusion about the stability of X needs some qualifications.

Maybe the testing procedure used is not sensitive enough to detect small but systematic net changes in X. More powerful tests of the presence of (curvi)linear trends in the marginal table TX may lead to different conclusions. Carrying out these tests is hindered by the absence of adequate computer programs. LCAG has no provisions for imposing (curvi)linear restrictions on the parameters. With Haberman's program NEWTON (curvi)linear restrictions on the two-variable log-linear effects T-X pose no problem, but NEWTON cannot easily handle modified path models. In other words, using NEWTON, (curvi)linear restrictions have to be applied to the relation T-X within the categories of the manifest variables L, C, S, and D, which is contrary to the logic of the relevant modified path model 1b in Figure 3.7. So we know in principle how to perform the desired more powerful tests, but for practical reasons it is impossible. However, given the test outcomes which do not provide indecisive borderline results, it is not unreasonable to uphold the hypothesis that there are no net changes in the underlying political orientation.

But test outcomes are not all that matters. An essential assumption in the above analyses is that the latent variable X has the same substantive meaning at all time points. Otherwise, it makes no sense to compare the conditional distributions of X over time. The meaning of latent variables is, however, mainly derived from the relations the latent variable has with the manifest variables. If these relations change over time, there is less reason to assume that the meaning of the latent variable remains the same. Therefore, the investigation of net changes in X is preferably carried out by means of models in which the relations between X and L, C, S, and D are the same for all time points.

In model $\{TXL, TXC, TXS, TXD\}$ used above, the relations between latent and manifest variables are allowed to change. In model $\{TX, XL, XC, XS, XD\}$ (model 3a in Figure 3.7), on the other hand, T has no direct relations with the manifest variables L, C, S, and D nor does T affect the relations between X and the manifest variables. But model $\{TX, XL, XC, XS, XD\}$ has to be rejected for the data in Table 6.8: $L^2 = 211.19$, df $= 153$, $p = .001$ (Pearson-$\chi^2 = 219.39$). Consequently, it does not make much sense to use this model to investigate the net changes in political orientation X. If model $\{TX, XL, XC, XS, XD\}$ had been accepted, more parsimonious models might have been defined assuming (curvi)linear relationships between T and X. The parameter estimates of such models can be found using NEWTON. Because T has no direct relationships with

L, C, S, and D, it is not necessary to apply the modified path analysis approach, as follows directly from the collapsibility theorem (Section 2.6).

Although the hypothesis that T has no direct influence whatever on the manifest variables has to be rejected, it is not necessary to conclude that the relations between X and the manifest variables have changed over time. Between models $\{TXL, TXC, TXS, TXD\}$ and $\{TX, XL, XC, XS, XD\}$ there is model $\{TX, XL, XC, XS, XD, TL, TC, TS, TD\}$ in which T directly influences the scores on the manifest variables, but does not affect the relations between the latent and the manifest variables (see model 2a, Figure 3.7). The parameters of this intermediate model can be estimated by means of LCAG, as explained in Chapter 3 (although it takes an enormous amount of computing time). The test results are rather favorable: $L^2 = 124.17$, df = 118, $p = .33$ (Pearson-$\chi^2 = 124.87$). Moreover, the intermediate model does not fit the data in Table 6.8 significantly worse than model $\{TXL, TXC, TXS, TXD\}$: $L^2 = 124.17 - 55.73 = 68.44$, df = 118 - 48 = 70, $p = .53$.

It is possible to test the significance of the net changes in X by postulating the intermediate model with the additional restriction that model $\{T, X\}$ holds in the marginal table TX, following the principles of the modified path analysis approach. The test results for the thus restricted intermediate model are $L^2 = 142.04$, df = 132, $p = .26$ (Pearson-$\chi^2 = 140.62$) and for the comparison of the restricted and unrestricted intermediate model $L^2 = 142.04 - 124.17 = 17.87$, df = 132 - 118 = 14, $p = .21$. There is no reason to reject the hypothesis that the basic political orientation is a stable characteristic in the sense that it does not show significant net changes over time. Of course, more powerful tests would be preferable, but because of the required modified path analysis approach, NEWTON can hardly be used to impose (curvi)linear restrictions on this intermediate model either. However, stable underlying political orientations do not strike one as odd. Thus, there are no substantive reasons to doubt the validity of the hypothesis of "no net change."

The parameter estimates for this restricted intermediate model as far as they concern the distribution of X and the relations between X and the manifest variables are very much the same as those presented in Table 6.9. The three latent classes have the same interpretations as discussed before. The effects of time on the indicators (not reported here) are generally not very large. Often they can be explained in terms of political incidents (actually the same kinds of events discussed in Section 6.3.1). The scores on the manifest variables, particularly the agreement with the political leaders, are determined by the underlying political orientation and by all sorts of political events. The major advantage of the intermediate model is that it separates these two influences and thus enables a clearer investigation of the net changes in X. So far only the "simple" analyses of Section 6.3.1 have been translated into the latent level, but translations of more complex analyses are possible. "Introduction of additional variables" (Section 6.3.2) into the "simple" analyses is possible by using the procedures outlined in Sections 3.3 and 3.7. Latent (quasi-)symmetry models (Section 4.4.3) can be used to investigate the similarity of the net changes in two or more latent characteristics (Section 6.4). However, the availability of the techniques does not guarantee that such analyses can actually be carried out. Lack of data, small sample sizes, lack of adequate computer programs, and other practical circumstances will often obstruct the application of latent class analyses of trend data.

Notes

1. In the next chapter on cohort analysis, Time or Period as a substitute for all kinds of events will be further discussed, particularly in relations with two other substitute variables, Age and Cohort. These latter two variables refer to the many events and processes that are connected with growing older and cohort succession respectively.

2. The appropriate design matrix was constructed (here and elsewhere in this book) using the program DESMAT, written and designed by Hummelman and the author. DESMAT uses a subroutine borrowed from Emerson (1965, 1968) which calculates the coefficients of orthonormal polynomials, if necessary taking unequal distances between the categories of a variable into account. Normalizing orthogonal polynomials enlarges the computational accuracy of particular parameter estimates.

3. The hypothesis underlying marginal homogeneity model 7 in Table 6.5 can also be tested by using nonconditional marginal homogeneity tests, for example, Stuart's Q-statistic (Section 5.2, Haberman, 1979: 499; Stuart, 1955). For the hypotheses underlying models 6 and 8, the available unconditional tests cannot be used. If S and C had been dichotomous variables, all marginal homogeneity models presented in Table 6.5 could have been unconditionally tested using log-linear (symmetry) models.

4. In the classic "linear" time-series analysis it is possible to investigate whether the changes in one characteristic are followed by similar changes in another characteristic with certain time lags (Gottman, 1981, Part VI; McCleary & Hay, 1980, Chapter 5). In a descriptive sense, especially with equally spaced time intervals, this is also possible within the log-linear framework proposed here: One just moves one of the two curves in Figure 6.2 one or more time points to the right. However, it is not clear how to carry out the necessary significance tests of the differences between such lagged curves.

5. In the log-linear time-series analysis proposed later the usual assumptions about the random shocks (errors) around the trend are made, especially the assumption that the

error terms are independent of each other. In other "linear" forms of time-series analysis, for example, ARIMA models, the random component can be divided into two parts: a "true" error part and a part that may be correlated with previous error terms in some systematic manner thus taking (partial) autocorrelation into account (Gottman, 1981; McCain & McCleary, 1979; McCleary & Hay, 1980).

6. To satisfy the restriction that all effects sum to zero, one should also express x_i in terms of deviations from the mean score x. If there are 20 pretests and 30 posttests, one can apply the score $x_i = -.6$ for all pretests and $x_i = .4$ for all posttests. As all other parameters in the model are usually expressed in terms of deviations from the mean and sum to zero, there certainly are advantages in applying this restriction to equation 6.10 as well. For reasons of simplicity of exposition, the dummy coding parametrization assigning the values zero and one to x_i has been chosen here.

7. There are 24 latent classes (8 × 3) consisting of the categories of the joint latent variable $T'X$, where T' is a quasi-latent variable perfectly connected with T. Independence of T' and X of course implies independence of T and X. For this model, no random initial estimates should be used. Otherwise, there is no guarantee that, for example, $X = 1$ has the same (left wing) meaning for each period and that independence of T and X makes any sense.

7. Cohort Analysis

7.1 Introduction

Cohort analyses are of prime interest whenever social change is studied from the angle of generation succession, that is, from the viewpoint that a society changes because its members are inevitably and continuously replaced by newcomers.

The term cohort analysis originates from demography, where cohort analysis is contrasted with period analysis (Pressat, 1972: 64-68). A (birth) cohort is a group of people born in the same period (the same year, the same decade, etc.). In demographic cohort analysis, data about vital events are obtained by following a particular cohort during its life-course. For example, age specific marriage rates are determined by noting how many people of a particular cohort marry when the cohort has reached the age 15-20, 20-25, etc. Period analysis on the other hand deals with data obtained at one particular moment of time. For example, age specific marriage rates are determined by registering at one particular point in time how many people marry among the age groups 15-20, 20-25, etc., where these age groups belong to different cohorts. The results of period and cohort analysis may have very different implications when they are used to predict demographic developments (Pressat, 1972: 64-68). For example, can the marriage behavior of a particular cohort of 20 year olds at age 25 be better predicted from the cohort's behavior at age 20 or from the marriage behavior of the 25 year olds of previous cohorts.

When interpreting the differences and similarities between the results of period and cohort analysis, three key variables appear to play a role:

Age, Period (Time), and Cohort. Not surprisingly, in the (other) social sciences, the term cohort analysis has come to refer to a technique by means of which one tries to unravel the separate influences of age, period, and cohort. Social scientists use cohort analysis to find out in what way processes of social change are related to the successive replacement of generations where each generation grows older in an ever changing historical context (Converse, 1976; Glenn, 1977; Hagenaars and Cobben, 1978; Mason & Fienberg, 1985; Riley, 1973; Riley & Foner, 1968; Riley, Johnson, & Foner, 1972; Riley, Riley, & Johnson, 1969; Ryder, 1951, 1965).

The cohort table in which several cohorts are followed over time is the nucleus of cohort analysis. Table 7.1 is such a cohort table containing the percentages of Dutch women that are not members of a religious denomination.

The columns of Table 7.1 pertain to the categories of the variable Time. Within the context of cohort analysis, Time is usually denoted as Period, a convention which will be followed in this chapter. Comparing the columns of Table 7.1 with one another informs us about the changes in religious membership over the different periods. The rows refer to the age categories, and rowwise comparisons tell us about the age differences in religious membership. What is peculiar about the cohort table is that there is a third way of comparing the cell percentages, namely, diagonalwise. Because the age intervals in Table 7.1 are the same as the intervals between the moments of observations (10 years), all cells belonging to a particular top-left bottom-right diagonal refer to people born in the same period, that is, to the same birth cohort.[1] Therefore, comparing these diagonals with one another tells us something about the differences between the cohorts.

This chapter deals with the log-linear analysis of a cohort table (Table 7.1). The expositions will differ from the presentations in the previous chapters. Because in the end a cohort study is nothing but a series of trend or panel studies, it is not necessary to introduce any new log-linear techniques or applications. In this chapter, it will be clarified what exactly is meant when it is said that in cohort analysis social change is studied from the viewpoints of age, period, and cohort. The main focus of this chapter will be on the logic, the pitfalls, and the potentialities of the cohort design, a survey of which may be found in the literature cited above.

TABLE 7.1 Estimated Percentages of Nonmembers of a Religious Denomination (N) According to Age (A), Period (P) and Cohort (C): Dutch Women Only; Between Brackets: Percentage Base

Age	Period							
	P_1: 1889	P_2: 1909	P_3: 1919	P_4: 1929	P_5: 1939	P_6: 1949	P_7: 1959	P_8: 1969
A_1: 20-30	C_6 1.90 (424073)	C_7 4.78 (477213)	C_8 7.27 (563766)	C_9 13.87 (678369)	C_{10} 16.00 (732955)	C_{11} 17.40 (771277)	C_{12} 18.01 (772604)	C_{13} 23.93 (972343)
A_2: 30-40	C_5 1.61 (329112)	C_6 4.22 (395255)	C_7 7.25 (456050)	C_8 14.14 (548317)	C_9 16.54 (630027)	C_{10} 18.20 (705662)	C_{11} 19.12 (763242)	C_{12} 22.28 (768865)
A_3: 40-50	C_4 1.13 (250609)	C_5 3.13 (304630)	C_6 5.61 (374580)	C_7 11.94 (434779)	C_8 14.88 (524677)	C_9 17.18 (610472)	C_{10} 18.61 (666165)	C_{11} 22.04 (744917)
A_4: 50-60	C_3 0.85 (202859)	C_4 2.46 (225263)	C_5 4.17 (279687)	C_6 9.10 (344436)	C_7 12.23 (407527)	C_8 15.05 (480889)	C_9 17.62 (596582)	C_{10} 21.39 (625616)
A_5: 60-70	C_2 0.65 (151520)	C_3 1.88 (163241)	C_4 3.11 (184501)	C_5 6.72 (23798)	C_6 9.25 (282921)	C_7 11.90 (341830)	C_8 15.63 (436410)	C_9 19.41 (545782)
A_6: 70+	C_1 0.47 (100398)	C_2 1.30 (115473)	C_3 2.12 (129306)	C_4 4.66 (149319)	C_5 6.36 (183925)	C_6 8.24 (230684)	C_7 11.25 (319868)	C_8 15.25 (458434)

NOTE: The estimated percentages are obtained through linear interpolation of the two nearest Census Data of 1899, 1909, 1920, 1930, 1947, 1960, and 1971.
SOURCE: Hagenaars and Cobben (1978)

7.2 Clarifying the Key Concepts Age, Period, and Cohort

Age, period, and cohort are ambiguous concepts whose complexities have given rise to most of the contentions about the usefulness of cohort analysis. *Age* is perhaps the least complex of the three key variables. In a strictly operational sense, age is simply the time that has elapsed between the date of birth and the moment of observation. But this operational definition is seldom or never of interest. Age is usually treated as a substitute variable, as an indicator of all kinds of processes and events associated with growing up and becoming old. For example, "getting older" refers to biological phenomena such as becoming sexually mature, getting tired faster, etc. Age is also used as a psychological variable and serves as a substitute for increase or decrease of intellectual capacities, development of personality, changing reactions in stress situations, etc. Age may also refer to all sorts of sociological phenomena: Not until a certain age is it permitted or appropriate to marry and have children; with changing age the family situation changes; age has to do with the position and the length of participation in particular social systems such as the occupational or the electoral system.

Although the labels "biological," "psychological," "sociological" are rather arbitrary, they make clear that when age is used as an explanation for a particular phenomenon, the age variable may refer to a large number of different variables which are the actual causes of the pertinent phenomenon. This makes the interpretation of age effects very difficult. For example, if small age differences with regard to religiosity have been found, the conclusion may be drawn that a particular "age part," a particular variable underlying the age concept, has only a weak effect on religiosity. In reality, however, the pertinent age part may exercise a large effect, but one that is compensated by the opposite effects of another age part one did not think of. Winsborough (1975: 203) comes across this problem in a cohort analysis of income, pointing out that growing older has a positive influence on income because of increasing work experience but also a negative one because of aging and obsolescence of training.

Finally, the fact that age effects may interact with period or cohort not only complicates the analysis of the cohort table, making it necessary to introduce higher order effects, but it also causes conceptual problems. As will be discussed below, cohort effects are sometimes interpreted as a special kind of interaction effect between age and period, namely, as effects of growing up under particular historical circumstances. This may make it difficult to attach separate meanings to the concepts age, period, and cohort.

Period, called Time in the previous chapter, is perhaps an even more complex variable. In a purely operational sense, period refers to the moments of observation. In practice, however, period effects are used as an indicator for the effects of all kinds of discrete events occurring at or between the moments of observation and for the influence of long term processes such as industrialization, urbanization, economic trends, gradual changes in educational standards, etc. Exactly what part of the complex variable Period must be held responsible for an observed period effect will be difficult to establish, the more so because one has to reckon with the facts that events have both short term and long term effects, that the effects of particular events may become visible only after some time, and that the effects of several events may interact with each other.

Finally, as with age, the possible interactions between period effects and age and cohort not only complicate the analysis but also cause conceptual problems as far as cohort effects are defined in terms of the idiosyncratic effects of the events people experience at a certain age.

Age and period appear to be very complex concepts, but their respective meanings can be separated from each other. As indicated above, the independent meaning of the complex concept *Cohort* is more troublesome. In strictly operational terms, cohort refers to date of birth. A cohort is a set of people born in the same period. Ryder (1965) gave a somewhat more general operational definition of cohort, namely, a set of people who have experienced a particular basic event in the same period. In addition to birth cohorts there are marriage cohorts (marrying as a basic event), labor market cohorts (entrance in the labor market as a fundamental event), etc. Age, as the time elapsed since the basic event, then refers to duration of marriage, of employment, etc. In this chapter only birth cohorts will be dealt with.

Ryder (1965) was one of the first to point out the usefulness of the cohort concept for social sciences other than demography and to recognize the complex meaning of cohort. In the first place, the variable Cohort may refer to differences in cohort composition. For example, cohorts may differ from each other in size, which may be of importance for their chances on the labor market, on the housing market, in the educational system, etc. There also may be relevant compositional differences with regard to background characteristics like the percentage of unmarried people, number of children, level of education; and also with regard to attitudinal variables such as Political Preference.

In the second place, a cohort can be seen as a bearer of the long term effects of particular events. Some cohorts will differ from each other because they have experienced different events before the first moment of observation. Other cohort differences are caused by the fact that cohorts are affected by the same events and trends but at a different age, and therefore with a different lasting impact. In this second sense, the meaning of cohort effects may become confounded with the meanings of age and period effects.

To elucidate the interconnected meanings of age, period, and cohort, the sociological concept *generation* is often brought to the fore. Sociologists use the term generation in several different ways, not all of which are relevant here. The genealogical definition is perhaps the oldest. The successive generations are grandparents, parents, children, etc. For cohort analysis, this meaning is the least relevant. It is impossible and would serve no useful purpose to describe the cohorts in Table 7.1 in terms of great-grandmothers, grandmothers, mothers, etc.

The structural-functionalist use of the concept generation is essentially an extension of the genealogical approach (Davis, 1940; Eisenstadt, 1964; Parsons, 1965). The generations, the age groups present at a particular moment, each have their own role in society. The process of cohort succession just means that the tasks of an older generation are taken over by a younger generation. The differences between the generations are not a source of social change. Generational differences originate only from the differences in social positions the generations occupy because of their being of different age. Once the younger people have taken over the places of the older ones, they become like these older people. Frictions in the form of generation conflicts may only temporarily arise when young people want to take over too quickly or older people do not want to give up their privileges (Bengtson, Furlong, & Laufer, 1974). In the end, the structural-functionalist approach has no need for the cohort concept as cohort effects are ultimately reduced to age effects.

The idea of generation as a vehicle of sociocultural change has been mainly developed by Mannheim (1928/1929, 1952 — see also Breitsamer, 1976). According to Mannheim, a generation is not simply a birth cohort. In the first instance, a birth cohort is just a generation location: Being born at the same time locates the cohort in history and delimits its possible experiences. Historical circumstances, in particular during the formative years (around 15), then determine whether such a potential generation develops a common view of the social world and a consciousness of itself as a sociocultural unity (Generation as Actuality, Generation Unit).

Various authors have discussed the relations between Mannheim's generation concept and the demographic cohort concept (Buchhofer, Friedrichs, & Lüdtke, 1970; Buss, 1974; Pfeil, 1967; Rosow, 1978). Although this relationship is not as simple as some other authors let us believe (Evan, 1959; Padioleau, 1973), bringing both concepts in contact with each other strengthens the meaning and usefulness of each. A sociocultural and historical analysis in line with Mannheim's views can suggest which potential generations have been actualized, what makes up their social and cultural identity, and what role they play in history. A firmer empirical base to these suggestions can be given by means of cohort analysis: Birth cohorts belonging to the same (actualized) generation should show similar cohort effects as far as these effects should reflect the various identities of the generations.

On the other hand, a "demographic" cohort analysis of Table 7.1, for example, may lead to suggestions about what events and processes caused changes in church membership, and what role various cohorts and "generations" played in this. These suggestions can then be used to guide a broader sociocultural and historical analysis of the secularization process.

How to separate the effects of age, period, and cohort both at the operational and at the conceptual level is discussed in Section 7.4 and illustrated in the sections thereafter by carrying out log-linear analyses for cohort Table 7.1. But before this, a few designs which are simpler than the age × period × cohort design will be considered to show why it is worth going to the bother of working with a complicated cohort table which involves the use of these three complex and partially overlapping concepts.

7.3 Alternative Designs

Schaie (1965) has indicated three data collection strategies, repeated application of which results in a cohort table. None of these three designs, the cross-sectional, the longitudinal, and the time-lag design, taken separately yield unbiased estimates of the effects of either of the three key variables Age, Period, and Cohort. Table 7.2 summarizes the pitfalls of each of these designs.

TABLE 7.2 Overview of the Pitfalls of the Cross-sectional, the Longitudinal, and the Time-lag Design

Data Collection	Threats to External Validity	Attributing the Differences to:	Implies the Fallacies (Threats to Internal Validity)
Cross-sectional (one period × all ages/cohorts)	period–centrism	age / cohort	life–course fallacy / cohort fallacy
longitudinal (one cohort × all ages/periods)	cohort-centrism	age / period	life–course fallacy / period fallacy
time-lag (one age group × all periods/ cohorts)	age-centrism	cohort / period	cohort fallacy / period fallacy

SOURCE: Hagenaars and Cobben (1978), partly based on Riley (1973)

In *cross-sectional* research, data are collected at just one point in time for all age groups or cohorts present at that moment (see Section 1.3). A particular column of Table 7.1 is the result of this type of data collection. Formulated in demographic terms: A period analysis is performed, the pitfalls of which become clear if we look at Column P_1 in Table 7.1. The differences in percentages of nonmembership might be completely explained in terms of effects of age: The older women get, the more religious they become. In that case, one commits what Riley (1973) called a "life-course fallacy." As inspection of column P_1 shows, each cell of column P_1 belongs not only to a different age category, but also to a different cohort. One might as well commit a "cohort fallacy" and attribute all percentage differences in nonmembership to the variable Cohort: Older generations (cohorts) are more religious than younger ones. Cohort and age effects are completely confounded and it is not possible to unravel the separate effects of age and cohort by means of cross-sectional research. Finally, the cross-sectional design involves the risk of period-centrism. If the age or cohort effects interact with period, the result obtained for P_1 cannot be generalized to other periods.

This can be illustrated by means of Figure 7.1 which depicts the effects of Age (A), Period (P), and Cohort (C) on Nonmembership of a Religious Denomination (N).

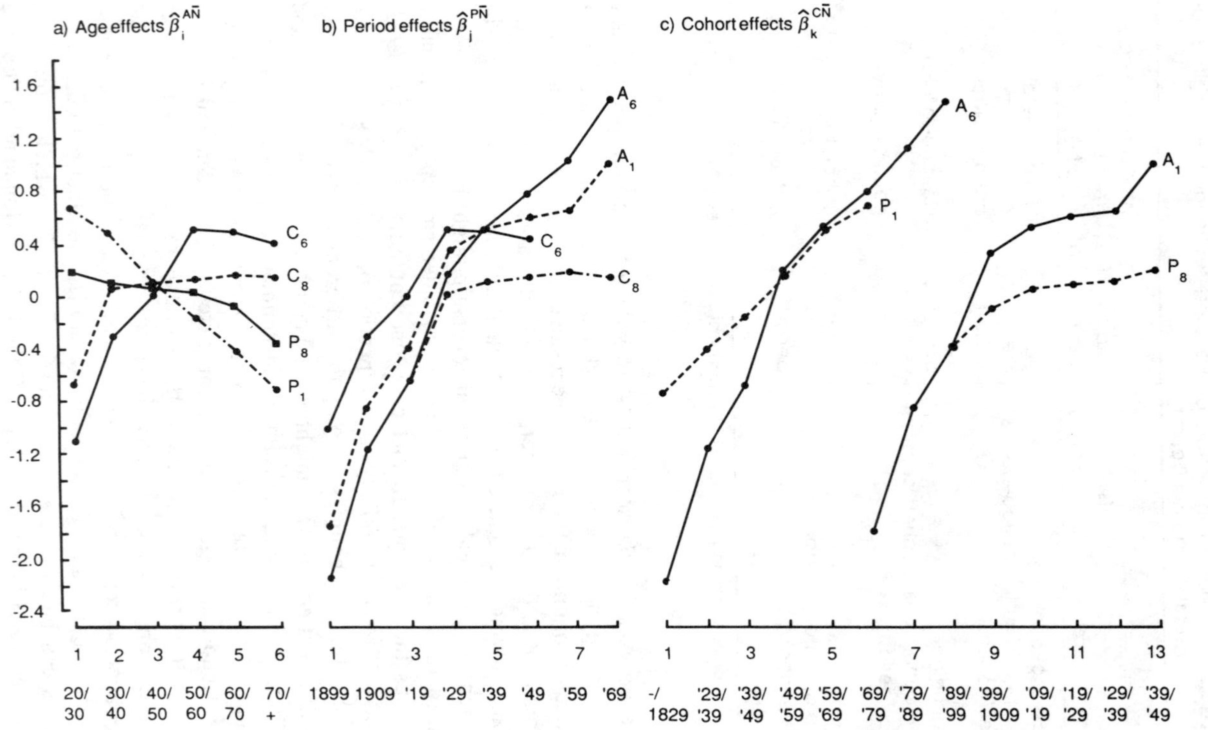

Figure 7.1. Log-Linear Effects $\hat{\beta}$ of Age (A), Period (P), and Cohort (C) on Nonmembership of a Religious Denomination (N); Data from Table 7.1

The curves P_1 and P_8 in Figure 7.1a represent the effects $\hat{\beta}_i^{A\bar{N}}$ for periods 1 and 8 and are obtained by estimating the parameters of the saturated model $\{AN\}$ for each of the columns P_1 and P_8 in Table 7.1.[2] Curve P_1 is practically linearly declining which implies that in 1899 the odds nonmember/member of a religious denomination decrease by a constant factor over the age categories in favor of the members. The risk of period-centrism appears from the different form curve P_8 has in Figure 7.1a. Curve P_8 is also declining, but, except for the older age groups, its slope is much less steep.

That these two curves may also be interpreted in terms of cohort effects appears from Figure 7.1c in which the same two curves P_1 and P_8 are found but now represent cohort effects. (Compared with the curves P_1 and P_8 in Figure 7.1a, the curves P_1 and P_8 in Figure 7.1c are "reversed" and show increasing effects, because at a particular point of time the younger age groups form the latest cohorts.)

The cross-sectional approach does not allow us to get unbiased estimates of age or cohort effects. The *longitudinal* data collection strategy in which one particular cohort is followed over time during all age phases confounds two other effects, namely, those of age and period. The data in diagonal C_6 in Table 7.1 might have been the result of longitudinal data collection. In Schaie's terminology, longitudinal research encompasses both panel and trend analysis: The cells of diagonal C_6 may contain the same respondents (as is the case here) or they may be the result of a new sample at each point in time.

Interpreting the percentage differences in diagonal C_6 solely in terms of age differences involves another form of committing a life-course fallacy: Cohort C_6 not only becomes older, but passes through several periods, experiences certain events. Attributing the percentage differences in diagonal C_6 solely to period involves a period fallacy. The possibility of cohort-centrism arises in the longitudinal design because the age and period effects may be different for the various cohorts.

These points can be demonstrated again by means of Figure 7.1. The age effects on the odds nonmember/member for cohort C_6 in Figure 7.1a are obtained by applying the saturated model $\{AN\}$ to diagonal C_6 in Table 7.1 (see note 2). These age effects are opposite to the age effects found in the cross-sectional analysis: As cohort C_6 grows older, nonmembership increases up to the age 50/60 after which it stabilizes. Depending on the (cross-sectional or longitudinal) research strategy used, one may conclude that women get more or get less religious as they grow older. The danger of cohort-centrism in longitudinal analysis becomes clear when the curve C_6 in Figure 7.1a is compared with the age effects indicated by curve C_8. In this latter curve, the odds nonmember/member have already stabilized at age 30/40.

The confounding of age and period effects in the longitudinal design appears when we compare Figure 7.1a with Figure 7.1b where the same two curves C_6 and C_8 are found but now represent period effects: Nonmembership increases up to 1929 after which it stabilizes.

In the third design, the *time-lag design*, data are collected for one particular age group over all periods or cohorts, resulting in row A_6 of Table 7.1 for example. The curve A_6 in Figure 7.1b shows the period effects on nonmembership for the oldest age category 70+. The time-lag period effects differ from the longitudinal period effects discussed above in that the stabilizing trend after 1929 has disappeared. The danger of age-centrism, inherent in the time-lag design, shows up in the comparison of the curves A_1 and A_6 in Figure 7.1b. The period effects for the youngest age group follow more or less the same pattern as those for the oldest age group, but the latter effects are much stronger than the former.

That the time-lag design confounds the effects of period and cohort, giving rise to period and cohort fallacy, follows from inspection of any particular row in Table 7.1 and from the fact that the curves A_1 and A_6 in Figure 7.1b show up again in Figure 7.1c, but now represent cohort effects.

Although it is impossible to obtain an unbiased estimate of the effects of even one of the three key variables by using the three data collection strategies mentioned in Table 7.2, Schaie (1965, 1970) expected that a repeated application of these strategies, resulting in what he called sequential designs, would yield unbiased estimates. An overview of these sequential designs and the resulting forms of the cohort table are presented in Figure 7.2.

In the time-sequential design, age and period are fully cross-classified; the cohort-sequential design yields a full cross-classification of age and cohort; the cross-sequential design yields a full cross-classification of cohort and period. None of these three more complete designs, however, provides an unbiased estimate of the effects of age, period, or cohort. For instance, in the time-sequential design, age and period effects are hopelessly confounded with cohort effects. The rectangle in Figure 7.2 shows that the cohorts are unequally represented in the rows or columns of this rectangle: Column '61 contains the cohorts 1 through 10 and column '65 the cohorts 5 through 14. the differences between these two columns with regard to the scores on the dependent variable may be

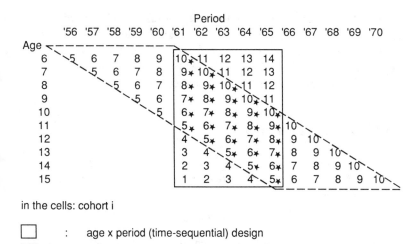

in the cells: cohort i

☐   :   age x period (time-sequential) design

◥＼
＼◢   :   age x cohort (cohort-sequential) design

✳✳✳
✳✳✳ :   cohort x period (cross-sequential) design

Figure 7.2. Schale's Sequential Designs
SOURCE: Hagenaars and Cobben (1978), adapted from Wohlwill (1970)

arbitrarily attributed to period effects or to cohort effects. Analogous remarks apply to the rows of this rectangle as well as to the other two sequential designs. Each sequential design provides "total effects," such as the differences between columns '61 and '65, which are *unknown* mixtures of the effects of two of the three key variables. Because of this, it is also not possible, as Schaie tried to do, to get unbiased estimates of the effects of age, period, and cohort by combining the outcomes of two or more sequential designs (Adam, 1978; Hagenaars & Cobben, 1978).

We seem to have arrived at a dead end. The main purpose of cohort analysis is to determine the importance of generation succession for processes of social change. In the first instance, the cohort table in which several generations or cohorts are followed over time, over their life-course, seemed to provide the means to realize this purpose, that is, if we could somehow separate the cohort effects from period and age effects. However, in each possible design which leads to a cohort table, the effects of cohort are confounded with those of age or period.

In the next section, a closer look will be taken at the source of this confounding of effects and at the several solutions that have been proposed to open up the dead end. These solutions will be evaluated against the background of what has been said in Section 7.2 about the substantive meanings of age, period, and cohort.

7.4 The Identification Problem

The main source of the problems with the analysis of a cohort table is that at the operational level there is an exact linear relationship among age, period, and cohort:

$$A = P - C \tag{7.1}$$

Age is exactly the difference between the moment of observation and date of birth. Once the scores on two of the three components A, P, or C are known, the score on the third variable is fixed. It is impossible to get a fully balanced cross-classification of age, period, and cohort: As shown in Figure 7.2, one variable is always nested in the other two. Expressed in experimental terms: It is impossible to let the third factor vary independently of the other two and to have at one particular point in time two persons who have the same age but are "assigned" to different cohorts.

This linear dependence among A, P, and C has some closely related consequences which have been partly dealt with in the previous section and can be further exemplified by means of Table 7.1, the age × period × cohort × nonmembership table.

Because of equation 7.1, Table 7.1 can be conceived in terms of an age × period design, and its frequencies f_{ijk+}^{APCN} and f_{ijkl}^{APCN} can be described exactly by means of the saturated models $\{AP\}$ and $\{APN\}$ respectively. But Table 7.1 may just as well be rearranged according to an age × cohort design and the same frequencies f_{ijk+}^{APCN} (f_{ijkl}^{APCN}) may be described exactly by means of model $\{AC\}$ (model $\{ACN\}$). It is also possible to rearrange Table 7.1 according to a cohort × period design and to describe exactly the same frequencies f_{ijk+}^{APCN} (f_{ijkl}^{APCN}) by means of model $\{PC\}$ (model $\{PCN\}$). All three representations fit the data equally well and it is impossible to prefer one to the other on empirical grounds.

Another consequence of equation 7.1 is that analyzing cohort Table 7.1 by means of models such as $\{APC\}$, $\{ACN\}$, and $\{PCN\}$ in which the effects of only two of the three key variables are taken into account yields biased estimates of the effects because the effects of the third, not explicitly included, variable are necessarily confounded with the effects of the other two. Analyses in which all three key variables are included cannot be carried out without further restrictions: The separate effects of A, P, and C are not identifiable. The parameters of the log-linear model $\{APCN\}$ are not identifiable because this model contains too many unknowns. But also the parameters of more restricted models such as $\{AP, AN, PN, CN\}$ (or, equivalently, $\{AC, AN, PN, CN\}$ or $\{PC, AN, PN, CN\}$) which have fewer unknowns than knowns are not identifiable because of restriction 7.1. There is perfect collinearity among A, P, and C and it is impossible to identify the separate effects without further assumptions. This problem is designated in cohort analysis as the "identification problem," more elaborated expositions of which are given by Fienberg and Mason (1985); Heckman and Robb (1985); Jagodzinski (1984); Palmore (1978); Pullum (1977, 1980); and Rodgers (1982a).[3]

Many proposals for the solution of the identification problem have been made, some of which failed for technical reasons and some of which were adequate from a technical point of view but lacked theoretical justification (Hagenaars & Cobben, 1978). A widely used example of a technically incorrect solution is the adjustment of the figures in a cohort table for the overall trend and the use of the adjusted figures to determine the age and cohort effects. Applied to Table 7.1, this adjustment procedure involves, in terms of the multiplicative model, the determination of the overall percentage of nonmembers at each point in time, dividing the cell percentages of nonmembers by the overall percentage in the pertinent column and fitting the log-linear model $\{ACN\}$ to the new cell entries. A simple examination of Table 7.1 and Figure 7.2 immediately makes clear that this procedure yields biased estimates of the effects of age and cohort. The overall trend in the column totals originates not only from period effects but is also caused by the unequal representation of the cohorts in the various columns and by the possible interactions of age and period. So the suggested correction procedure confounds at least period and cohort effects. Analogous criticisms apply to attempts to adjust for age or cohort effects by adjusting the cell figures for the overall age group or cohort differences.

It is also wrong to conclude the existence of cohort effects on religiosity in Table 7.1 from the existence of age-period interaction effects on religiosity and to use the age-period interaction effects as estimates of the cohort effects. Although a cohort effect can be seen as a special kind of interaction effect in the $A \times P$ table (comparable with the "diagonal" effects in the diagonal models discussed in Section 4.2.4 — see Goodman, 1972c, 1984; Hout, 1983, Section 3), the presence of age-period interaction effects does not logically imply the presence of cohort effects, nor does the absence of age-period interaction effects necessarily mean the absence of cohort effects. The presence of age-period interaction effects on religiosity in Table 7.1 might mean just that: The effects of period on nonmembership vary with age, and may not point to any "independent" cohort effects. Furthermore, if there are cohort effects on nonmembership, the main effects of age and period will absorb the linear component of the relationship between cohort and nonmembership because of equation 7.1. Therefore, the absence of age-period interactions does not exclude the existence of very strong (linear) cohort effects.

The "trend correction" and the "interaction approach," although often used, fail to solve the identification problem. Baltes disentangled the Gordian knot in the classical manner: Drastically, effectively, but cutting the rope forever (Baltes, 1968; Baltes, Cornelius, & Nesselroade, 1979; Baltes, Reese, & Nesselroade, 1977; see also Labouvie & Nesselroade, 1985). He proposed working with just two of the three key variables. For him, age, period, and cohort are only ways of ordering subjects on the time dimension where, once the order on two dimensions is known, the order on the third dimension does not provide any new information. With this proposal, Baltes solved the identification problem by reducing cohort analysis to purely descriptive analysis and treating age, period, and cohort as purely operational terms. However, it is hard to imagine how the variation in a particular phenomenon, for example, religiosity, can be described in terms of two of the three key variables, for example, age and cohort, without any reference whatsoever to what age and cohort stand for, that is, without any attempt to explain the variation in religiosity in terms of the variables of which age and cohort are the substitutes. In explanatory analyses, the meanings of the three key factors do matter and age, period, and cohort cannot be treated as purely descriptive operational terms, one of which can be arbitrarily deleted. We need other arguments to delete one of the three key factors.

It is not possible to base the decision to delete one of the key factors on purely empirical grounds. For example, all two-factor models $\{APN\}$, $\{ACN\}$, and $\{PCN\}$ fit the data in Table 7.1 perfectly and there are no empirical reasons to choose one above the other. Models $\{AP, AN, PN\}$, $\{AC, AN, CN\}$, and $\{PC, PN, CN\}$ do not necessarily yield the same expected frequencies. But even if only one of these three unsaturated models fits the data, for example, model {*AP, AN, PN*}, this does not mean that the factor cohort can be deleted. Only when there are truly no higher order interactions in any of the three two-factor designs and only when cohort is not linearly related to the dependent variable is the omission of the cohort factor justified. Otherwise, the effects of age and period in model {*AP, AN, PN*} are confounded with the cohort effects.

Pullum (1977) nevertheless advocates this empirical strategy and extends it to the three-factor design. If one of the two-factor models {*AP, AN, PN*}, {*AC, AN, CN*}, and {*PC, PN, CN*} fits the data, that model should be chosen. If all three show a significant lack of fit, the three-factor model {*AP, AN, PN, CN*} (or, equivalently, {*AC, AN, PN, CN*} or {*PC, AN, PN, CN*}) is set up. The unidentifiability of the parameters in this three-factor model is solved by estimating the effects of age, period, and cohort on nonreligiosity in such a way that they resemble the corresponding effects in the two-factor models as closely as possible, where a model carries more weight as it fits the data better. Though this is a technically correct solution for the identification problem, it is hard to justify it theoretically. After all, why should the effects of age, period, and cohort in the three-factor model resemble the corresponding effects in the two-factor models when in each of the two-factor designs the effects are confounded with the effects of the deleted factor?

The identification problem should first of all be solved on the basis of theoretical considerations by explicating the possible meanings of the three key variables and their possible effects on the dependent variable. The cause of the problem, equality restriction 7.1, exists only at the operational level. If we look at what is meant by the operational variables Age, Period, and Cohort (Section 7.2) at the theoretical level, no such simple linear relationship exists among the three key variables. The difficulty is, however, that the effects of the theoretically relevant variables have to be found by means of a design which uses the operational age × period × cohort categorization. So the identification problem has to be solved, but in a way which is in agreement with the theoretical meanings of age, period, and cohort.

Sometimes, the conceptual clarification of the three key concepts may lead to the conclusion that only two are relevant. To give an example: In a cohort analysis on level of education among 30 years old, age effects can generally be ignored. This solves the identification problem as Baltes wanted, but for different reasons.

In other situations, the theoretical analysis may point out that what is meant by one of the key factors, for example, period, is just one particular variable which can be measured directly, for example, economic prosperity. Replacing Period by some Prosperity Index generally solves the identification problem as there is in principle no simple linear relation among Age, Cohort, and the Economic Prosperity Index.[4]

Even if none of the key factors can be deleted or replaced by other variables, theoretical considerations might be helpful by pointing out that it is reasonable to assume that particular age groups, periods, or cohorts will have the same effects on the dependent variable. For example, substantive arguments might lead to the conclusion that religiosity does not change between the ages 35 and 45 and hence that for Table 7.1 the effects $\beta_2^{A\overline{N}}$ and $\beta_3^{A\overline{N}}$ in model $\{AP, AN, PN, CN\}$ are equal. Setting the effects $\beta_2^{A\overline{N}}$ and $\beta_3^{A\overline{N}}$ equal to each other corresponds to collapsing the categories A_2 and A_3 in Table 7.1. In this way the strict equality 7.1 is broken and all parameters in the three-factor model $\{AP, AN, PN, CN\}$ are identifiable. Building upon the work of Graybill (1961) and Mason, Mason, Winsborough, and Poole (1973), Fienberg and Mason (1978) formally showed that the introduction of one appropriate extra linear restriction, for example, an equality restriction, on the effects of A, P, or C on the dependent variable is sufficient to identify all effects in a three-factor model such as model $\{AP, AN, PN, CN\}$ for Table 7.1. The possibilities and limitations of this approach are discussed more extensively in Section 7.5.5.

If the substantive arguments are not precise enough to deduce strict equality restrictions on the effect parameters, it may be possible to infer weaker inequality constraints such as period effects (on nonmembership) are larger than age effects, period effects follow a monotonically increasing pattern, differences between the age groups are larger at a younger age than at an older age, etc. Mathematical programming techniques (Feiring, 1986; Kim, 1984a; Land & Felson, 1981) can be used to obtain parameter estimates for three-factor models subjected to the inequality constraints. Usually, the inequality constraints will not be restrictive enough to get unique point estimates of the parameters, but they might give the maximum and minimum values the parameter estimates can attain. If these intervals are small enough, they might give the relevant information. Although as yet mathematical programming techniques have not been used to attack the identification problem in cohort analysis, the applications of mathematical programming techniques by Land and Felson and Kim to social science research problems suggest that it may be well worth trying (see Mason & Smith, 1985, notes 12, 16 for a somewhat opposite point of view).

Solving the identification problem should be part of a strategy in which cohort analysis is used in an exploratory way to find the true causes of variation in the dependent variable which hide behind the operational terms age, period, and cohort. This task is the more difficult because, as we saw in Section 7.2, the substantive meanings of the three central concepts may partly overlap. Riley (1973) formulated the strategy to be followed.

> The researcher can circumvent the identification fallacy by recognizing age and date for what they are — powerful tools for exploratory research on social dynamics, but blunt instruments for precise measurement or analysis. After all, age and date are mere surrogates for other variables. . . . Fuller understanding will develop as, through skillful exploration we can specify and index these more meaningful variables. (p. 47)

Excellent examples of this approach were presented by Mason and Fienberg (1985). For the cohort analysis of Table 7.1, the approach advocated here involves starting with a clarification of the possible substantive meanings of Age, Period, and Cohort in the context of an exploratory investigation into the phenomenon of religiosity. With these meanings in mind, Table 7.1 is analyzed from different perspectives by means of two- and three-factor models. Thus Table 7.1 is analyzed from the age × period viewpoint using models such as $\{APN\}$ and $\{AP, PN\}$, from the age × cohort viewpoint with models like $\{ACN\}$ and $\{AC, AN\}$, and from the period × cohort viewpoint with models like $\{PCN\}$ and $\{PC, PN\}$. Moreover, the available theoretical information is used to set up three-factor models like $\{AP, AN, PN, CN\}$ in which different reasonable (equality) restrictions are imposed to make the model identifiable. Different kinds of restrictions are preferable to determine the sensitivity of the outcomes to the kinds of restrictions actually made.

By looking at the same data from different points of view, the researcher may discover certain regularities which may put him on the track of the underlying, theoretically relevant variables which otherwise may have gone unnoticed. A nice example of this can be found in Figure 7.1. If we compare the curves C_6 and C_8 in Figure 7.1a resulting from the age × cohort model $\{ACN\}$, we see an interaction effect of age and cohort on nonmembership. In cohort C_6 the odds nonmember/member increase from age group 20/30 to age group 50/60, but are constant after age 60. In cohort C_8 the odds also increase initially but stabilize at an earlier age, namely, at age 40. Had we drawn the age curves for the other cohorts in Figure 7.1a, we would have seen the same initial increase in nonmembership for all cohorts but with "stabilizing points" that occur at different ages for different cohorts. A simple explanation of this complex interaction is not readily available in terms of the age × cohort design.

If we look at Table 7.1 from the period × cohort perspective and inspect the "same" curves C_6 and C_8 in Figure 7.1b, a simple pattern arises: For both cohorts the odds nonmembership/membership increase up to 1929 after which they stabilize. If the period curves for the other cohorts had been drawn, the same pattern would have been found. This suggests that the complex interaction in Figure 7.1a is just a simple period effect. The search for the underlying explanatory variables is directed away from explaining the complex age-cohort interaction effect toward investigating what makes the period 1899-1929 different from the period 1929-1969 with regard to events and processes that influenced religiosity.

In the next section, it will be shown how to carry out the several two- and three-factor log-linear analyses for Table 7.1.

7.5 Log-Linear Analysis of a Cohort Table

7.5.1 Introduction: Obtaining the Data and Clarifying the Meaning of Age, Period, and Cohort

The data to be analyzed in Section 7.5 are presented in Table 7.3, columns Women, and are essentially the same as the data in Table 7.1.[5]

The format of Table 7.3 (or 7.1) is largely determined by the kind of data that were available. Although in the Dutch Census the age of the respondents is recorded exactly, the published sources the secondary investigator has to use present cross-classifications of age (A) and non-membership of a religious denomination (N) in which age is sometimes classified in categories of five years, and sometimes in categories of ten years. Therefore, it was necessary to use 10-year age intervals in Table 7.3 (Table 7.1).

This created a problem as the censuses did not take place exactly every ten years, but in 1899, 1909, 1920, 1930, 1947, 1960, and 1971. If one sets up a period × age table in which age is divided into categories of ten years and the actual census dates are used for period, it is not

TABLE 7.3 Period × Age × Cohort × Nonmembership of a Religious Denomination

			1. Women		2. Men	
Period	Age	Cohort	1. Nonmember	2. Member	1. Nonmember	2. Member
1. 1889	1. 20-30	6	806	41,602	1,151	39,387
	2. 30-40	5	530	32,381	904	30,721
	3. 40-50	4	283	24,778	544	24,088
	4. 50-60	3	172	20,114	298	19,066
	5. 60-70	2	99	15,054	158	13,598
	6. 70+	1	48	9,992	70	8,178
2. 1909	1. 20-30	7	2,281	45,440	3,191	43,052
	2. 30-40	6	1,668	37,858	2,621	35,041
	3. 40-50	5	954	29,510	1,541	27,924
	4. 50-60	4	554	21,972	857	20,894
	5. 60-70	3	307	16,017	430	14,579
	6. 70+	2	150	11,397	189	9,514
3. 1919	1. 20-30	8	4,099	52,278	5,315	49,875
	2. 30-40	7	3,306	42,299	4,700	39,269
	3. 40-50	6	2,102	35,357	3,106	32,965
	4. 50-60	5	1,166	26,802	1,732	25,126
	5. 60-70	4	574	17,876	824	16,406
	6. 70+	3	274	12,657	331	10,712
4. 1929	1. 20-30	9	9,409	58,428	10,602	55,374
	2. 30-40	8	7,753	47,079	9,660	42,982
	3. 40-50	7	5,191	38,287	6,865	35,198
	4. 50-60	6	3,134	31,309	4,323	28,857
	5. 60-70	5	1,564	21,715	2,121	19,899
	6. 70+	4	697	14,235	841	12,385

(continued)

TABLE 7.3 Continued

Period	Age	Cohort	1. Women		2. Men	
			1. Nonmember	2. Member	1. Nonmember	2. Member
5. 1939	1. 20-30	10	11,727	61,568	12,807	59,181
	2. 30-40	9	10,421	52,582	12,174	48,515
	3. 40-50	8	7,807	44,661	9,599	40,579
	4. 50-60	7	4,984	35,769	6,650	32,352
	5. 60-70	6	2,617	26,575	3,574	23,217
	6. 70+	5	1,170	17,222	1,462	14,908
6. 1949	1. 20-30	11	13,420	63,708	14,357	62,459
	2. 30-40	10	12,843	57,723	14,291	53,990
	3. 40-50	9	10,488	50,559	12,149	45,896
	4. 50-60	8	7,237	40,852	9,185	36,422
	5. 60-70	7	4,068	30,115	5,390	26,634
	6. 70+	6	1,900	21,168	2,393	18,064
7. 1959	1. 20-30	12	13,915	63,346	14,670	64,457
	2. 30-40	11	14,593	61,731	15,841	58,599
	3. 40-50	10	12,397	54,219	13,491	50,448
	4. 50-60	9	10,512	49,146	11,600	44,035
	5. 60-70	8	6,821	36,820	7,891	31,346
	6. 70+	7	3,600	28,387	4,378	23,545
8. 1969	1. 20-30	13	23,268	73,966	26,173	77,075
	2. 30-40	12	17,130	59,756	19,069	61,596
	3. 40-50	11	16,418	58,074	17,942	54,785
	4. 50-60	10	13,382	49,180	15,223	46,084
	5. 60-70	9	10,594	43,985	11,169	35,858
	6. 70+	8	6,993	38,850	7,258	28,054
Total			285,426	1,824,399	331,110	1,693,189

SOURCE: See Table 7.1

334

possible to assign each cell of the table unambiguously to one particular cohort.

Fienberg and Mason (1978) dealt extensively with the case where the length of the age intervals does not correspond with the intervals between the moments of observation, but their exposition mainly pertained to the situation that two or more age categories fit exactly within the period categories. For example, the observations take place every ten years and age is coded in categories of five years. In such cases, the cells of the age × period table may still be assigned unambiguously to the cohorts. Cases with irregularly spaced periods or irregular age groups are still too difficult to handle.

Therefore, in Table 7.1, the observation periods have been adjusted to intervals of exactly ten years. Starting with the data of 1899 and 1909, estimates have been made of the number of nonmembers and members of a religious denomination for 10-year period intervals. These estimates were obtained by means of linear interpolation between the two nearest censuses per age category, using an idea developed by Price (1974).

The results of such an interpolation are somewhat arbitrary. Different forms of (non)linear interpolation might yield somewhat different results. In retrospect, it might perhaps have been better to estimate the numbers of (non)members for the years 1900, 1910, 1920, . . . , 1970, leaving out 1940. In Tables 7.1 and 7.3, the numbers of (non)members for the four periods 1929, 1939, 1949, and 1959 are estimated on the basis of only three censuses: 1930, 1947, and 1960.

Interpolation in general, but this one in particular, creates dependence among the observations which will affect the outcomes of statistical tests. The test statistics L^2 and Pearson-χ^2 will therefore be treated as descriptive measures of fit relying mainly on w and $\hat{\delta}$ (equations 2.32 and 2.35) to judge the size of the discrepancy between "observed" and estimated expected frequencies.

Another reason for treating the test statistics as descriptive measures is that at each point in time, the whole Dutch population was investigated and not a sample of it. Even if the population at each particular point in time is regarded as a random sample from some superpopulation, it still makes no sense to perform statistical tests because given the enormous "sample" size − the total number of women in Table 7.3 is 2,109,825 − even very small, substantively insignificant, effects are statistically significant, leading to the rejection of all nonsaturated models.

There is, moreover, one other problem with the use of statistical testing procedures. Because the whole population is interviewed in each census, in each particular cohort the same people have been followed during their life-course and for each cohort we have panel data. To test the (net) changes in nonmembership within the cohorts, the marginal homogeneity models described in Chapter 4 have to be used. But then we need the multiwave turnover tables of nonmembership, tables that are not available from the published sources. Therefore, one is forced to treat the data in Table 7.3 (or 7.1) as if they were trend instead of panel data which results in less than optimal statistical testing procedures (see Section 5.2). For all these reasons, no significance levels will be employed in the cohort analyses discussed below.

These first analyses pertain only to the data for women. The data for men will be analyzed later on in Section 7.6.2. Separate analyses for men and women serve the purpose of diminishing the possibly confounding effect of the (demographic) factor mortality. According to most investigations, including the present one, Dutch women are somewhat more religious than men. Because on average women also live longer than men, a cohort table based on the total population instead of the subpopulation women (or men) will show (partly) spurious age effects, according to which older people get more religious.

In general, the observed changes in nonmembership in Table 7.3 may be caused by the demographic phenomena mortality, birth, and migration. If immigrants are more religious than emigrants or if religious people have more children or live longer than nonreligious people, the percentages of nonmembers will show particular changes over time that may be erroneously attributed to age, period, or cohort. The best thing to do is to correct the frequencies in a cohort table for these demographic factors before performing a cohort analysis. For the period 1947-1960, the Dutch Central Bureau of Statistics (Centraal Bureau voor de Statistiek) tried to estimate the effects of the demographic factors on the proportion of nonmembers and concluded that the net effect of these factors is probably negligible (CBS, 1968). Insufficient data are available to carry out comparable analyses for the entire period 1899 - 1971.

The most conspicuous feature of Table 7.1 is that the percentage of nonmembers has increased spectacularly over time from 0.47% for the oldest cohort C_1 (at age 70+) to 23.93% for the youngest cohort C_{13} (at age 20-30). Explanations of this trend are usually given in terms of long term processes of rationalization and secularization, the beginnings of which are marked by the French Revolution and the Industrial Revolution. Period effects should be understood in terms of these processes but also in terms of events that occurred between 1899 and 1969 such as the two World Wars, the Great Depression in the 1930s, and the high level of economic prosperity in the 1960s. Exact statements on how these events and processes are related to membership of a religious denomination cannot be found in the literature.

Cohort effects can probably be interpreted best in terms of changes in religious education at home and at school. Because of the secularization trend, religious values and the importance of membership of a religious denomination are less and less emphasized in the education of children. In this way, each new cohort starts out at a higher level of nonmembership and bears with it through time the spirit of the period in which it has been brought up.

Thus both period and cohort effects are expected to reflect the ongoing secularization process. When period and cohort are considered simultaneously, cohort effects perhaps better represent continuously ongoing processes, whereas the influence of discrete events like the World Wars will be better reflected in period effects.

Bahr (1970) has presented an overview of the relation between "aging and religious affiliation" as found in the literature. According to some studies, age is positively related to religious affiliation; according to others there is a negative relationship, a curvilinear relationship, or no relation at all. All relationships are "explainable" in the following terms: At different ages people have different responsibilities with regard to children or to their position on the labor market; older people withdraw from society (and its religious organizations); or — the opposite — older people become more religious facing death.

There is empirical evidence for all these (opposite) relationships. The reason for this becomes clear when one looks at Figure 7.1 in which the age effects in the age × period design are opposite to the age effects in the age × cohort design and in which the dangers of cohort- or period-centrism are illustrated. This confusion about the nature of the age effects is perhaps sufficient reason to carry out cohort analyses despite the problems connected with them.

7.5.2 The Age × Period Design

Analyzing Table 7.3 from the viewpoint of the age × period design involves no special difficulties: Given the way Table 7.3 has been set up,

TABLE 7.4 Test Statistics for the Age × Period Design

Model	L^2	Pearson-χ^2	df	w	$\hat{\delta}_{r/u}$	(r, u) [a]
1. $\{AP, N\}$	98,526.25	82,214.75	47	0.22	—	
2. $\{AP, AN\}$	92,440.31	76,786.56	42	0.21	-0.05	$(r = 1, u = 2)$
3. $\{AP, PN\}$	12,182.86	11,111.64	40	0.08	0.85	$(r = 1, u = 3)$
4. $\{AP, AN, PN\}$	2,325.38	2,271.26	35	0.03	0.97	$(r = 1, u = 4)$
					0.78	$(r = 3, u = 4)$

a. Comparing restricted model r with unrestricted model u.
SOURCE: Table 7.4, columns Women

the age × period design is a regular balanced design in which all age groups are cross-classified with all periods.

Effect (logit) model $\{AP, N\}$ in which neither age nor period influences nonmembership yields an extremely large value of the test statistics L^2 and Pearson-χ^2 (Table 7.4, model 1), as was to be expected given the total number of women. But the value of w is also rather large, at least large enough to look for better fitting models.

The introduction of age effects on nonmembership (Table 7.4, model 2) hardly changes these large values. In terms of $\hat{\delta}$, model 2 fits the data even worse than model 1. An important improvement occurs when period effects are taken into account. For model $\{AP, PN\}$ (model 3), L^2 and w are much smaller than for the previous two models and also, according to $\hat{\delta}$, model 3 has to be preferred to models 1 and 2. Model $\{AP, AN, PN\}$ (model 4) in which both age and period effects are taken into account performs even better: w is very low, and $\hat{\delta}_{1/4}$ is almost one which means that the main effects of age and period on nonmembership explain almost totally the discrepancies between the observed frequencies and the estimated expected frequencies in model 1. According to $\hat{\delta}_{3/4}$, model 4 has also to be preferred to model 3 in which only period effects are present. Given the favorable test results for model 4, it does not seem necessary to set up the saturated model. Nevertheless, the saturated model will be briefly discussed below, after examination of the age and period effects on nonmembership as found in model 4 and depicted in Figure 7.3a,b, curves $A \times P$.

The effects of age are much smaller than the period effects. In terms of the maximum γ parameters of the multiplicative effect model (equation 2.47): $\hat{\gamma}^*_{max} = (\hat{\tau}^*_{max})^2 \approx 2$ for the age effects on nonmembership, and $\hat{\gamma}^*_{max} \approx 21$ for the period effects. The effects of age are such that the odds

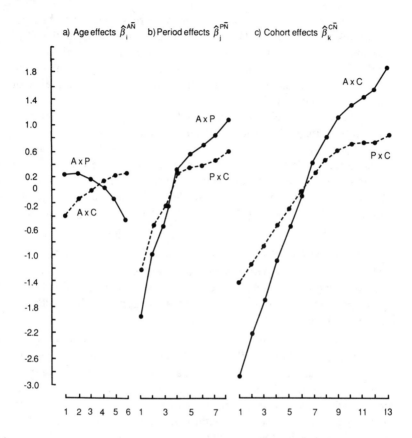

a) Age effects $\hat{\beta}_i^{A\bar{N}}$ b) Period effects $\hat{\beta}_j^{P\bar{N}}$ c) Cohort effects $\hat{\beta}_k^{C\bar{N}}$

Figure 7.3. Log-Linear Effects $\hat{\beta}$ on Nonmembership of a Religious Denomination (N); Two-Factor Designs A (Age) $\times$ P (Period) $\times$ C (Cohort), and P (Period) $\times$ C (Cohort); Models 4 in Tables 7.4, 7.5, and 7.6; Data from Table 7.3, Columns Women

nonmember/member are the same for the two youngest age categories, but, from age 40 on, the odds get smaller at an increasing rate. Ultimately, the odds nonmember/member are about twice as large in the younger age groups than in the oldest one. In terms of Bahr's (1970) life-course effects, religion gains more and more importance after people have acquired a certain social position, as one's children become more and more independent, and as the end of life approaches.

The odds nonmember/member strongly increase over the entire period of observation, albeit at a somewhat decreasing rate as of 1929 (P_4). It seems as if the Great Depression and the World Wars slowed down the secularization trend. In this respect, it is noteworthy that in the period 1959-1969 the period curve accelerates again. Perhaps the influence of World War II had waned.

According to model 4, that is, model $\{AP, AN, PN\}$, age effects on nonmembership are the same for all periods, and period effects the same for all age groups. In the discussion of Figure 7.1, mention was made of the dangers of period- and age-centrism which arise when age and period have interacting effects on nonmembership. The nature of these possible interactions can be found by means of the saturated model $\{APN\}$. The parameters $\hat{\beta}_i^{A\overline{N}}$, $\hat{\beta}_j^{P\overline{N}}$, and $\hat{\beta}_{ij}^{AP\overline{N}}$ of model $\{APN\}$ can be used to compute the conditional age effects $\hat{\beta}_i^{A\overline{N}/P}$ and the conditional period effects $\hat{\beta}_j^{P\overline{N}/A}$. If these conditional effects are depicted, it turns out that the shape of the age curves for the successive periods gradually change from the shape of curve P_1 in Figure 7.1a to the form of curve P_8 in that figure and the period curves for the successive age categories gradually change from curve A_1 to curve A_6 in Figure 7.1b.

The three-variable parameters $\hat{\beta}_{ij}^{AP\overline{N}}$ (not reported here) show a remarkable pattern: They are all positive for the cells pertaining to cohorts C_7, C_8, and C_9 and negative for almost all other cells. This is in agreement with the adjusted residuals of model $\{AP, AN, PN\}$ (not reported here) which indicate that on the basis of model 4 the percentages of nonmembers are underestimated in the cells belonging to cohorts C_7, C_8, and C_9, but overestimated for the cells pertaining to cohorts C_1 through C_5 and C_{10} through C_{13}. These outcomes suggest that it may be worthwhile to investigate what makes cohorts C_7, C_8, and C_9 comparatively less religious, but most of all, they indicate once more that, in the age × period design, cohort plays a confounding role.

7.5.3 The Age × Cohort Design

The data in Table 7.3 (or 7.1) were gathered by means of a time-sequential design (Figure 7.2). Consequently, for Table 7.3, the age × period design is a balanced design, but the age × cohort design is a nested design in which most cohorts occur in combination with just a few of the age categories and none of the age categories is cross-classified with all cohorts. Because age has 6, cohort has 13, and nonmembership has 2 categories, the full age × cohort × nonmembership table has 6 × 13 × 2 =

78 × 2 categories. As appears from Table 7.1 (7.3), only 6 × 8 × 2 = 48 × 2 of these 78 × 2 cells contain data. The presence of structural zero cells in table ACN requires some special provisions. When using a program such as ECTA in which the iterative proportional fitting procedure has been implemented, the full 6 × 13 × 2 table ACN has to be set up. The cells for which no data are available are filled with structural zeros. ECTA provides the estimated expected frequencies (which are set to zero for the structurally empty cells) and the test results, but not the parameter estimates. Parameter estimates can be obtained by means of standard regression programs using the estimated expected frequencies as values of the dependent variable and dummy or effect codes for the independent variables Age, Cohort, and Nonmembership. Employing a program such as FREQ, there is no need to set up the fully cross-classified table ACN and the parameter estimates are obtained directly. The only difficulty is setting up the appropriate design matrix.

The fact that most cohorts are observed for just a few age categories may have consequences for the interpretation of the overall effect parameter $\beta^{\bar{N}}$, depending on the kinds of identifying restrictions imposed on the parameters. For example, if in effect model {AC, AN, CN} the usual identifying restrictions that the effects are zero after summation over the appropriate subscripts, $\Sigma_i \beta_i^{A\bar{N}} = \Sigma_k \beta_k^{C\bar{N}} = 0$, are imposed, then $\beta_i^{A\bar{N}}$ and $\beta_k^{C\bar{N}}$ can be given the usual interpretations in terms of deviations from the overall effect $\beta^{\bar{N}}$, but $\beta^{\bar{N}}$ itself is no longer equal to the mean log-odds nonmember/member in Table 7.3. Imposing the identifying restriction that the weighted cohort effects sum to zero, where the weights are equal to the number of age categories that are observed for a particular cohort, restores the original interpretation of $\beta^{\bar{N}}$ (Fienberg & Mason, 1978, Note 3). As $\beta^{\bar{N}}$ is just a scale factor, a nuisance parameter, such a weighting will be omitted here.

Identifying restrictions are also problematic for the three-variable interaction parameters ACN. Defining higher order interaction terms in nested designs is often cumbersome as several of these terms are logically impossible. The easiest way out is to estimate the parameters of the saturated model {ACN} by performing a set of separate analyses on Table 7.3, one for each age category in order to estimate the cohort effects and one for each cohort to determine the age effects (see Figure 7.1). In this way, the three-variable parameters β_{ik}^{ACN} in model {ACN} are not equal to zero after summation over $i = 1, \ldots, I$ or over $k = 1, \ldots, K$, but their sum is zero for each row and each column of table AC. For example:

$$\sum_{k=6}^{13} \beta_{1k}^{A\overline{CN}} = \sum_{k=1}^{8} \beta_{6k}^{A\overline{CN}} = \sum_{i=5}^{6} \beta_{i2}^{A\overline{CN}} = \sum_{i=1}^{6} \beta_{i6}^{A\overline{CN}} \qquad (7.2)$$

The interaction effects $\beta_{ij}^{A\overline{CN}}$ of age and cohorts C_1 and C_{13} on nonmembership are by definition zero, as these cohorts are observed for only one age category; the observed frequencies $f_{113l}^{AC\,N}$ and f_{61l}^{ACN} have already been reproduced by the estimated expected frequencies of model $\{AC, AN, CN\}$.

For nested designs and tables with structural zero cells, determination of the number of degrees of freedom may be difficult. Fienberg and Mason (1978, Table 6) present a table in which the degrees of freedom are given for most models that are relevant for the analysis of a cohort table. In the analyses discussed below, the number of degrees of freedom will be simply determined by comparing the age × cohort models with the age × period models.

In model $\{AC, N\}$ (model 1 in Table 7.5) it is assumed that the odds nonmember/member are the same for all combinations of age and cohort, that is, for all cells in Table 7.3. Because this is the same assumption that was made in model $\{AP, N\}$ (model 1 in Table 7.4) and because table AC is just a rearranged table AP, models 1 in Tables 7.5 and 7.4 have the same estimated expected frequencies and the same test statistics and degrees of freedom. For similar reasons, model $\{AC, AN\}$ equals model $\{AP, AN\}$. Given the test outcomes, introducing age effects does not improve the fit of model 1.

Adding cohort effects does make a difference. Model $\{AC, CN\}$ (model 3, Table 7.5), which is not equivalent to any of the models in Table 7.4, fits the data better than models 1 and 2. Model 3 has 35 degrees of freedom, because, compared with model 1, 12 independent extra cohort effects on nonmembership have to be estimated. Model 4 in which both age and cohort effects are present appears to be the best model. The number of degrees of freedom of model 4 is five less than for model 3 because model 4 contains five extra independent age effects on nonmembership in comparison with model 3. Comparison of the test results in Tables 7.4 and 7.5 suggests that the role of cohort in the age × cohort design is similar to the role of period in the age × period design and that age does not play a very important role in either design.

TABLE 7.5 Test Statistics for the Age × Cohort Design

Model	L^2	Pearson-χ^2	df	w	$\delta_{r/u}$	$(r, u)^a$
1. $\{AC,N\}$	98,526.25	82,214.75	47	0.22	—	
2. $\{AC,AN\}$	92,440.31	76,786.56	42	0.21	-0.05	$(r = 1, u = 2)$
3. $\{AC,CN\}$	12,708.53	11,825.14	35	0.08	0.83	$(r = 1, u = 3)$
4. $\{AC,AN,CN\}$	4,511.11	4,356.91	30	0.05	0.93	$(r = 1, u = 4)$
					0.59	$(r = 3, u = 4)$

a. Comparing restricted model r with unrestricted model u.
SOURCE: Table 7.4, columns Women

The nature of the age effects on nonmembership in model $\{AC, AN, CN\}$ are very different from the age effects in model $\{AP, AN, PN\}$. As the A × C age curve in Figure 7.3a shows, the maximum age effects $\hat{\gamma}^*_{max}$ are about the same in the age × period and the age × cohort design, but the direction of the effects is opposite to each other. Now acquiring a certain social position, having the children leave home, and getting old appear to go hand in hand with becoming less religious.

The cohort effects on nonmembership according to model $\{AC, AN, CN\}$ are presented in Figure 7.3c, curve A × C. Although these effects have to be treated with some caution — the younger and the older cohorts are observed over at most a few age categories — each following cohort appears to have a higher chance of being a nonmember. The cohort effects are very large: $\hat{\gamma}^*_{max} \approx 125$.

Some outcomes of the saturated model $\{ACN\}$ were briefly dealt with when the effects in Figure 7.1 were discussed. In particular, attention was paid to the age-cohort interaction effects on nonmembership (which could be interpreted more easily as period effects). The three-variable effects ACN as well as the adjusted residuals of model $\{AC, AN, CN\}$ are most easily interpreted in terms of the omitted factor Period: for all cells pertaining to periods P_4, P_5, and P_6, the chances of being a nonmember are underestimated by model 4, whereas for the remaining cells (and periods) these chances are usually overestimated.

7.5.4 The Period × Cohort Design

If the period × cohort design is applied to Table 7.3 (7.1), we again have a nested design that has to be handled in a manner similar to the age × cohort design.

TABLE 7.6 Test Statistics for the Period × Cohort Design

Model	L^2	Pearson-χ^2	df	w	$\hat{\delta}_{r/u}$	$(r, u)^a$
1. {PC,N}	98,526.25	82,214.75	47	0.22	—	
2. {PC,PN}	12,182.86	11,111.64	40	0.08	0.85	$(r = 1, u = 2)$
3. {PC,CN}	12,708.53	11,825.14	35	0.08	0.83	$(r = 1, u = 3)$
4. {PC,PN,CN}	859.04	846.89	28	0.02	0.99	$(r = 1, u = 4)$
					0.90	$(r = 2, u = 4)$
					0.92	$(r = 3, u = 4)$

a. Comparing restricted model r with unrestricted model u.
SOURCE: Table 7.4, columns Women

The test statistics for the several models are presented in Table 7.6. Models 1, 2, and 3 were already discussed above (see Tables 7.4 and 7.5). New is model {PC, PN, CN} (Table 7.6, model 4) which fits the data excellently in all respects. The residual frequencies for model 4 are smaller than for any other unsaturated model considered so far and they do not show an immediately recognizable pattern. To the extent that a systematic pattern is present, the percentage of nonmembers tends to be somewhat overestimated in the cells pertaining to age groups A_1 and A_6 and for the cells in age groups A_2, A_3, and A_4 there is a slight tendency toward underestimation.

The period and cohort effects on nonmembership found in model 4 are depicted in Figure 7.3, curves $P \times C$. The form of the period curve $P \times C$ is about the same as the form of the period curve $A \times P$, only the size of the period effects are now smaller: $\hat{\gamma}^*_{max} \approx 7.5$ as opposed to 21 in the age × period design. Similar remarks apply to the cohort curves $P \times C$ and $A \times C$ ($\hat{\gamma}^*_{max} \approx 10$ as opposed to 125 in the age × cohort design). It looks as if in model {AP, AN, PN} the period factor partly absorbs the omitted cohort effects and the cohort factor partly absorbs the omitted period effects in model {AC, AN, CN}. The independent effects of age and period become visible in model {PC, PN, CN} (to the extent that age effects may be ignored).

The two-factor analyses carried out so far suggest the following about the influence of age, period, and cohort on nonmembership. In the first place, age is a less important cause of nonmembership than period or cohort. This is not surprising as the enormous increase in nonmembership over time that is obvious from Table 7.1 cannot be caused by age as such. Nevertheless, the age effects cannot be overlooked completely. We are uncertain about the kind of relation age has with nonmembership.

There is empirical evidence for both a positive and a negative relationship. This is understandable as we realize that, in the one case, age effects are confounded with period effects and, in the other case, with cohort effects. A three-factor design is needed to discover the true effects of age.

Both cohort and period effects reflect the increase in nonmembership over time. The period effects, moreover, suggest a stabilization or slowing down of this trend between 1929 and 1959 (see Figures 7.1b and 7.3b). Period and cohort effects cannot be reduced to one another; they seem to have independent effects. Deleting either the effects of period or the effects of cohort on nonmembership from model {PC, PN, CN} results in a worse fitting model (Table 7.6). Moreover, the interaction effects of age and period on nonmembership in model {APN} are best understood as cohort effects and the interaction effects of age and cohort on nonmembership in model {ACN} point to period effects. So, a three-factor design in which the effects of age, period, and cohort are considered simultaneously is called for.

7.5.5 The Three-Factor Design: Age × Period × Cohort

Given the at least partially nonoverlapping meanings rendered to the concepts age, period, and cohort in Section 7.5.1 and given the outcomes of the two-factor analyses in Sections 7.5.2 through 7.5.4, it makes sense to try to estimate simultaneously the influence of the three factors, age, period, and cohort on nonmembership. The point of departure will be model {APC, AN, PN, CN}, in which the term APC represents at most 48 independent parameters (including the overall effect) and could be replaced without altering the results by any of the terms AP, AC, or PC and with the annotation that certain higher order interaction effects on nonmembership may be added to the model, a point to which we will return below.

Applying logit model {APC, AN, PN, CN} to the data in Table 7.3 yields unique estimates of the expected frequencies, but not of the effects of A, P, and C on N because of the linear dependence among A, P, and C (equation 7.1).[6] Imposing appropriate linear restrictions on the effect parameters β_i^{AN}, β_j^{PN}, or β_k^{CN} renders the model identifiable (Fienberg & Mason, 1978; Graybill, 1961; Mason et al., 1973). If the age and period intervals are equal, one linear restriction suffices, otherwise more restrictions are needed (Fienberg & Mason, 1978). The linear restriction used most often is the equality restriction. Setting two age, two period, or two cohort effects on nonmembership equal to each other is sufficient to make all effects in model $\{APC, AN, PN, CN\}$ identifiable. This may be clarified in an intuitive way. If an ordinary log-linear model, such as model $\{AP, AN, PN\}$ is applied to Table 7.1 without any (identifying) restrictions on the parameters, the estimated expected frequencies can be uniquely estimated but the parameter estimates $\hat{\beta}_i^{A\bar{N}}$, $\hat{\beta}_j^{P\bar{N}}$, and $\hat{\beta}_k^{C\bar{N}}$ are not identifiable (Section 2.3). However, the differences among the $\hat{\beta}_i^{A\bar{N}}$ parameters, among the $\hat{\beta}_j^{P\bar{N}}$ parameters, and among the $\hat{\beta}_k^{C\bar{N}}$ parameters are uniquely identified. By imposing a linear restriction on each of the three parameter sets, for example, setting a particular parameter equal to zero (dummy coding) or setting the sum of the parameters equal to zero (effect coding — equation 2.11), the parameters themselves are made identifiable. Whatever identifying restriction is chosen, the estimated expected frequencies remain the same. The parameter estimates vary according to the identification restriction that has been chosen, but these differences are trivial because, given an appropriate interpretation of the parameters, the same substantive conclusions will be reached.

When model $\{APC, AN, PN, CN\}$ is applied to Table 7.3 without any restrictions, not even the differences within the parameter sets $\beta_i^{A\bar{N}}$, $\beta_j^{P\bar{N}}$, and $\beta_k^{C\bar{N}}$ are identified but the differences among the differences in each parameter set are. For example, $(\beta_1^{A\bar{N}} - \beta_2^{A\bar{N}}) - (\beta_3^{A\bar{N}} - \beta_4^{A\bar{N}}) = c$, where c is a known constant. By making the effects for age categories 1 and 2 equal to each other, that is, by setting the difference $(\beta_1^{A\bar{N}} - \beta_2^{A\bar{N}})$ equal to zero, the difference $\beta_3^{A\bar{N}} - \beta_4^{A\bar{N}}$ is also identified (equal to $-c$) and so are all other parameter differences within each parameter set.[7] The parameter estimates themselves can then be obtained by applying the usual identifying restrictions such as setting the sum of the effects equal to zero. Whatever single (equality) restriction is used to make the parameter differences identifiable, the estimated expected frequencies remain the same and are identical to the estimated expected frequencies obtained when model $\{APN, AN, PN, CN\}$ is applied without an extra (equality) restriction. However, the estimated parameter values do depend on the kind of (equality) restriction chosen, and the kinds of conclusions that are drawn about the effects of age, period, and cohort may vary when different (equality) restrictions are made.

Mason et al. (1973) recommended imposing more than one (equality) restriction. These extra restrictions are not "arbitrary" parametrization of the differences between the parameters which yield the same estimated expected frequencies, but imply assumptions about the data which may or may not be true. Different sets of two or more equality restrictions will generally result in different estimated expected frequencies and test statistics. Thus, it is possible to test which set of restrictions agrees best with the data.

Imposing more than one restriction has another advantage. Equating two effects in model $\{APN, AN, PN, CN\}$, for example, $\beta_1^{AN} = \beta_2^{AN}$, is equivalent to collapsing the two categories concerned, for example, A_1 and A_2. Because of this, the strict linear dependence among A, P, and C is broken, but A, P, and C will still be very strongly correlated. Multicollinearity, which particularly affects the stability of the estimates, poses a serious problem. Although the multicollinearity problem will never completely disappear from cohort analysis, imposing more than one restriction reduces the degree of multicollinearity.

Many objections have been raised against Mason et al.'s solution of the identification problem. The principal objections come from those that argue that it makes no sense to investigate the effects of *three* separate variables given the strict linear dependence among age, period, and cohort (equation 7.1). Essentially, this is the argument put forward by Baltes that has been discussed (and rejected) in Section 7.4. If it is impossible to give at least partly independent meanings to the three central factors, there is naturally no reason to try to estimate the parameters of three-factor models. But if separate meanings exist, three-factor models can be used to find in an exploratory way the possibly underlying independent variables of which age, period, and cohort are the inadequate, but mostly the only, available operationalizations.

A second set of objections focuses on the particular kind of model that is generally used to separate the effects of age, period, and cohort, namely model $\{APC, AN, PN, CN\}$. In this effect model only the main effects of A, P, and C (on nonmembership) are taken into account and possible interaction effects on nonmembership are not reckoned with (Glenn, 1976, 1977; with reactions of Fienberg and Mason, 1978; Knoke and Hout, 1976; Mason, Mason, & Winsborough, 1976). One might argue that the literature is full of standard linear models (regression, factor, LISREL models) in which no interaction effects occur but which seem to provide satisfactory explanations of the phenomena studied. Why must there be higher order interaction effects when analyzing a cohort table?

There is, however, a more principal side to this objection: Adding unrestricted higher order interactions to model $\{APC, AN, PN, CN\}$

always results in an unidentifiable model, even after imposing extra equality restrictions on the main effects. A model such as {*APC, APN, PN, CN*} in which age and period have interacting effects on nonmembership has too many unknowns to be able to identify the parameters. So it seems logically impossible to take higher order interactions effects into account. But this is not completely true. As Fienberg and Mason (1985) pointed out, model {*APC, AN, PN, CN*} has degrees of freedom which can be used to introduce certain restricted higher order interaction terms.

The third set of objections concerns the impossibility of objectively choosing the right linear (equality) restriction (see Pullum, 1977; Rodgers, 1982a, 1982b; and, for a reaction, Smith, Mason, & Fienberg, 1982). As stated above, different single equality restrictions may lead to completely different parameter estimates but always yield the same estimated expected frequencies. So there is no way to choose among the different sets of parameter estimates on empirical grounds. Imposing sets of two or more equality constraints yields different values for the test statistics, but this means only that the selection of one set of restrictions instead of another can be made on empirical grounds. It does not imply that the right set of restrictions, the set that corresponds with the true state of affairs in the population, is chosen, or even that the true set is among the sets tested. Moreover, from some (unpublished) monte carlo experiments carried out by the author it became clear that, even when the true equality restrictions are made, rather small measurement errors in the dependent variable may lead to completely wrong conclusions about the effects of the three key factors, a finding which probably has much to do with the high multicollinearity among the three key factors.

So we have a serious problem here. But the situation need not be hopeless if the investigator has some substantive theoretical ideas about the possible effects of age, period, and cohort. Sometimes, these theoretical ideas may be precise enough to determine what kinds of linear (equality) restrictions are appropriate. But in any case, the investigator should be able to infer from these theoretical ideas which empirical outcomes are nonsensical and which at least plausible. Whether available "side information" (Converse, 1976: 20) can solve the identification problem is not to be decided a priori, but within a particular concrete research setting.

Taking stock of what we expected and found so far about the effects of age, period, and cohort on nonmembership, the following appears to be relevant for the application of the three-factor design. Table 7.1 showed a huge increase in nonmembership over time, a trend which can hardly be explained in terms of age effects. So we may expect large positive effects of period and/or cohort on nonmembership and relatively small effects of age (the nature of these effects not being known). The outcomes of the two-factor designs discussed above make it rather certain that the effects of age are indeed rather small. Although none of the effects in any of the three two-factor design could be estimated unbiasedly, it seems impossible to explain the outcomes in Figures 7.1 and 7.3 in any meaningful way other than that age effects are much smaller than period and/or cohort effects. Therefore, it is more plausible (and safer) to impose equality restrictions on the age effects than on the period or cohort effects. Given the interpretation of the possible age effects in Section 7.5.1 and Bahr's (1970) summary of the effects of age on religiosity, particularly restriction $\beta_3^{A\overline{N}} = \beta_4^{A\overline{N}}$ or $\beta_4^{A\overline{N}} = \beta_5^{A\overline{N}}$ seems appropriate.

How the increase in nonmembership over time will be distributed among period and cohort effects is uncertain, but the two-factor analyses have made it at least plausible that between 1929 and 1959, the years of the Great Depression and the two World Wars, the upward trend has been slowed down or even halted in terms of period effects. It is hard to see how else to explain the outcomes in Figures 7.1a,b (certainly after adding the other curves for the other age groups and cohorts) and 7.3b. Therefore, other plausible equality restrictions might be $\beta_4^{P\overline{N}} = \beta_5^{P\overline{N}}$ or $\beta_5^{P\overline{N}} = \beta_6^{P\overline{N}}$.

Armed with this knowledge, the identification problem can be attacked in order to obtain the parameter estimates of model $\{APC, AN, PN, CN\}$. Imposing the suggested equality restrictions on the parameters of effect model $\{APC, AN, PN, CN\}$ is possible in a straightforward manner by means of a design matrix. How to impose equality restrictions when the iterative proportional fitting procedure is used was shown by Fienberg and Mason (1978). In short, their procedure amounts to collapsing those categories of age, period, or cohort whose effects are supposed to be equal to each other. After computation of the estimated expected frequencies for the collapsed table, this table has to be unfolded again in order to determine the values of the test statistics.

The test statistics for all models $\{APC, AN, PN, CN\}$ with just one equality restriction are reported in Table 7.7, model 1. The number of degrees of freedom can be found by comparing model $\{APC, AN, PN, CN\}$ with the two-factor models discussed above, for example, with model $\{PC, PN, CN\}$ (model 4 in Table 7.6). Model $\{PC, PN, CN\}$ has

TABLE 7.7 Test Statistics for the Three-Factor Age × Period × Cohort Design

Model {APC, AN, PN, CN}	L^2	Pearson-χ^2	df	w	$\hat{\delta}_{r/u}$	(r, u)[a]
1. One (varying) restriction	62.75	62.12	24	0.01	0.91	(r = model 4, Table 6.7, u=1)
2. $\beta_1^{A\overline{N}} = \beta_2^{A\overline{N}} = \beta_3^{A\overline{N}}$	171.71	171.44	25	0.01	0.62	(r = 2, u= 1)
3. $\beta_2^{A\overline{N}} = \beta_3^{A\overline{N}} = \beta_4^{A\overline{N}}$	64.03	63.28	25	0.01	−0.02	(r = 3, u= 1)
4. $\beta_3^{A\overline{N}} = \beta_4^{A\overline{N}} = \beta_5^{A\overline{N}}$	68.66	67.91	25	0.01	0.05	(r = 3, u= 1)
5. $\beta_4^{A\overline{N}} = \beta_5^{A\overline{N}} = \beta_6^{A\overline{N}}$	136.86	136.22	25	0.01	0.52	(r = 3, u= 1)
6. $\beta_2^{A\overline{N}} = \beta_3^{A\overline{N}} = \beta_4^{A\overline{N}} = \beta_5^{A\overline{N}}$	80.69	79.77	26	0.01	0.16	(r = 3, u= 1)
7. $\beta_4^{A\overline{N}} = \beta_5^{A\overline{N}} = \beta_4^{A\overline{N}} = \beta_5^{A\overline{N}}$	66.19	65.61	25	0.01	0.01	(r = 3, u= 1)

a. Comparing restricted model r with unrestricted model u.
SOURCE: Table 7.4, columns Women

28 degrees of freedom. The term APC in the three-factor model $\{APC,$ $AN, PN, CN\}$ refers to just as many independent parameters as the term PC in two-factor model $\{PC, PN, CN\}$. Moreover, both models contain the same period and cohort effects. However, the age effects that are present in model $\{APC, AN, PN, CN\}$ cannot be found in model $\{PC,$ $PN, CN\}$. Because age has six categories, there are five additional independent parameters $\beta_i^{A\overline{N}}$ to be estimated in the three-factor model. Nevertheless, the number of degrees of freedom of model 1 in Table 7.7 is not $28 - 5 = 23$, but $23 + 1 = 24$, as one extra equality restriction was made to make the model identifiable.

Given the more than two million cases, the value of the test statistic for model 1 is incredibly low: $L^2 = 62.75$. Also in terms of w (= .005), the three-factor model fits perfectly. As appears from the large value of $\hat{\delta}$, it also has to be preferred to the best model obtained so far (model 4, Table 7.6). Moreover, the adjusted residuals for model 1 are very small — most of them are not even "significant" — and do not show a conspicuous pattern. There is no need to complicate model $\{APC, AN,$ $PN, CN\}$ with higher order interaction effects on nonmembership.

A large number of models with only one equality restriction has been tried out, particularly models in which the effects of two consecutive age categories on nonmembership were set equal. The parameter estimates for the model with restriction $\beta_2^{A\overline{N}} = \beta_3^{A\overline{N}}$ differed very little from the parameter estimates in the model with $\beta_3^{A\overline{N}} = \beta_4^{A\overline{N}}$ or with $\beta_4^{A\overline{N}} = \beta_5^{A\overline{N}}$ (and resembled rather closely the curves "$A_4 = A_5, P_4 = P_5$" in Figure 7.4). Very much the same results were obtained for the model with restriction $\beta_4^{P\overline{N}} = \beta_5^{P\overline{N}}$. The similarities of these findings certainly corroborate the plausibility of the ideas behind the selected equality restrictions.

Such similarities are by no means a matter of course. If restrictions are applied that were thought to be unlikely, for example, $\beta_1^{P\overline{N}} = \beta_2^{P\overline{N}}$ and $\beta_2^{C\overline{N}} = \beta_3^{C\overline{N}}$, totally different results are obtained, as shown in Figure 7.4. It seems impossible to give a meaningful interpretation to the parameter estimates resulting from these unlikely restrictions.

Likewise the model with restriction $\beta_5^{A\overline{N}} = \beta_6^{A\overline{N}}$ and to an even greater extent the model with restriction $\beta_1^{A\overline{N}} = \beta_2^{A\overline{N}}$ yielded parameter estimates that differed from the parameter estimates in the models with restrictions $\beta_2^{A\overline{N}} = \beta_3^{A\overline{N}}$, $\beta_3^{A\overline{N}} = \beta_4^{A\overline{N}}$, or $\beta_4^{P\overline{N}} = \beta_5^{P\overline{N}}$. Given the plausibility of these latter restrictions, we have an indication that the restrictions $\beta_1^{A\overline{N}} = \beta_2^{A\overline{N}}$ and $\beta_5^{A\overline{N}} = \beta_6^{A\overline{N}}$ are not valid. Further evidence for this is obtained from the

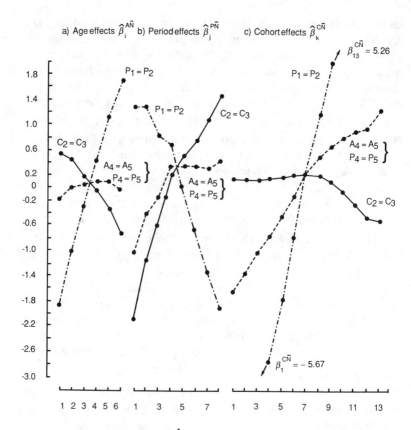

Figure 7.4. Log-Linear Effects $\hat{\beta}$ on Nonmembership of a Religious Denomination (N); Three-Factor Design A (Age) × P (Period) × C (Cohort); Models 1 and 7 in Table 7.8; Data from Table 7.3, Columns Women

test outcomes for models 2 through 6. These models contain more than one equality restriction and have different estimated expected frequencies. The test statistics for models 3, 4, and 6 in which the effects of the middle age categories were equated are more or less the same as for model 1, but the test statistics for models 2 and 5 in which the effect of an extreme age category was set equal to the effects of the neighboring middle age categories are less favorable than for model 1.

Model 7 in which a plausible age restriction is combined with a plausible period restriction was chosen as the final model. In all respects, this model fits the data as well as model 1. The curves "$A_4 = A_5$, $P_4 = P_5$" depict the parameter estimates of model 7.

According to model 7, cohort effects are the main determinants of the variation in nonmembership: $\hat{\gamma}^*_{max} = 17.67$, whereas for period $\hat{\gamma}^*_{max} = 4.47$ and for age only 1.32. Each cohort has a higher level of nonmembership than the previous one, although the differences between the successive cohorts become somewhat less from cohort C_9 on (except for the very last cohort). Period effects cause an increase in the odds nonmember/member except for the period 1929-1959, the years of the Great Depression and the two World Wars. The cohort effects may represent the ongoing secularization trend mentioned before, not interrupted by the 1929-1959 events, although the slowing down of the cohort curve after cohort C_9 (born around 1905) can perhaps be interpreted in terms of the lagged effects of the period 1929-1959. The increase in nonmembership for cohort C_{13} may point to the influence of the "protest generation" of the 1960s, but it is very dangerous to attach too much meaning to the outcomes of the earliest and latest cohorts because they are observed for just one or two periods and age categories. More reliable in this respect is the increase of nonmembership in the decade 1959-1969 after 30 years of stabilization, an effect which affects all cohorts and age categories equally.

Given that the equality restrictions imposed are valid, we also have unbiased estimates of the age effects, which is especially important given the completely different outcomes in the two-factor designs $A \times P$ and $A \times C$ (see Figure 7.3a). The age curve more or less supports Bahr's (1970) Traditional Model. Nonmembership increases as people leave their parental home, look for a job, and start raising a family of their own. A relatively high level of nonmembership is maintained as long as people are actively engaged at home, on the labor market, etc. With the withdrawing from active life and with approaching death, nonmembership declines somewhat.

These outcomes are partly in agreement with the outcomes of the two-factor designs and partly different from them. The main conclusions are that there is evidence for an important, continuously ongoing, secularization trend between 1929 and 1969, that people have sought relief in religion from the miserable circumstances of the Great Depression and the World Wars and that age has a weak curvilinear relationship with nonmembership. None of these conclusions can be proven definitely but they have been made plausible. Only cohort analysis, contrary to cross-sectional or panel and trend analysis, could reveal these effects.

7.6 Extending the Cohort Table

7.6.1 Introducing Additional Variables

The above analyses have been kept simple in that the dependent variable Nonmembership was a dichotomy and the only independent variables were Age, Period, and Cohort. Extension to a polytomous dependent variable is straightforward (see Section 2.7.2). The extension of the dependent "variable" from a univariate to a multivariate frequency distribution may look more complicated, but this extension also turns out to be simple. For example, an investigator may be interested in the effects of education (E) on nonmembership and in particular in the question whether the effects of E on N depend on age, period, or cohort. The bivariate conditional distributions EN, in particular the effects of education on nonmembership, are then the dependent "variable" in the cohort analysis. By comparing model $\{APCE, AN, PN, CN, EN\}$ with model $\{APCE, AEN, PEN, CEN\}$ (and after application of the necessary [equality] restriction(s) on the effects of A, P, or C on N), the investigator can determine whether and how age, period, and cohort influence the effects of E on N.

The same example (and models) can be viewed as an instance of introducing an extra independent variable E into cohort analysis. An explanation of a particular phenomenon such as nonmembership will seldom take place purely in terms of age, period, and cohort effects. Where possible, extra independent variables will be introduced. In some cases, these independent variables are regarded as components of one of the key factors. For example, one might be of the opinion that the cohort effects on nonmembership have to be interpreted in terms of educational differences between the cohorts. By introducing education into the analysis, it can be tested to what extent cohort effects are indeed educational effects. If the cohort effects disappear after the introduction of education, the only relevant variable underlying cohort is education.

In this example, it is assumed that the data on education are available at the individual level or at least that the complete cross-classification $APCEN$ is available. This will often not be the case. But if the average level of education or the marginal distribution of education is known for each cohort, it is still possible to test whether cohort differences are educational differences. All persons within a certain cohort then receive a "contextual" educational score that is equal to the average educational level of their cohort or to the percentage of people in their cohort that have at least a high school education. If not every cohort has a unique score on this contextual educational variable, the simultaneous influence of cohort and "education" on nonmembership can be determined (see note 4). If the cohort effects disappear after the introduction of "education," cohort effects may be interpreted as educational effects in the sense of: The higher is the average educational level of the cohort to which an individual belongs, the larger (or smaller) are the odds nonmember/member. If such contextual variables are employed, one should beware of the ecological fallacy. In the above example, it would be wrong to conclude purely on the basis of the effects of the contextual education variable that the higher is a person's educational level, the larger are the odds nonmember/member. Individual data are needed to determine the relations at the individual level.

7.6.2 An Example: Introducing Gender

Table 7.4 contains not only the data for women, but also for men. By introducing the variable Gender (G) into cohort analyses, the following questions may be answered: What is the influence of gender on nonmembership? Is this influence the same for all ages, periods, and cohorts? Are the effects of age, period, and cohort the same for men and women?

The point of departure for all following analyses is model 7 in Table 7.7, that is, model $\{APC, AN, PN, CN\}$ with the restrictions $\beta_4^{AN} = \beta_5^{AN}$ and $\beta_4^{P\bar{N}} = \beta_5^{P\bar{N}}$. Application of this model to Table 7.3, columns Men, yields the following test statistics: $L^2 = 114.31$, Pearson-$\chi^2 = 113.63$, df = 25, $w = 0.007$. These test statistics are slightly less favorable than the corresponding test statistics obtained for women. But this had to be expected as the model was partly selected by inspection of the data for women. But given the total number of cases (2,024,299), the model fits the data for men very well too. The test statistics for the assumption that the model holds for both men and women simultaneously are reported in Table 7.8, model 9. The L^2 statistic and df for model 9 are the sum of the separate L^2 values and df for men and women.

Model 9 is a very unrestricted model in that it implies that the same model is valid for men and women but without restricting any of the parameters to being the same for men and women. The most restrictive model is model 1 in Table 7.8. This logit model indicates that gender has no effect on nonmembership whatsoever.

TABLE 7.8 Test Statistics for the Age × Period × Cohort × Gender Design; Restrictions: $A_4 = A_5$, $P_4 = P_5$

Model	L^2	Pearson-χ^2	df	w	$\hat{\delta}_{r/u}$	$(r, u)^a$
1. {APCG, AN, PN, CN}	8,382.27	8,382.34	73	0.06	—	
2. {APCG, AN, PN, CN, GN}	2,159.99	2,160.70	72	0.03	0.74	$(r = 1, u = 2)$
3. {APCG, AGN, PN, CN}	1,430.90	1,426.42	68	0.03	0.30	$(r = 2, u = 3)$
4. {APCG, AN, PGN, CN}	1,312.93	1,313.83	66	0.02	0.34	$(r = 2, u = 4)$
5. {APCG, AN, PN, CGN}	270.03	268.99	60	0.01	0.85	$(r = 2, u = 5)$
6. {APCG, AGN, PGN, CN}	335.20	334.78	62	0.01	0.82	$(r = 2, u = 6)$
7. {APCG, AGN, PN, CGN}	229.54	228.35	56	0.01	0.86	$(r = 2, u = 7)$
8. {APCG, AN, PGN, CGN}	210.78	209.43	54	0.01	0.87	$(r = 2, u = 8)$
9. {APCG, AGN, PGN, CGN}	180.50	179.23	50	0.01	0.88	$(r = 2, u = 8)$

a. Comparing restricted model r with unrestricted model u.
SOURCE: Table 7.4

Model 2 allows for a direct effect of gender on nonmembership but in such a way that this effect is the same for all age groups, periods, and cohorts. Given the test outcomes, model 2 must be preferred to model 1. The odds nonmember/member are different for men and women. In the further analyses, model 2 has been chosen as the baseline model.

The three-variable interaction effects AGN, PGN, CGN are added to the baseline model in models 3, 4, and 5 respectively. The test statistics for these models clearly show that in particular the introduction of gender-cohort interaction effects on nonmembership greatly improves the fit of model 2.

Models 6, 7, and 8 each contain two of the three three-variable interaction terms AGN, PGN, CGN. Model 9 contains all three. The goodness-of-fit of the five last models in Table 7.8 is about the same. For reasons of parsimony, model 5, which only contains the CGN three-variable interaction term, has to be preferred. This preference is confirmed when the size of the three-variable interaction effects AGN, PGN, CGN are examined in model 9. All three-variable effects are rather small. The interaction effects of age and gender can be ignored, the largest effect being $\hat{\beta}_{11}^{AG\overline{N}} = 0.03$. Period-gender interaction effects all have about the same strength as the age-gender interactions with just one "exception": $\hat{\beta}_{11}^{PG\overline{N}} = 0.07$. The cohort-gender interaction effects are somewhat larger, several of them 0.05, the largest being $\hat{\beta}_{121}^{CG\overline{N}} = 0.09$.

The main effect of gender on nonmembership in model 5 is $\hat{\beta}_{1}^{G\overline{N}} = -0.16$. The strength of the gender effect on nonmembership is about the same as that of age: $\hat{\gamma}_{max}^{*} = 1.38$. The odds that a women is a nonmember of a religious denomination are somewhat smaller than for men. Because there are no interaction effects on nonmembership between age and gender and between period and gender, this difference in nonmembership between men and women remains the same throughout the life-course and for all periods. Although the cohort-gender interaction effects are small, they show an interesting pattern. Disregarding the extreme (oldest and youngest) cohorts, the differences between men and women with regard to nonmembership as measured by the conditional effects $\hat{\beta}_{1\ k}^{G\overline{N}/C}$ are stable for the older cohorts but become smaller and smaller as of cohort C_7 (although women remain somewhat more religious than men).

The period and age effects on nonmembership found in model 5 are much the same as depicted in Figure 7.4, curves "$A_4 = A_5$, $P_4 = P_5$" (and are the same for men and women). Totally different effects for men and women, particularly different period effects, would have cast doubt on the validity of the previous findings and of the imposed equality restrictions, as there is no theoretical reason why men and women would be totally differently affected by the kinds of period effects mentioned.

Cohort effects are largely the same for men and women and similar to the outcomes in Figure 7.4. The interaction effects of cohort and gender on nonmembership are such that the slowing down of the upward secularization trend as of cohort C_9 is somewhat stronger for men than for women (again disregarding the extreme cohorts). Interpreting, as before, cohort effects in terms of changes in ways of bringing up children, a possible explanation of this interaction might be that religious values are more strongly stressed for girls than for boys, but that this difference is waning.

Whether or not this suggestion is pure speculation, this seems to be the most important use of cohort analysis: To find by means of the operational age, period, and cohort effects suggestions for the underlying real causes of the phenomena studied. Because cohort analysis can be used to disentangle the confounding effects of age, period, and cohort, it is in this respect better suited to explore the data than most other designs. Whether or not the suggestions are true is a matter for further research in which the underlying variables are measured directly.

Notes

1. Fienberg and Mason (1978: 37-42) discussed cohort tables in which the age intervals and the distances between the moments of observation were not the same. See also Section 7.5.1.

2. Actually, the results in Figure 7.1 are not based on the data in Table 7.1, but on Table 7.3. However, the differences between these two tables are for all practical purposes negligible.

3. This kind of perfect linear relationship among three key variables and all the problems it causes are by no means unique to cohort analysis (Blalock, 1982: 226-234; Smith et al., 1982). It occurs in panel research (Kessler & Greenberg, 1981), in research on social mobility (Hope, 1975, 1981), in the investigation of status inconsistency (Hornung, 1980; Lenski, 1954; Wilson, 1979), in contextual analysis (Burstein, 1980), in research on interviewer effects (Hagenaars & Heinen, 1982), etc. In all these researches, there is a first "key" variable (a first wave variable, father's occupation, a first status component, an individual characteristic, a respondent characteristic), a second key variable (a second wave variable, son's occupation, a second status component, a group characteristic, an interviewer characteristic), and a third key variable which is nothing but the difference between the first and the second key variables. In all these fields,

determining the effects of these three key variables on a particular dependent variable gives rise to the same kind of identification problem that arises in cohort analysis.

4. If each period has a unique score on the prosperity index and if no additional restrictions are introduced, the effects of age, cohort, and period (= prosperity) on, for example, nonmembership, can still not be separated from each other. To be identifiable, at least two periods should have the same prosperity score or particular constraints should be imposed on the effects of prosperity on religiosity such as the restriction that the relationship between prosperity and religiosity is linear (see the discussion below about imposing extra [equality] constraints on the effect parameters).

5. In contrast to Table 7.1, all frequencies in Table 7.3 have been divided by 10 and rounded to the nearest integers. This does not noticeably affect the parameter estimates or the descriptive goodness-of-fit measures $\hat{\delta}$ and w, nor will there be any effect on the degrees of freedom. However, the reported test statistics L^2 and Pearson-χ^2 have to be multiplied by 10 to obtain the values of these test statistics in terms of the original frequencies in Table 7.1. Because, for reasons outlined below, L^2 and Pearson-χ^2 are used here as descriptive measures, this multiplication was omitted.

6. As Fienberg and Mason (1985) showed, the main "linear" effects of A, P, and C on N are not identified, but the effects of higher order terms like A^2 or product terms like AP may be identified for particular models.

7. This may be verified by setting up a simple cohort table with three age categories, three periods, and five cohorts, writing out the equations for logit model {APN, AN, PN, CN} and performing the necessary calculations by hand.

8. A Summary View

Log-linear models are excellent tools for analyzing long term and short term changes reflected in categorical data obtained from panel, trend, or cohort studies. Not only can the nature of the net and gross changes be described in an insightful manner by means of log-linear models, but these models can also be used to find and test explanations of the changes in terms of "external" variables. Moreover, imperfections in the data, such as those connected with unreliable measurements, nonresponse, test-retest effects, etc., can be remedied by applying special kinds of log-linear models, in particular log-linear models with latent variables.

Facilities to carry out ordinary log-linear models without latent variables are available in statistical packages such as SPSS^X, SAS, BMDP, and GLIM. Special purpose programs like Goodman and Fay's ECTA or Haberman's FREQ can be used once the multivariate frequency table is available. Log-linear models with latent variables can be handled by Clogg's MLLSA, Haberman's LAT or NEWTON, and Hagenaars' LCAG. PC versions of MLLSA, ECTA, and FREQ have been implemented in Eliason's CDAS, using one standard, easy-to-use input language. So, log-linear models not only have an enormous potential for the analysis of social change, but this potential can actually be mobilized using widely available computer programs.

Nevertheless, it is not all roses. Despite the fact that never before in history has so much data on as many different subjects been accessible, the investigator may still not acquire the data he or she needs. The cohort analysis reported in Chapter 7 is a case in point. Although it is remarkable that one is able to follow patterns of (non)membership of a religious denomination for several cohorts over a period as long as 1899-1969, it

360

is not what the investigators originally had in mind. They intended to carry out a cohort analysis on "frequency of church going" and not on (non)membership. Furthermore, they wanted to specify the cohort analysis not only according to gender but also marital status and education or income. These theoretically based wishes were obstructed by lack of data. The pertinent data were available from national surveys, but only for a limited period of time (from 1960 on), making it impossible to follow any of the cohorts over the entire life-course. It was therefore decided to use census data which covered a longer period of time but concerned nonmembership instead of "frequency of church going" and did not provide information on marital status. These kinds of dilemmas will often be encountered when using secondary data.

But even when all relevant data are available, there may still be problems because of small sample size. Particularly in panel studies, a large sample is required to perform adequately the desired statistical tests. But even in trend analyses, the introduction of a moderately large number of variables may cause problems because of too sparse tables. At the one extreme, the investigator may decide to employ just a few, coarsely categorized, variables leading to adequate statistical testing outcomes. At the other extreme, all relevant variables may be introduced keeping their original categorization with the result that the outcomes of the statistical tests cannot be trusted. It is often difficult to find a sensible middle-of-the-road solution.

Finally, even when the right kind of data is available and the sample size is large enough, lack of appropriate software may preclude actually carrying out the planned analyses. For example, as shown in Section 6.6, the algorithm for carrying out latent class analyses over time where T (Time) is directly related to the manifest variables and moreover has a direct (curvi)linear relationship with the latent variable is well known, but at this time there is no software package available in which such an algorithm has been implemented.

Another example is provided by the tests on marginal homogeneity that play an important role in panel and trend analyses (Sections 4.3 and 6.4). Using the conditional marginal homogeneity tests, many kinds of hypotheses about social change can be investigated, but at the same time the usefulness of these conditional tests is lessened by the fact that they do not provide estimated expected frequencies if the pertinent hypothesis is valid. Extending the procedures of Bishop et al. (1975, Section 8.2.4), and Haber and Brown (1986), appropriate unconditional test procedures which do provide estimated expected frequencies can be developed but they are likewise not available at this moment.

The study of social change by means of secondary data is full of exciting potentialities but also of inevitable pitfalls. The spirit in which the investigation of processes of social change has to be carried out (and in which this book has been written) has been eloquently expressed by Nunnally (1973):

> Interesting ideas outstrip the available methodology for their investigation; measurement techniques are crucially important, but difficult to develop in many cases; research strategies either are doomed to produce equivocal results or are unfeasible in the light of available resources and technical developments; but despite all the difficulties, the questions are intriguing and the search for answers is fun. (p. 109)

References

Achen, C.H. (1986). *The statistical analysis of quasi-experiments*. Berkeley: University of California Press.

Adam, J. (1978). Sequential strategies and the separation of age, cohort, and time-of-measurement contributions to developmental data. *Psychological Bulletin, 85*, 1309-1316.

Agresti, A. (1984). *Analysis of categorical data*. New York: John Wiley.

Agresti, A. (1989). Tutorial on modeling ordered categorical response data. *Psychological Bulletin, 105*, 290-301.

Agresti, A., Chuang, C., & Kezouh, A. (1987). Order-restricted score parameters in association models for contingency tables. *Journal of the American Statistical Association, 82*, 619-623.

Aitkin, M. (1980). A note on the selection of log-linear models. *Biometrics, 36*, 173-178.

Akaike, H. (1973). Information theory and an extension of the maximum likelihood principle. In B.N. Petrov & B.F. Csaki (Eds.), *Second International Symposium on Information Theory* (pp. 267-281). Budapest, Hungary: Academiai Kiado.

Akaike, H. (1987). Factor analysis and AIC. *Psychometrika, 52*, 317-332.

Alba, R.D. (1987). Interpreting the parameters of log-linear models. *Sociological Methods and Research, 16*, 45-77.

Allison, P.D. (1980). Analyzing collapsed contingency tables without actually collapsing. *American Sociological Review, 45*, 123-130.

Allison, P.D. (1984). *Sage university paper 46. Event history analysis: Regression for longitudinal event data*. Beverly Hills, CA: Sage.

Allison, P.D. (1987) Estimation of linear models with incomplete data. In C.C. Clogg (Ed.), *Sociological Methodology 1987* (Vol. 17, pp. 71-104). Washington DC: American Sociological Association.

Allison, H.E., Zwick, C.J., & Brinser, A. (1958). Recruiting and maintaining a consumer panel. *Journal of Marketing, 22*, 377-390.

Anderson, A.B., Basilevsky, A., & Hum, D.P. (1983). Missing data. In P.H. Rossi, J.D. Wright, & A.B. Anderson (Eds.), *Handbook of survey research* (pp. 415-494). New York: Academic Press.

Andrich, D. (1985). An elaboration of Guttman scaling with Rasch models for measurement. In N.B. Tuma (Ed.), *Sociological Methodology 1985* (pp. 33-80). San Francisco: Jossey Bass.

Andrich, D. (1988). *Sage university paper 68. Rasch models for measurement.* Newbury Park, CA: Sage.

Arminger, G. (1976). Anlage und Auswertung von Paneluntersuchungen. In K. Holm (Ed.). *Die Befragung 4* (pp. 134-235). Munich, W. Germany: Francke Verlag.

Bahr, H.M. (1970). Aging and religious disaffiliation. *Social Forces, 49,* 60-71.

Bailar, B.A. (1975). The effects of rotation group bias on estimates from panel surveys. *Journal of the American Statistical Association, 70,* 23-30.

Bailar, B.A., Bailey, L., & Corby, C. (1978). A comparison of some adjustment and weighting procedures for survey data. In K. Namboodiri (Ed.), *Survey sampling and measurement* (pp. 175-198). New York: Academic Press.

Bailey, K.D. (1988). The conceptualization of validity: Current perspectives. *Social Science Research, 17,* 117-136.

Baltes, P.B. (1968). Longitudinal and cross-sectional sequences in the study of age and generation effects. *Human Development, 11,* 145-171.

Baltes, P.B., Cornelius, S.W., & Nesselroade, J.R. (1979). Cohort effects in developmental psychology. In J.R. Nesselroade & P.B. Baltes (Eds.), *Longitudinal research in the study of behavior and development* (pp. 61-88). New York: Academic Press.

Baltes, P.B., Reese, H.W., & Nesselroade, J.R. (1977). *Life-span developmental psychology: Introduction to research methods.* Belmont, CA: Brooks/Cole.

Barnes, S.H., & Kaase, M. (1979). *Political action.* Beverly Hills, CA: Sage.

Beck, P.A. (1975). Models for analyzing panel data: A comparative review. *Political Methodology, 1,* 357-380.

Benedetti, J.K., & Brown, M.B. (1978). Strategies for the selection of log-linear models. *Biometrics, 34,* 680-686.

Bengtson, V.L., Furlong, M.J., & Laufer, R.S. (1974). Time, aging, and the continuity of social structure: Themes and issues in generational analysis. *Journal of Social Issues, 30,* 1-30.

Bennett, C.A., & Lumsdaine, A.A. (Eds.). (1975). *Evaluation and experiment.* New York: Academic Press.

Berger, M.P. (1985). *Some aspects of the application of the generalized multivariate analysis of variance model.* Unpublished doctoral dissertation, Tilburg University, Tilburg, The Netherlands.

Bhapkar, V.P. (1973). On the comparison of proportions in matched samples. *Sankhya.* Series A, *35,* 341-365.

Bishop, Y.M.M., Fienberg, S.E., & Holland, P.W. (1975). *Discrete multivariate analysis: Theory and practice.* Cambridge, MA: MIT Press.

Blalock, H.M., Jr. (1968). The measurement problem: A gap between the languages of theory and research. In H.M. Blalock & A.B. Blalock (Eds.), *Methodology in social research* (pp. 5-27). New York: McGraw-Hill.

Blalock, H.M., Jr. (1982). *Conceptualization and measurement in the social sciences.* Beverly Hills, CA: Sage.

Bock, R.D. (1975). *Multivariate statistical methods in behavioral research.* New York: McGraw-Hill.

Bohrnstedt, G.W. (1969). Observations on the measurement of change. In E.F. Borgatta (Ed.), *Sociological Methodology 1969* (pp. 113-136). San Francisco: Jossey-Bass.

Bohrnstedt, G.W. (1983). Measurement. In P.H. Rossi, J.D. Wright, & A.B. Anderson (Eds.), *Handbook of survey research* (pp. 70-121). New York: Academic Press.

Bollen, K.A. (1989). A new incremental fit index for general structural equation models. *Sociological Methods and Research, 17*, 303-316.

Bonett, D.G., & Bentler, P.M. (1983). Goodness-of-fit procedures for the evaluation and selection of log-linear models. *Psychological Bulletin, 93*, 149-166.

Bozdogan, H. (1987). Model selection and Akaike's information criterion (AIC): The general theory and its analytical extensions. *Psychometrika, 52*, 345-370.

Bradburn, N.M. (1983). Response effects. In P. Rossi et al. (Eds.), Handbook of survey research (pp. 284-328). New York: Academic Press.

Breiger, R.L. (1981). The social class structure of occupational mobility. *American Journal of Sociology, 87*, 578-611.

Breiger, R.L. (1982). A structural analysis of occupational mobility. In P.V. Marsden & N. Lin (Eds.), *Social structure and network analysis* (pp.17-32). Beverly Hills, CA: Sage.

Breitsamer, J. (1976). Ein Versuch zum "Problem der Generationen". *Kölner Zeitschrift für Soziologie und Sozial-Psychologie, 28*, 451-478.

Bridge, R.G., Reeder, L.G., Kanouse, D., Kinder, D.R., Nagy, V.T., & Judd, C.M. (1977). Interviewing changes attitudes — sometimes. *Public Opinion Quarterly, 41*, 56-64.

Brier, S.S. (1978). The utility of systems of simultaneous logistic response equations. In K.F. Schuessler (Ed.), *Sociological methodology 1979* (pp. 119-129). San Francisco: Jossey-Bass.

Buchhofer, B., Friedrichs, J., & Lüdtke, H. (1970). Alter, Generationsdynamik und Soziale Differenzierung. Zur Revision des Generationsbegriffs als Analytisches Konzept. *Kölner Zeitschrift für Soziologie und Sozial-Psychologie, 22*, 300-334.

Buck, S.F., Fairclough, E.H., Jephcott, J.S.G., & Ringer, D.W.C. (1977). Conditioning and bias in consumer panels — some new results. *Journal of the Market Research Society, 19*, 59-75.

Burstein, L. (1980). *Review of research in education: Vol. 8. The analysis of multilevel data in educational research and evaluation*. Washington, DC: American Educational Research Association.

Buss, A.R. (1974). Generational analysis: Description, explanation and theory. *Journal of Social Issues, 30*, 55-71.

Butler, D., & Stokes, D. (1969). *Political change in Britain*. London, England: Macmillan.

Bye, B.V., & Schechter, E.S. (1986). A latent Markov model approach to the estimation of response errors in multiwave panel data. *Journal of the American Statistical Association, 81*, 375-380.

Caillez, F., & Pagès, J.P. (1976). *Introduction à l'Analyse des Données*. Paris, France: SMASH.

Campbell, D.T. (1963). From description to experimentation: Interpreting trends as quasi-experiments. In C.W. Harris (Ed.), *Problems in measuring change* (pp. 212-242). Madison: University of Wisconsin Press.

Campbell, D.T. (1969). Reforms as experiments. *American Psychologist, 24,* 409-429.

Campbell, D.T., & Boruch, R.F. (1975). Making the case for randomized assignment to treatments by considering the alternatives: Six ways in which quasi-experimental evaluations in compensatory education tend to underestimate effects. In C.A. Bennett & A.A. Lumsdaine (Eds.), *Evaluation and experiment* (pp. 195-269). New York: Academic Press.

Campbell, D.T., & Clayton, K.N. (1961). Avoiding regression effects in panel studies of communication impact. *Studies in Public Communication* (No. 3, pp. 99-118). Indianapolis: Bobbs-Merill. (Reprint, No. S 353).

Campbell, D.T., & Erlebacher, A. (1975). How regression artifacts in quasi experimental evaluations can mistakenly make compensatory education look harmful. In E.L. Struening & M. Guttentag (Eds.), *Handbook of evaluation research* (Vol. 1, pp. 597-620). Beverly Hills, CA: Sage.

Campbell, D.T., & Stanley, J.S. (1966). *Experimental and quasi-experimental designs for research.* Chicago: Rand McNally.

Carlsson, G. (1972). Lagged structures and cross-sectional methods. *Acta Sociologica, 15,* 323-341.

Centraal Bureau voor de Statistiek. (1968). *Kerkelijke Gezindte, A: Algemene Inleiding.* 13e Algemene Volkstelling, 31 mei 1960, Deel 7, Staatsuitgeverij. The Hague, The Netherlands: Author.

Chaffee, S.H., & McLeod, J.M. (1968). Sensitization in panel design: A coorientational experiment. *Journalism Quarterly, 45,* 661-669.

Chapman, D.W. (1976). A survey of nonresponse imputation procedures. *Proceedings of the Social Statistics Section of the American Statistical Association* (pp. 245-251). Washington, DC: American Statistical Association.

Chen, T. (1979). Log-linear models for categorized data with misclassifications and double sampling. *Journal of the American Statistical Association, 74,* 481-488.

Chen, T., & Fienberg, S. (1974). Two-dimensional contingency tables with both completely and partially cross-classified data. *Biometrics, 30,* 629-642.

Clarridge, B.R., Sheehy, L.L., & Hansen, T.S. (1977). Tracing members of a panel: A 17-year follow-up. In K.F. Schuessler (Ed.), *Sociological methodology 1978* (pp. 185-203). San Francisco: Jossey-Bass.

Clogg, C.C. (1979a). *Measuring underemployment.* New York: Academic Press.

Clogg, C.C. (1979b). Some latent structure models for the analysis of Likert type data. *Social Science Research, 8,* 287-301.

Clogg, C.C. (1981a). New developments in latent structure analysis. In D.J. Jackson & E.F. Borgatta (Eds.), *Factor analysis and measurement in sociological research* (pp. 215-246). Sage Studies in International Sociology 21. Beverly Hills, CA: Sage.

Clogg, C.C. (1981b). Latent structure models of mobility. *American Journal of Sociology, 86,* 836-868.

Clogg, C.C. (1982a). Some models for the analysis of association in multiway cross-classifications having ordered categories. *Journal of the American Statistical Association, 77,* 803-815.

Clogg, C.C. (1982b). Using association models in sociological research: Some examples. *American Journal of Sociology, 88,* 114-135.

Clogg, C.C., & Eliason, S.R. (1987). Some common problems in log-linear analysis. *Sociological Methods and Research, 16*, 8-44.

Clogg, C.C., & Goodman, L.A. (1984). Latent structure analysis of a set of multidimensional contingency tables. *Journal of the American Statistical Association, 79*, 762-771.

Clogg, C.C., & Goodman, L.A. (1985). Simultaneous latent structure analysis in several groups. In N.B. Tuma (Ed.), *Sociological methodology 1985* (pp. 18-110). San Francisco: Jossey-Bass.

Clogg, C.C., & Goodman, L.A. (1986). On scaling models applied to data from several groups. *Psychometrika, 51*, 123-136.

Clogg, C.C., Rubin, D.B., Schenker, N., Schultz, B., & Weidman, L. (1986). *Simple Bayesian methods for logistic regression*. Paper presented at the American Statistical Association Annual Meeting, Chicago.

Clogg, C.C., & Sawyer, D.O. (1981). A comparison of alternative models for analyzing the scalability of response patterns. In S. Leinhardt (Ed.), *Sociological methodology 1981* (pp. 240-280). San Francisco: Jossey-Bass.

Cochran, W.G. (1950). The comparison of percentages in matched samples. *Biometrika, 37*, 256-266.

Cohen, J. (1977). *Statistical power analysis for the behavioral sciences*. New York: Academic Press.

Coleman, J.S. (1964). *Introduction to mathematical sociology*. London, England: Free Press.

Coleman, J.S. (1984). Interdependence among qualitative attributes. *Journal of Mathematical Sociology, 10*, 29-50.

Converse, P.E. (1976). *The dynamics of party support: Cohort-analyzing party identification*. Beverly Hills, CA: Sage.

Cook, T.D., & Campbell, D.T. (1979). *Quasi-experimentation: Design and analysis issues for field settings*. Chicago: Rand McNally.

Cox, D.R. (1977). *The analysis of binary data*. London, England: Chapman and Hall.

Crider, D.M., & Willits, F.K. (1973). Respondent retrieval bias in a longitudinal survey. *Sociology and Social Research, 58*, 56-65.

Crider, D.M., Willits, F.K., & Bealer, R.C. (1971). Tracking respondents in longitudinal surveys. *Public Opinion Quarterly, 35*, 613-620.

Crider, D.M., Willits, F.K., & Bealer, R.C. (1973). Panel studies: Some practical problems. *Sociological Methods and Research, 2*, 3-19.

Croon, M. (1988, December 18-21). *Latent class analysis with ordered latent classes*. Paper presented at the SMABS Conference, organized by the Society for Multivariate Analysis in the Behavioral Sciences, Groningen.

Cuttance, P., & Ecob, R. (Eds.). (1987). *Structural modelling by example*. Cambridge, England: Cambridge University Press.

Daniel, W.W. (1975). Nonresponse in sociological surveys: A review of some methods for handling the problem. *Sociological Methods and Research, 3*, 291-307.

Darroch, J.N., & Ratcliff, D. (1972). Generalized iterative scaling for loglinear models. *Annals of Mathematical Statistics, 43*, 1470-1480.

Davis, J.A. (1975a). Analyzing contingency tables with linear flow graphs: D systems. In D.R. Heise (Ed.), *Sociological Methodology 1976* (pp. 111-145). San Francisco: Jossey-Bass.

Davis, J.A. (1975b). The log linear analysis of survey replications. In K.C. Land & S. Spilerman (Eds.), *Social indicator models* (pp. 75-104). New York: Russell Sage.

Davis, J.A. (1978). Studying categorical data over time. *Social Science Research, 7,* 151-179.

Davis, K. (1940). The sociology of parent-youth conflict. *American Sociological Review, 5,* 523-535.

Dayton, C.M., & McReady, G.B. (1983). Latent structure analysis of repeated classifications with dichotomous data. *British Journal of Mathematical and Statistical Psychology, 36,* 189-201.

De Leeuw, J. (1973). *Canonical analysis of categorical data.* Leyden, The Netherlands: DSWO Press.

Deming, W.E., & Stephan, F.F. (1940). On a least squares adjustment of a sampled frequency table when the expected marginal totals are known. *Annals of Mathematical Statistics, 11,* 427-444.

Dempster, A.P., Laird, N.M., & Rubin, D.B. (1977). Maximum likelihood from incomplete data via the EM algorithm. *Journal of the Royal Statistical Society, Series B, 39,* 1-38.

Droege, R.C. (1971). Effectiveness of follow-up techniques in large-scale longitudinal research. *Developmental Psychology, 5,* 27-31.

Duncan, O.D. (1969). Some linear models for two-wave, two-variable panel analysis. *Psychological Bulletin, 72,* 177-182.

Duncan, O.D. (1975a). Partitioning polytomous variables in multiway contingency analysis. *Social Science Research, 4,* 167-182.

Duncan, O.D. (1975b). Some linear models for two-wave, two-variable panel analysis, with one-way causation and measurement error. In H.M. Blalock, A. Aganbegian, F.M. Borodkin, R. Boudon & V. Capecchi (Eds.), *Quantitative sociology: International perspectives on mathematical and statistical modelling* (pp. 285-306). New York: Academic Press.

Duncan, O.D. (1979a). Testing key hypotheses in panel analysis. In K.F. Schuessler (Ed.), *Sociological methodology 1980* (pp. 279-289). San Francisco: Jossey-Bass.

Duncan, O.D. (1979b). Indicators of sex typing: Traditional and egalitarian, situational and ideological response. *American Journal of Sociology, 85,* 251-260.

Duncan, O.D. (1981). Two faces of panel analysis: Parallels with comparative cross-sectional analysis and time-lagged association. In S. Leinhardt (Ed.), *Sociological Methodology 1981* (pp. 281-318). San Francisco: Jossey-Bass.

Duncan, O.D. (1985). New light on the 16-fold table. *American Journal of Sociology, 91,* 88-128.

Duncan, O.D., Sloane, D.M., & Brody, C. (1982). Latent classes inferred from response-consistency effects. In K.G. Jöreskog & H. Wold (Eds.), *Systems under indirect observation* (part I, pp. 19-64). Amsterdam, The Netherlands: North Holland.

Dykstra, R.L., & Lemke, J.H. (1988). Duality of I projections and maximum likelihood estimation for log-linear models under cone constraint. *Journal of the American Statistical Association, 83,* 546-554.

Eckland, B.K. (1968). Retrieving mobile cases in longitudinal surveys. *Public Opinion Quarterly, 32,* 51-64.

Ehrenberg, A.S.C. (1960). A study of some potential biases in the operation of a consumer panel. *Applied Statistics, 9,* 20-27.

Eisenstadt, S.N. (1964). *From generation to generation: Age groups and social structure.* New York: Free Press.

Emerson, P.L. (1965). A Fortran generator of polynomials orthonormal over unequally spaced and weighted abscissas. *Educational and Psychological Measurement, 25,* 867-871.

Emerson, P.L. (1968). Numerical construction of orthogonal polynomials from a general recurrence formula. *Biometrics, 28,* 695-701.

Erdfelder, E. (1984). Zur Bedeutung und Kontrolle des ß-Fehlers bei der Inferenzstatistischen Prüfung Log-Linearer Modelle. *Zeitschrift für Sozialpsychologie, 15,* 18-32.

Evan, W.M. (1959). Cohort analysis of survey data: A procedure for studying long-term opinion change. *Public Opinion Quarterly, 13,* 63-72.

Everitt, B.S. (1984a). A note on parameter estimation for Lazarsfeld's latent class model using the EM-algorithm. *Multivariate Behavioral Research, 19,* 79-89.

Everitt, B.S. (1984b). *An introduction to latent variable models.* London, England: Chapman and Hall.

Evers, M., & Namboodiri, N.K. (1978). On the design matrix strategy in the analysis of categorical data. In K.F. Schuessler (Ed.), *Sociological methodology 1979* (pp. 86-111). San Francisco: Jossey-Bass.

Fay, R.E. (1986). Causal models for patterns of nonresponse. *Journal of the American Statistical Association, 81,* 354-365.

Feiring, B.R. (1986). *Sage university paper 60. Linear programming: An introduction.* Beverly Hills, CA: Sage.

Fenech, A.P., & Westfall, P.H. (1988). The power function of conditional log-linear model tests. *Journal of the American Statistical Association, 83,* 198-203.

Ferber, R. (1953). Observations on a consumer panel operation. *Journal of Marketing, 17,* 246-259.

Ferber, R. (1964). Does a panel operation increase the reliability of survey data: The case of consumer savings. *Proceedings of the Social Statistics Section of the American Statistical Association* (pp. 210-211). Washington, DC: American Statistical Association.

Ferber, R. (1966). *The reliability of consumer reports of financial assets and debts.* Urbana: University of Illinois.

Fienberg, S.E. (1977, 1980). *The analysis of cross-classified categorical data.* Cambridge, MA: MIT Press.

Fienberg, S.E. (1978). A note on fitting and interpreting parameters in models for categorical data. In K.F. Schuessler (Ed.), *Sociological methodology 1979* (pp. 112-118). San Francisco: Jossey-Bass.

Fienberg, S.E., & Mason, W.M. (1978). Identification and estimation of age-period-cohort models in the analysis of discrete archival data. In K.F. Schuessler (Ed.), *Sociological methodology 1979* (pp. 1-67). San Francisco: Jossey-Bass.

Fienberg, S.E., & Mason, W.M. (1985) Specification and implementation of age, period and cohort models. In W.M. Mason & S.E. Fienberg (Eds.), *Cohort analysis in social research* (pp. 45-88). New York/Berlin: Springer Verlag.

Fitzgerald, R., & Fuller, L. (1982). I hear you knocking but you can't come in. *Sociological Methods and Research, 11*, 3-32.

Freedman, D., Thornton, A., & Camburn, D. (1980). Maintaining response rates in longitudinal studies. *Sociological Methods and Research, 9*, 87-98.

Fuchs, C. (1982). Maximum likelihood estimation and model selection in contingency tables with missing data. *Journal of the American Statistical Association, 77*, 270-278.

Gart, J.J., Pettigrew, H.M., & Thomas, D.G. (1985). The effect of bias variance estimation, skewness and kurtosis of the empirical logit on weighted least squares analysis. *Biometrika, 72*, 179-190.

Gart, J.J., & Zweifel, J.R. (1967). On the bias of various estimators of the logit and its variance with application to quantal bioassay. *Biometrika, 54*, 181-187.

Gifi, A. (1981). *Non-linear multivariate analysis*. Leyden, The Netherlands: DSWO Press.

Glenn, N.D. (1976). Cohort analysts' futile quest: Statistical attempts to separate age, period and cohort effects. *American Sociological Review, 41*, 900-904.

Glenn, N.D. (1977). *Sage university paper 5. Cohort analysis*. Beverly Hills, CA: Sage.

Glock, C.Y. (1952). *Participation bias and re-interview effect in panel studies*. University Micro Films No. 4185. Ann Arbor, MI.

Glock, C.Y. (1955). Some applications of the panel method to the study of change. In P.F. Lazarsfeld & M. Rosenberg, *The Language of social research* (pp. 242-250). New York: Free Press.

Gokhale, D.V., & Kullback, S. (1978). *The information in contingency tables*. New York: Marcel Dekker.

Goldstein, H. (1979). *The design and analysis of longitudinal studies: Their role in the measurement of change*. London, England: Academic Press.

Goldthorpe, J.H. (1980). *Social mobility and class structure in modern Britain*. Oxford, England: Oxford University Press.

Goodman, L.A. (1968). The analysis of cross-classified data: Independence, quasi-independence, and interaction in contingency tables with or without missing cells. *Journal of the American Statistical Association, 63*, 1019-1131.

Goodman, L.A. (1969). How to ransack social mobility tables and other kinds of cross-classified tables. *American Journal of Sociology, 75*, 1-40.

Goodman, L.A. (1971a). The analysis of multidimensional contingency tables: Stepwise procedures and direct estimation methods for building models for multiple classification. *Technometrics, 13*, 33-61.

Goodman, L.A. (1971b). Partitioning of chi-square, analysis of marginal contingency tables, and estimation of expected frequencies in multidimensional contingency tables. *Journal of the American Statistical Association, 66*, 339-344.

Goodman, L.A. (1972a). A modified multiple regression approach to the analysis of dichotomous variables. *American Sociological Review, 37*, 28-46.

Goodman, L.A. (1972b). A general model for the analysis of surveys. *American Journal of Sociology, 77*, 1035-1086.

Goodman, L.A. (1972c). Some multiplicative models for the analysis of cross-classified data. In L. LeCam, J. Neyman, & E.L. Scott (Eds.), *Proceedings of the Sixth Berkeley Symposium on Mathematical Statistics and Probability* (pp. 649-696). Berkeley: University of California Press.

Goodman, L.A. (1973a). Causal analysis of panel studies and other kinds of surveys. *American Journal of Sociology, 78*, 1135-1191.

Goodman, L.A. (1973b). The analysis of multidimensional contingency tables when some variables are posterior to others: A modified path analysis approach. *Biometrika, 60*, 179-192.

Goodman, L.A. (1973c). Guided and unguided methods for the selection of models for a set of T multidimensional contingency tables. *Journal of the American Statistical Association, 68*, 165-175.

Goodman, L.A. (1974a). The analysis of systems of qualitative variables when some of the variables are unobservable. Part I − A modified latent structure approach. *American Journal of Sociology, 79*, 1179-1259.

Goodman, L.A. (1974b). Exploratory latent structure analysis using both identifiable and unidentifiable models. *Biometrika, 61*, 215-231.

Goodman, L.A. (1975). The relationship between modified and usual multiple regression approaches to the analysis of dichotomous variables. In D.R. Heise (Ed.), *Sociological methodology 1976* (pp. 83-110). San Francisco: Jossey-Bass.

Goodman, L.A. (1981a). Three elementary views of log linear models for the analysis of cross-classifications having ordered categories. In S. Leinhardt (Ed.), *Sociological Methodology 1981* (pp. 193-239). San Francisco: Jossey-Bass.

Goodman, L.A. (1981b). Criteria for determining whether certain categories in a cross-classification table should be combined with special reference to occupational categories in an occupational mobility table. *American Journal of Sociology, 87*, 612-650.

Goodman, L.A. (1983). The analysis of dependence in cross-classifications having ordered categories, using log-linear models for frequencies and loglinear models for odds. *Biometrics, 39*, 149-160.

Goodman, L.A. (1984). *The analysis of cross-classified data having ordered categories.* Cambridge, MA: Harvard University Press.

Goodman, L.A. (1985) The analysis of cross-classified data, having ordered and/or unordered categories: Association models, correlation models, and asymmetry models for contingency tables with or without missing entries. *Annals of Statistics, 13*, 10-69.

Goodman, L.A. (1987). New methods for analyzing the intrinsic character of qualitative variables using cross-classified data. *American Journal of Sociology, 93*, 529-583.

Goodman, L.A., & W.H. Kruskal (1954). Measures of association for cross-classifications. *Journal of the American Statistical Association, 49*, 732-764.

Gottman, J.M. (1981). *Time-series analysis: A comprehensive introduction for social scientists.* Cambridge, England: Cambridge University Press.

Goudy, W.J. (1976). Nonresponse effects on relationships between variables. *Public Opinion Quarterly, 40*, 360-369.

Graybill, F.A. (1961). *An introduction to linear statistical models*, Vol. I. New York: McGrawHill.

Greenberg, D.F., & Kessler, R.C. (1982). Equilibrium and identification in linear panel models. *Sociological Methods and Research, 10*, 435-451.

Grizzle, J.E., Starmer, C.F., & Koch, G.C. (1969). Analysis of categorical data by linear models. *Biometrics, 25*, 137-156.

Grover, R. (1987). Estimation and use of standard errors of latent class model parameters. *Journal of Marketing Research, 24*, 298-304.

Haber, M., & Brown, M.B. (1986). Maximum likelihood methods for log-linear models when expected frequencies are subject to linear constraints. *Journal of the American Statistical Association, 81*, 477-482.

Haberman, S.J. (1974). *The analysis of frequency data*. Chicago: University of Chicago Press.

Haberman, S.J. (1976). Iterative scaling procedures for log-linear models for frequency data derived by indirect observation. *Proceedings of the American Statistical Association 1975: Statistical Computing Section*, pp. 45-50.

Haberman, S.J. (1977). Product models for frequency tables involving indirect observation. *Annals of Statistics, 5*, 1124-1147.

Haberman, S.J. (1978). *Analysis of qualitative data. Vol. 1. Introductory Topics*. Academic Press, New York.

Haberman, S.J. (1979). *Analysis of qualitative data. Vol. 2. New Developments*. New York: Academic Press.

Haberman, S.J. (1982). Analysis of dispersion of multinomial responses. *Journal of the American Statistical Association, 77*, 568-580.

Haberman, S.J. (1988a). A warning on the use of chi-squared statistics with frequency tables with small expected cell counts. *Journal of the American Statistical Association, 83*, 555-560.

Haberman, S.J. (1988b). A stabilized Newton-Raphson algorithm for log-linear models for frequency tables derived by indirect observation. In C.C. Clogg (Ed.), *Sociological methodology 1988* (Vol. 18, pp. 193-212). American Sociological Association, Washington, DC.

Hagenaars, J.A. (1978). Latent probability models with direct effects between indicators. *Quality and Quantity, 12*, 205-221.

Hagenaars, J.A. (1983). De beantwoording van "typische panelvragen" met behulp van loglineaire modellen. *Sociale Wetenschappen, 26*, 147-186.

Hagenaars, J.A. (1984, 3-5 October). *LCAG-Loglinear Modelling with Latent Variables: A modified Lisrel Approach*. Paper presented at the International Conference on Methodological Research, Amsterdam, The Netherlands.

Hagenaars, J.A. (1985). *Loglineaire analyse van herhaalde surveys: Panel-, trend- en cohortonderzoek*. Unpublished doctoral dissertation. Tilburg, The Netherlands: Tilburg University Press.

Hagenaars, J.A. (1986a). Symmetry, quasi-symmetry, and marginal homogeneity on the latent level. *Social Science Research, 15*, 241-255.

Hagenaars, J.A. (1986b). Ontbrekende gegevens bij loglineaire analyses met en zonder latente variabelen. In J.H. Segers & E.J. Bijnen (Eds.), *Onderzoeken: Reflecteren en Meten* (pp. 57-77). Tilburg, The Netherlands: Tilburg University Press.

Hagenaars, J.A. (1987). Establishing causal order among variables in delinquency research: The failure of panel analysis. In J. Janssen, F. Marcotorchino, & J.M. Proth (Eds.), *Data analysis: The ins and outs of solving real problems* (pp. 331-344). New York: Plenum.

Hagenaars, J.A. (1988a). LCAG — Loglinear modelling with latent variables: A modified LISREL approach. In W.E. Saris & I.N. Gallhofer (Eds.), *Sociometric research: Volume 2. Data analysis* (pp. 111-130). London, England: Macmillan.

Hagenaars, J.A. (1988b). Latent structure models with direct effects between indicators: Local dependence models. *Sociological Methods and Research, 16*, 379-405.

Hagenaars, J.A. (1988c, May/June). *Log-linear analysis with latent variables and missing data*. Paper presented at the International Conference on Social Science Methodology, ISA, RC#33, Dubrovnik, Yugoslavia.

Hagenaars, J.A., & Bijnen, E.J. (1974). Models to determine test-retest reliability with the aid of path analysis with an application to background variables. *The Netherlands Journal of Sociology* [Sociologia Neerlandica], *10*, 244-257.

Hagenaars, J.A., & Cobben, N.P. (1978). Age, cohort, and period: A general model for the analysis of social change. *The Netherlands Journal of Sociology* [Sociologia Neerlandica], *14*, 59-92.

Hagenaars, J.A., & Halman, L.C. (1989). Searching for ideal types: The potentialities of latent class analysis. *European Sociological Review, 5*, 81-96.

Hagenaars, J.A., & Heinen, A.G. (1980). Analyse van contingentietabellen met behulp van het loglineaire model. In J.H.G. Segers & J.A.P. Hagenaars (Eds.), *Sociologische Onderzoeksmethoden: deel II, Technieken van Causale Analyse* (Chapter 6, pp. 185-258). Assen, The Netherlands: Van Gorcum.

Hagenaars, J.A., & Heinen, A.G. (1982). Effects of role-independent interviewer characteristics on responses. In W. Dijkstra & J. van der Zouwen (Eds.), *Response behaviour in the Survey-Interview* (pp. 91-130). London: Academic Press.

Hagenaars, J.A., Heinen, A.G.J., & Hamers, P.A.M. (1980). Causale modellen met diskrete latente variabelen: een variant op de lisrel-benadering. *MDN, 5*, 38-54.

Hagenaars, J., & Luijkx, R. (1987). *LCAG: Latent class analysis models and other loglinear models with latent variables: Manual LCAG*. Working Paper Series # 17, Department of Sociology, Tilburg University, Tilburg, The Netherlands.

Hagenaars, J., Stouthard, P., & Wauters, F. (1990). Panel attrition. In M.K. Jennings & J.W. van Deth (Eds.), *Continuities in political action* (Appendix B). Frankfurt, W. Germany: DeGruyter.

Hakim, C. (1982). *Secondary analysis in social research*. London, England: Unwin Hyman.

Hamdan, M.A., Pirie, W.R., & Arnold, J.C. (1975). Simultaneous testing of McNemar's problem for several populations. *Psychometrika, 40*, 153-161.

Hannan, M.T., Robinson, R., & Warren, J.T. (1974). The causal approach to measurement error in panel analysis: Some further contingencies. In H.M. Blalock (Ed.), *Measurement in the social sciences* (pp. 293-324). Hawthorne, NY: Aldine.

Harder, T. (1973). *Dynamische Modelle in der Empirischen Sozialforschung.* Stuttgart, W. Germany: Teubner.

Hauck, W.W., Anderson, S., & Leaky, F.J., III. (1982). Finite-sample properties of some old and some new estimators of a common odds ratio from multiple 2 × 2 tables. *Journal of the American Statistical Association, 77,* 145-151.

Hauser, R.M. (1979). Some exploratory methods for modelling mobility tables and other cross-classified data. In K.F. Schuessler (Ed.), *Sociological methodology 1980* (pp. 13-458). San Francisco: Jossey-Bass.

Hays, W.L. (1981). *Statistics.* New York: Holt, Rinehart & Winston.

Heckman, J., & Robb, R. (1985). Using longitudinal data to estimate age, period, and cohort effects in earnings equations. In W.M. Mason & S.E. Fienberg (Eds.), *Cohort analysis in social research* (pp. 137-150). New York/Berlin: Springer Verlag.

Heinen, A.G., Hagenaars, J.A., & Croon, M.A. (1988, December 18-21). *Latent trait models in LCA perspective.* Paper presented at the SMABS Conference, organized by the Society for Multivariate Analysis in the Behavioral Sciences, Groningen, The Netherlands.

Heise, D.R. (1970). Causal inference from panel data. In E.F. Borgatta & G.W. Bohrnstedt (Eds.), *Sociological methodology 1970* (pp. 3-27). San Francisco: Jossey-Bass.

Heise, D.R. (1975). *Causal Analysis.* New York: John Wiley.

Hibbs, D.A., Jr. (1977). On analyzing the effects of policy interventions: Box-Jenkins and Box-Tiao versus structural equation models. In D.R. Heise (Ed.), *Sociological Methodology 1977* (pp. 137-179). San Francisco: Jossey-Bass.

Hildebrand, D.K., Laing, J.D., & Rosenthal, H. (1977). *Prediction analysis of cross classifications.* New York: John Wiley.

Hocking, R., & Oxspring, H. (1971). Maximum likelihood estimation with incomplete multinomial data. *Journal of the American Statistical Association, 66,* 65-70.

Hocking, R., & Oxspring, H. (1974). The analysis of partially categorized contingency data. *Biometrics, 30,* 469-483.

Holm, K. (1979). Das Allgemeine Lineare Modell. In K. Holm (Ed.), *Die Befragung 6* (UTB 436, pp. 11-214). Munich, W. Germany: Francke Verlag.

Hope, K. (1975). Models of status inconsistency and social mobility effects. *American Sociological Review, 40,* 322-343.

Hope, K. (1981). Constancies in the analysis of social stratification: A replication study. *Quality and Quantity, 15,* 481-503.

Hornung, C.A. (1980). Status inconsistency, achievement motivation and psychological stress. *Social Science Research, 9,* 362-180.

Hout, M. (1983). *Sage university paper 31. Mobility tables.* Beverly Hills, CA: Sage.

Hout, M., Duncan, O.D., & Sobel, M.E. (1987). Association and heterogeneity: Structural models of similarities and differences. In C.C. Clogg (Ed.), *Sociological methodology 1987* (Vol 17, pp. 145-184). Washington, DC: The American Sociological Association.

Hsiao, C. (1986). *Analysis of panel data*. Cambridge, England: Cambridge University Press.

Huckfeldt, R.R., Kohfeld, C.Q., & Likens, T.W. (1982). *Sage university paper 27. Dynamic modelling: An introduction*. Beverly Hills, CA: Sage.

Hyman, H.H. (1954). *Interviewing in social research*. Chicago: University of Chicago Press.

Hyman, H.H. (1955). *Survey design and analysis*, New York: Free Press.

Hyman, H.H. (1972). *Secondary analysis of sample surveys: Principles, procedures and potentialities*. New York: John Wiley.

Inglehart, R. (1977). *The Silent revolution: Changing values and political styles among Western publics*. Princeton, NJ: Princeton University Press.

Inglehart, R. (1979). Value priorities and socioeconomic change. In S.H. Barnes & M. Kaase (Eds.), *Political action* (pp. 305-342). Beverly Hills, CA: Sage.

Jacob, H. (1984). *Sage university paper 42. Using published data: Errors and remedies*. Beverly Hills, CA: Sage.

Jagodzinski, W. (1984). Identification of parameters in cohort models. *Sociological Methods and Research, 12*, 375-398.

Jagodzinski, W., Kühnel, S.M., & Schmidt, P. (1987). Is there a "Socratic effect" in nonexperimental panel studies. *Sociological Methods and Research, 15*, 259-302.

Johnston, J. (1972). *Econometric methods*. Tokyo, Japan: McGraw-Hill.

Jöreskog, K.G., & Sörbom, D. (1979). *Advances in factor analysis and structural equation models*. Cambridge, MA: Abt.

Judd, C.M., & Kenny, D.A. (1981). *Estimating the effects of social interventions*. Cambridge, England: Cambridge University Press.

Kaplan, D. (1988). The impact of specification error on the estimation, testing, and improvement of structural equation models. *Multivariate Behavioral Research, 23*, 69-86.

Kaufman, R.L., & Schervish, P.G. (1986). Using adjusted crosstabulations to interpret log-linear relationships. *American Sociological Review, 51*, 717-733.

Kaufman, R.L., & Schervish, P.G. (1987). Variations on a theme: More uses of odds ratios to interpret log-linear parameters. *Sociological Methods and Research, 16*, 218-255.

Kawasaki, S. (1982). An analysis of multivariate ordinal polytomous data using the log-linear probability model. *Journal of Mathematical Sociology, 9*, 15-31.

Kawasaki, S., & Zimmermann, K.F. (1981). Measuring relationships in the loglinear model by some compact measures of Association. *Statistische Hefte, 22*, 94-121.

Kendall, P. (1954). *Conflict and mood: Factors affecting stability and response*. New York: Free Press.

Kenny, D.A., & Harackiewicz, J.M. (1979). Cross-lagged panel correlation: Practice and promise. *Journal of Applied Psychology, 64*, 372-379.

Kerlinger, F.N., & Pedhazur, E.J. (1973). *Multiple regression in behavioral research*. New York: Holt, Rinehart & Winston.

Kessler, R.C. (1977). Rethinking the 16-fold table problem. *Social Science Research, 16*, 84-107.

Kessler, R.C., & Greenberg, D.F. (1981). *Linear panel analysis: Models of quantitative change*. New York: Academic Press.

Kiecolt, K.J., & Nathan, L.E. (1985). *Sage university paper 53. Secondary analysis of survey data*. Beverly Hills, CA: Sage.

Kiiveri, H., & Speed, T.P. (1982). Structural analysis of multivariate data: A review. In S. Leinhardt (Ed.), *Sociological methodology 1982* (pp. 209-289). San Francisco: Jossey-Bass.

Kim, J.O. (1984a). Sensitivity analysis in sociological research. *American Sociological Review, 49*, 272-282.

Kim, J.O. (1984b). PRU-measures of association for contingency tables analysis. *Sociological Methods and Research, 13*, 3-44.

Kimberly, J.R. (1976). Issues in the design of longitudinal organizational research. *Sociological Methods and Research, 4*, 321-347.

Knoke, D., & Burke, P.J. (1980). *Sage university paper 20. Log-linear models*. Beverly Hills, CA: Sage.

Knoke, D., & Hout, M. (1976). Reply to Glenn. *American Sociological Review, 41*, 905-908.

Koehler, K.J. (1986). Goodness-of-fit tests for log-linear models in sparse contingency tables. *Journal of the American Statistical Association, 81*, 483-493.

Kohfeld, C.W., & Salert, B. (1982). Representations of dynamic models. *Political Methodology, 8*, 1-32.

Krippendorf, K. (1986). *Sage university paper 62. Information theory: Structural models for qualitative data*. Sage Publications, Beverly Hills.

Kritzer, H.M. (1977). Analyzing measures of association derived from contingency tables. *Sociological Methods and Research, 5*, 387-417.

Kritzer, H.M. (1983). The identification problem in cohort analysis. *Political Methodology, 9*, 35-50.

Labouvie, E.W., & Nesselroade, J.R. (1985). Age, period and cohort analysis and the study of individual development and social change. In J.R. Nesselroade & E. von Eye (Eds.), *Individual development and social change* (pp. 189-212). New York: Academic Press.

Lana, R.E. (1969). Pretest sensitization. In R. Rosenthal, & R.L. Rosnow (Eds.), *Artifact in behavioral research* (pp. 119-141). New York: Academic Press.

Land, K.C., & Felson, M. (1981). Sensitivity analysis of arbitrarily identified simultaneous-equation models. In P.V. Marsden (Ed.), *Linear models in social research* (pp. 155-177). Beverly Hills, CA: Sage.

Landis, J.R., & Koch, G.G. (1979). The analysis of categorical data in longitudinal studies of behavioral development. In J.R. Nesselroade & P.B. Baltes (Eds.), *Longitudinal research in the study of behavior and development* (pp. 233-262). New York: Academic Press.

Langeheine, R., & Rost, J. (Eds.). (1988). *Latent trait and latent class models*. New York: Plenum.

Larntz, K. (1978). Small sample comparisons of exact levels for chi-square goodness-of-fit statistics. *Journal of the American Statistical Association, 73*, 253-263.

Lawal, H.B. (1984). Comparisons for the X^2, Y^2, Freeman-Tukey and William's improved G^2 test statistic in small samples of one-way multinomials. *Biometrika, 71*, 415-418.

Lazarsfeld, P.F. (1948). The use of panels in social research. *Proceedings of the American Philosophical Society, 92*, 405-410.

Lazarsfeld, P.F. (1950a). The logical and mathematical foundation of latent structure analysis. In S. Stouffer (Ed.), *Measurement and prediction* (pp. 362-412), Princeton, NJ: Princeton University Press.

Lazarsfeld, P.F. (1950b). The interpretation and mathematical foundation of latent structure analysis. In S. Stouffer (Ed.), *Measurement and prediction* (pp. 413-472). Princeton, NJ: Princeton University Press.

Lazarsfeld, P.F. (1955). Interpretation of statistical relations as a research operation. In P.F. Lazarsfeld & M. Rosenberg (Eds.), *The language of social research* (pp. 115-125). New York: Free Press.

Lazarsfeld, P.F. (1972a). The problem of measuring turnover. In P.F. Lazarsfeld, A.K. Pasanella, & M. Rosenberg (Eds.), *Continuities in the language of social research* (pp. 358-362). New York: Free Press.

Lazarsfeld, P.F. (1972b). Mutual effects of statistical variables. In P.F. Lazarsfeld, A.K. Pasanella, & M. Rosenberg (Eds.), *Continuities in the language of social research* (pp. 388-398). New York: Free Press.

Lazarsfeld, P.F., Berelson, B., & Gaudet, H. (1944). *The people's choice.* New York: Duell, Sloan and Pearce.

Lazarsfeld, P.F., & Fiske, M. (1938). The "panel" as a new tool for measuring opinion. *Public Opinion Quarterly, 2*, 596-612.

Lazarsfeld, P.F., & Henry, N.W. (1968). *Latent Structure Analysis.* Boston: Houghton Mifflin.

Lazarsfeld, P.F., Pasanella, A.K., & Rosenberg, M. (Eds.). (1972). *Continuities in the language of social research.* New York: Free Press.

Lee, E.S., Forthofer, R.N., & Lorimor, R.J. (1989). *Sage university paper 71. Analyzing complex survey data.* Newbury Park, CA: Sage.

Lenski, G.E. (1954). Status crystallization: A non-vertical dimension of social status. *American Sociological Review, 9*, 405-413.

Little, R.J. (1982). Models for nonresponse in sample surveys. *Journal of the American Statistical Association, 77*, 237-250.

Little, R.J., & Rubin, D.B. (1987). *Statistical analysis with missing data.* New York: John Wiley.

Long, J.S. (1983a). *Sage university paper 33. Confirmatory factor analysis.* Beverly Hills, CA: Sage.

Long, J.S. (1983b). *Sage university paper 34. Covariance structure models: An introduction to LISREL.* Beverly Hills, CA: Sage.

Long, J.S. (1984). Estimable functions in log-linear models. *Sociological Methods and Research, 12*, 399-432.

Long, J.S. (Ed.) (1988). *Common problems/ proper solutions: Avoiding error in quantitative research.* Newbury Park, CA: Sage.

Lösel, F., & Wüstendörfer, E. (1974). Zum Problem Unvollständiger Datenmatrizen in der Empirischen Sozialforschung. *Kölner Zeitschrift für Soziologie und Sozial-Psychologie, 26,* 342-357.

Luijben, T. (1989). Statistical guidance for model modification in covariance structure analysis. Amsterdam, The Netherlands: Sociometric Research Foundation.

Maccoby, E.M. (1956). Pitfalls in the analysis of panel data: A research note on some technical aspects of voting. *American Journal of Sociology, 61,* 359-362.

Macdonald, K.I. (1983). On the interpretation of a structural model of the mobility table. *Quality and Quantity, 17,* 203-224.

Madden, T.J., & Dillon, W.R. (1982). Causal analysis and latent class models: An application to a communication hierarchy of effects model. *Journal of Marketing Research, 19,* 472-490.

Madow, W.G. (Ed.) (1983). *Incomplete data in sample surveys,* (Vols. 1, 2, 3), New York: Academic Press.

Magidson, J. (1981). Qualitative variance, entropy and correlation ratios for nominal dependent variables. *Social Science Research, 10,* 177-194.

Magidson, J., Swan, J., & Berk, R. (1981). Estimating nonhierarchical and nested log-linear models. *Sociological Methods and Research, 10,* 3-49.

Mannheim, K. (1928/1929). Das Problem der Generationen. *Kölner Vierteljahreshefte für Soziologie, 7,* 157-185, 309-330.

Mannheim, K. (1952). The problem of generations. In P. Kecskemeti (Ed.), *Essays on the sociology of knowledge* (pp. 276-322). London, England: Routledge & Kegan Paul.

Marascuilo, L.A., & Serlin, R.C. (1979). Tests and contrasts for comparing change parameters for a multiple McNemar data model. *British Journal of Mathematical and Statistical Psychology, 32,* 105-112.

Margolin, B.H., & Light, R.J. (1974). An analysis of variance for categorical data, II: Small sample comparisons with chi-square and other competitors. *Journal of the American Statistical Association, 69,* 755-764.

Marini, M.M., Olsen, A.R., & Rubin, D.B. (1979). Maximum-likelihood estimation in panel studies with missing data. In K.F. Schuessler (Ed.), *Sociological methodology 1980* (pp. 314-357). San Francisco: Jossey-Bass.

Marsden, P.V. (1985). Latent structure models for relationally defined social classes. *American Journal of Sociology, 90,* 1002-1021.

Mason, W.M., & Fienberg, S.E. (Eds.). (1985). *Cohort analysis in social research: Beyond the identification problem.* New York: Springer Verlag.

Mason, W.M., Mason, K.O., & Winsborough, H.H. (1976). Reply to Glenn. *American Sociological Review, 41,* 904-905.

Mason, K.O., Mason, W.M., Winsborough, H.H., & Poole, W.K. (1973). Some methodological issues in cohort analysis of archival data. *American Sociological Review, 38,* 242-257.

Mason, W.M., & Smith, H.L. (1985). Age-period-cohort analysis and the study of deaths from pulmonary tuberculosis. In W.M. Mason & S.E. Fienberg (Eds.), *Cohort analysis in social research: Beyond the identification problem.* New York: Springer.

Matsueda, R.L., & Bielby, W.T. (1986). Statistical power in covariance structure models. In N.B. Tuma (Ed.), *Sociological Methodology 1986* (Vol. 16, pp. 120-158). American Sociological Association, Washington, DC.

Mayer, L.S., & Carroll, S.S. (1987). Testing for lagged, cotemporal, and total dependence in cross-lagged panel analysis. *Sociological Methods and Research, 16*, 187-217.

Mayer, L.S., & Carroll, S.S. (1988). Measures of dependence for cross-lagged panel models. *Sociological Methods and Research, 17*, 93-119.

McAllister, R.J., Butler, E.W., & Goe, S.J. (1973). Evolution of a strategy for the retrieval of cases in longitudinal survey research. *Sociology and Social Research, 58*, 37-47.

McAllister, R.J., Goe, S.J., & Butler, E.W. (1973). Tracking respondents in longitudinal surveys: Some preliminary considerations. *Public Opinion Quarterly, 37*, 413-416.

McCain, L.J., & McCleary, R. (1979). The statistical analysis of the simple interrupted time-series quasi-experiment. In T.D. Cook & D.T. Campbell (Eds.), *Quasi-experimentation: Design and analysis issues for field settings* (pp. 233-294). Chicago: Rand McNally.

MacCallum, R. (1986). Specification searches in covariance structure modeling. *Psychological Bulletin, 100*, 107-120.

McCleary, R., & Hay, R.A., Jr. (1980). *Applied time series analysis for the social sciences*. Beverly Hills, CA: Sage.

McCullagh, P., & Nelder, J.A. (1983). *Generalized linear models*. London, England: Chapman and Hall.

McCullough, B.C. (1978). Effects of variables using panel data: A review of techniques. *Public Opinion Quarterly, 42*, 199-220.

McCutcheon, A.L. (1987). *Sage university paper 64. Latent class analysis*. Newbury Park, CA: Sage.

McGuire, W.J. (1960). Cognitive consistency and attitude change. *Journal of Abnormal and Social Psychology, 3*, 345-353.

McNemar, Q. (1947). Note on the sampling error of the difference between correlated proportions or percentages. *Psychometrika, 17*, 153-157.

Meilijson, I. (1989). A fast improvement to the EM algorithm on its own terms. *Journal of the Royal Statistical Society: Series B, 51*, 127-138.

Milligan, G.W. (1980). Factors that effect type I and type II error rates in the analysis of multidimensional contingency tables. *Psychological Bulletin, 87*, 238-244.

Mohr, L.B. (1982). On rescuing the nonequivalent-control-group design. *Sociological Methods and Research, 11*, 53-80.

Mood, A.M., Graybill, F.A., & Boes, D.C. (1974). *Introduction to the Theory of Statistics*. Tokyo, Japan: McGraw-Hill.

Mooijaart, A. (1982). Latent structure analysis for categorical variables. In K.G. Jöreskog & H. Wold (Eds.), *Systems under indirect observations. Part I*. (pp. 1-18). Amsterdam, The Netherlands: North Holland.

Mooijaart, A. (1978). *Latent structure models*. Unpublished doctoral dissertation, Leyden, The Netherlands.

Mulaik, S.A., James, L.R., Van Alstine, J., Bennett, N., Lenni, S., & Stilwell, C.D. (1989). Evaluation of goodness-of-fit indices for structural equation models. *Psychological Bulletin, 105*, 430-445.

Muthén, B., Kaplan, D., & Hollis, M. (1987). On structural equation modeling with data that are not missing completely at random. *Psychometrika, 52*, 431-462.

Nesselroade, J.R., & Baltes, P.N. (Eds.). (1979). *Longitudinal research in the study of behavior and development*, New York: Academic Press.

Neter, J., Maynes, E.S., & Ramanathan, R. (1965). The effect of mismatching on the measurement of response errors. *Journal of the American Statistical Association, 60*, 1005-1027.

Neter, J., & Waksberg, J. (1964). Conditioning effects from repeated household interviews. *Journal of Marketing, 28*, 51-56.

Nosanchuk, T.A. (1970). Pretesting effects: An inductive model. *Sociometry, 33*, 12-19.

Nunnally, J.C. (1973). Research strategies and measurement methods for investigating human development. In J.R. Nesselroade & H.W. Reese (Eds.), *Life-Span developmental psychology: Methodological issues* (pp. 87-109). New York: Academic Press.

O'Grady, K.E., & Medoff, D.R. (1988). Categorical variables in multiple regression: Some cautions. *Multivariate Behavioral Research, 23*, 243-260

Padioleau, J.G. (1973). L'analyse par cohortes appliquée aux enquêtes par sondages. *Revue Française de Sociologie, 14*, 513-528.

Palmore, E. (1978). When can age, period and cohort be separated? *Social Forces, 57*, 282-295.

Parsons, T. (1965). Youth in the context of American society. In E.H. Erikson (Ed.), *The Challenge of Youth* (pp. 110-141). New York: Doubleday.

Pelz, D.C., & Andrews, F.M. (1964). Detecting causal priorities in panel study data. *American Sociological Review, 29*, 836-848.

Pelz, D.C., & Lew, R.A. (1970). Heise's causal model applied. In E.F. Borgatta & G.W. Bohrnstedt (Eds.), *Sociological methodology 1970* (pp. 28-37). San Francisco: Jossey-Bass.

Pfeil, E. (1967). Der Kohorten-Ansatz in der Soziologie: Ein Zugang zum Generationsproblem? *Kölner Zeitschrift für Soziologie und Sozial-Psychologie, 19*, 645-657.

Plewis, I. (1981). A comparison of approaches to the analysis of longitudinal categoric data. *British Journal of Mathematical and Statistical Psychology, 34*, 118-123.

Plewis, I. (1985). *Analysing change: Measurement and explanation using longitudinal data*. Chichester, England: John Wiley.

Popper K.R. (1959). *The logic of scientific discovery*. London, England: Hutchinson.

Pressat, R. (1972). *Demographic analysis*. Chicago: Aldine Atherton.

Price, D.O. (1974). Constructing cohort data from discrepant age intervals and irregular reporting periods. *Social Science Quarterly, 55*, 167-174.

Pullum, T.W. (1977). Parametrizing age, period, and cohort effects: An application to U.S. delinquency rates, 1964-1973. In K.F. Schuessler (Ed.), *Sociological methodology 1978* (pp. 116-140). San Francisco: Jossey-Bass.

Pullum, T.W. (1980). Separating age, period, and cohort effects in white U.S. fertility, 1920-1970. *Social Science Research, 9*, 225-244.

Raftery, A.E. (1986a). Choosing models for cross-classifications. *American Sociological Review, 51,* 145-146.

Raftery, A.E. (1986b). A note on Bayes factors for log-linear contingency table models with vague prior information. *Journal of the Royal Statistical Society. Series B. 48,* 249-250.

Rao, J.N., & Thomas, D.R. (1988). The analysis of cross-classified categorical data from complex sample surveys. In C.C. Clogg (Ed.), *Sociological Methodology 1988* (pp. 213-270). American Sociological Association, Washington D.C.

Read, T.R., & Cressie, N.A. (1988). *Goodness-of-fit statistics for discrete multivariate data.* New York/Berlin: Springer Verlag.

Reichardt, C.S. (1979). The statistical analysis of data from nonequivalent group designs. In T.D. Cook & D.T. Campbell (Eds.), *Quasi-experimentation: Design and analysis issues for field settings* (pp. 147-206). Chicago: Rand McNally.

Reiser, M. (1981). Latent trait modelling of attitude items. In G.W. Bornstedt & E.F. Borgatta (Eds.), *Social measurement: Current issues* (pp. 117-144). Beverly Hills, CA: Sage.

Reynolds, H.T. (1977). *The analysis of cross-classifications.* New York: Free Press.

Riley, M.W. (1973). Aging and cohort succession: Interpretations and misinterpretations. *Public Opinion Quarterly, 37,* 35-49.

Riley, M.W., & Foner, A. (Eds.). (1968). *Aging and Society. Vol. 1. An inventory of research findings.* New York: Russell Sage.

Riley, M.W., Johnson, M., & Foner, A. (Eds.) (1972). *Aging and society. Vol. 3. A sociology of age stratification.* New York: Russell Sage.

Riley, M.W., Riley, J.W., Jr., & Johnson, M.E. (Eds.). (1969). *Aging and society. Vol. 2. Aging and the professions.* New York: Russell Sage.

Rindskopf, D. (1983). A general framework for using latent class analysis to test hierarchical and nonhierarchical learning models. *Psychometrika, 48,* 85-97.

Rodgers, W.L. (1982a). Estimable functions of age, period, and cohort effects. *American Sociological Review, 47,* 774-787.

Rodgers, W.L. (1982b). Reply to comment by Smith, Mason & Fienberg. *American Sociological Review, 47,* 793-796.

Rosenberg, M. (1968). *The logic of survey analysis.* New York: Basic Books.

Rosenberg, M., Thielens, W., & Lazarsfeld, P.F. (1951). The panel study. In C. Selltiz, M. Jahoda, M. Deutsch & S.W. Cook (Eds.), *Research methods in social relations. Part Two: Selected Techniques* (Chapter 18). New York: Dryden Press.

Rosow, I. (1978). What is a cohort and why? *Human Development, 21,* 65-75.

Rozelle, R.M., & Campbell, D.T. (1969). More plausible rival hypotheses in the cross-lagged panel correlation technique. *Psychological Bulletin, 71,* 74-80.

Rubin, D.B. (1974). Characterizing the estimation of parameters in incomplete-data problems. *Journal of the American Statistical Association, 69,* 467-474.

Ryder, N.B. (1951). *The cohort approach: Essays in the measurement of temporal variations in demographic behavior.* Unpublished doctoral dissertation, Princeton University, Princeton, NJ.

Ryder, N.B. (1965). The cohort as a concept in the study of social change. *American Sociological Review, 30,* 843-681.

Saris, W.E., & Satorra, A. (1988). Characteristics of structural equation models which affect the power of the likelihood ratio test. In W.E. Saris & I.N. Gallhofer (Eds.), *Sociometric research. Vol. 2. Data analysis* (pp. 220-236). London, England: Macmillan.

Saris, W.E., Satorra, A., & Sörbom, D. (1987). The detection and correction of specification errors in structural equation models. In C.C. Clogg (Ed.), *Sociological Methodology 1987* (Vol. 17, pp. 105-130). The American Sociological Association, Washington, DC.

Saris, W.E., & Stronkhorst, H. (1984). *Causal modelling in nonexperimental research,* Amsterdam, The Netherlands: Sociometric Research Foundation.

Satorra, A., & Saris, W. (1985). The power of the likelihood ratio test in covariance structure analysis. *Psychometrika, 50,* 83-90.

Schaie, K.W. (1965). A general model for the study of developmental problems. *Psychological Bulletin, 64,* 92-107.

Schaie, K.W. (1970). Age related changes in cognitive structure and functioning. In L.R. Goulet & P.B. Baltes (Eds.), *Life-span developmental psychology: Research and theory* (pp. 485-507). New York: Academic Press.

Scheuren, F., & Lock, O. (1975). Fiddling around with nonmatches and mismatches. *Proceedings of the Social Statistics Section of the American Statistical Association* (pp. 627-633). Washington, DC: American Statistical Association.

Schoenberg, R. (1977). Dynamic models and cross-sectional data: The consequences of dynamic misspecification. *Social Science Research, 6,* 133-144.

Schuman, H., & Presser, S. (1981). *Questions and answers in attitude surveys: Experiments on question form, wording, and context.* New York: Academic Press.

Schwartz, J.E. (1985). The neglected problem of measurement error in categorical data. *Sociological Methods and Research, 13,* 435-466.

Schwartz, J.E. (1986). A general reliability model for categorical data applied to Guttman scales and current-status data. In N.B. Tuma (Ed.), *Sociological methodology 1986* (Vol. 16, pp. 79-119). American Sociological Association, Washington, D.C.

Schwarz, G. (1978). Estimating the dimensions of a model. *Annals of Statistics, 6,* 461-464.

Sclove, S.L. (1987). Application of model-selection criteria to some problems in multivariate analysis. *Psychometrika, 52,* 333-343.

Sewell, W. H., & Hauser, R.M. (1975). *Education, occupation, and earnings.* New York: Academic Press.

Simonton, D.K. (1977). Cross-sectional time-series experiments: Some suggested statistical analyses. *Psychological Bulletin, 84,* 489-502.

Smith, H.L., Mason, W.M., & Fienberg, S. (1982). More chimeras of the age-period-cohort accounting framework: Comment on Rodgers. *American Sociological Review, 47,* 787-793.

Smith, K.W. (1976). Marginal standardization and table shrinking: Aids in the traditional analysis of contingency tables. *Social Forces, 54,* 669-793.

Smith, T.W. (1983). The hidden 25 percent: An analysis on the 1980 general social survey. *Public Opinion Quarterly, 47,* 386-404.

Sobel, M.E. (1988). Some models for the multiway contingency table with a one-to-one correspondence among categories. In C.C. Clogg (Eds.), *Sociological methodology 1988* (pp. 165-192). American Sociological Association, Washington, D.C.

Sobel, M.E., & Arminger, G. (1986). Platonic and operational true scores in covariance structure analysis. *Sociological Methods and Research, 15,* 44-58.

Sobol, M.G. (1959). Panel mortality and panel bias. *Journal of the American Statistical Association, 54,* 52-68.

Spencer, N. (1974). Panels or polls. *Journal of the Market Research Society, 16,* 283-286.

Steeh, C.G. (1981). Trends in nonresponse rates 1952-1979. *Public Opinion Quarterly, 45,* 40-57.

Steiger, J.H. (1979a). The relationship between external variables and common factors. *Psychometrika, 44,* 93-97.

Steiger, J.H. (1979b). Factor indeterminacy in the 1930's and the 1970's: Some interesting parallels. *Psychometrika, 44,* 157-167.

Steiger, J.H., & Schönemann, P.H. (1978). A history of factor indeterminacy. In S. Shye (Ed.), *Theory construction and data analysis in the behavioral sciences.* (pp. 136-178). San Francisco: Jossey-Bass.

Struening, E.L., & Guttentag, M. (1975). *Handbook of evaluation research* (Vol. 1). Beverly Hills, CA: Sage.

Stuart, A. (1955). A test for homogeneity of the marginal distributions in a two-way classification. *Biometrika, 42,* 412-416.

Sutcliffe, J.P. (1965a). A probability model for errors of classification. I. General considerations. *Psychometrika, 30,* 73-96.

Sutcliffe, J.P. (1965b). A probability model for errors of classification. II. Particular cases. *Psychometrika, 30,* 129-155.

Suyapa, E., Silvia, M., & MacCallum, R.C. (1988). Some factors affecting the success of specification searches in covariance structure modeling. *Multivariate Behavioral Research, 23,* 297-326.

Swafford, M. (1980). Parametric techniques for contingency table analysis. *American Sociological Review, 45,* 664-690.

Swaminathan, H., & Algina, J. (1977). Analysis of quasi-experimental time-series designs. *Journal of Multivariate Behavioral Research, 12,* 111-131.

Thomas, K. (1973). *Religion and the decline of magic: Studies in popular beliefs in sixteenth- and seventeenth-century England.* Harmondsworth, England: Penguin.

Thornton, A., Freedman, D.S., & Camburn, D. (1982). Obtaining respondent cooperation in family panel studies. *Sociological Methods and Research, 11,* 33-51.

Tuma, N.B., & M.T. Hannan (1984). *Social dynamics: Models and methods.* New York: Academic Press.

Turner, C.F., & Martin, E. (Eds.). (1984). *Surveying subjective phenomena* (Vols. 1, 2). New York: Russell Sage.

Van de Pol, F., & De Leeuw, J. (1986). A latent Markov model to correct for measurement error. *Sociological Methods and Research, 15,* 118-141.

Van der Eijk, C. (Ed.). (1980). *Longitudinaal Enquete Onderzoek: Mogelijkheden en Problemen.* (Mededelingen 6/7, FSW-A). Amsterdam, The Netherlands: Universiteit van Amsterdam.

Van der Heijden, P.G.M. (1987). *Correspondence analysis of longitudinal categorical data.* Leyden, The Netherlands: DSWO Press.

Visser, R. (1985). *Analysis of longitudinal data in behavioural and social research.* Leyden, The Netherlands: DSWO Press.

Wheaton, B. (1987). Assessment of fit in overidentified models with latent variables. *Sociological Methods and Research, 16,* 118-154.

Wiggins, L.M. (1955). *Mathematical models for the analysis of multi-wave panels.* Doctoral Dissertation Series No. 12.481. Ann Arbor, MI.

Wiggins, L.M. (1973). *Panel analysis: Latent probability models for attitude and behavior processes.* Amsterdam, The Netherlands: Elsevier.

Williams, W.H. (1969). The systematic bias effects of incomplete responses in rotation samples. *Public Opinion Quarterly, 44,* 593-602.

Williams, W.H., & Mallows, C.L. (1970). Systematic biases in panel surveys due to differential non-response. *Journal of the American Statistical Association, 65,* 1338-1349.

Wilson, K.L. (1979). Status inconsistency and the hope technique I. *Social Forces, 57,* 1228-1248.

Winsborough, H.H. (1975). Age, period, cohort, and education effects on earnings by race. In K.C. Land & S. Spilerman (Eds.), *Social Indicator Models* (pp. 201-218). New York: Russell Sage.

Winship, C., & Mare, R.D. (1989). Loglinear models with missing data: A latent class approach. In C.C. Clogg (Ed.), *Sociological Methods 1989* (pp. 331-369). Washington, D.C.: American Sociological Association.

Wohlwill, J.F. (1970). Methodology and research strategy in the study of developmental change. In L.R. Goulet & P.B. Baltes (Eds.), *Life-span developmental psychology: Research and theory.* (pp. 149-191). New York: Academic Press.

Yamane, T. (1973). *Statistics: An introductory analysis.* New York: Harper & Row.

Yee, A.H., & Gage, N.L. (1968). Techniques for estimating the source and direction of causal influence in panel data. *Psychological Bulletin, 70,* 115-126.

Zahn, D.A., & Fein, S.B. (1979). Large contingency tables with large cell frequencies: A model search algorithm and alternative measures of fit. *Psychological Bulletin, 86,* 1189-1200.

Zeisel, H. (1968). *Say it with figures.* New York: Harper & Row.

Zeller, R.A., & Carmines, E. (1980). *Measurement in the social sciences.* Cambridge, England: Cambridge University Press.

Zelterman, D. (1987). Goodness-of-fit tests for large sparse multinomial distributions. *Journal of the American Statistical Association, 82,* 624-629.

Author Index

Achen, C. H. 229
Adam, J. 325
Agresti, A. 83, 84
Aitkin, M. 61, 91
Akaike, H. 66, 67
Alba, R. D. 36
Algina, J. 303
Allison, H. E. 270
Allison, P. D. 16, 253, 287
Anderson, A. B. 88, 250
Andrews, F. M. 240
Andrich, D. 145
Arminger, G. 201, 270
Arnold, J. C. 210

Bahr, H. M. 337, 339, 349
Bailar, B. A. 270
Bailey, K. D. 201
Bailey, L. 270
Baltes, P. B. 21, 328, 329, 347
Barnes, S. H. 99, 128, 136, 149, 150
Basilevsky, A. 250
Bealer, R. C. 248, 270
Beck, P. A. 183
Benedetti, J. K. 61
Bengtson, V. L. 319
Bennett, C. A. 269
Bentler, P. M. 63, 65, 66, 91
Berelson, B. 147
Berger, M. P. 206
Berk, R. 51
Bhapkar, V. P. 210
Bielby, W. T. 91
Bijnen, E. J. 182
Bishop, Y. M. M. 21, 30, 44, 45, 50, 52, 54, 61, 68, 88, 90, 91, 146, 149, 156, 157, 158, 161, 162, 175, 181, 201, 211, 270, 300, 361
Blalock, H. M. Jr. 93, 358
Bock, R. D. 24
Boes, D. C. 50
Bohrnstedt, G. W. 164, 185, 201, 269
Bollen, K. A. 64
Bonett, D. G. 63, 65, 66, 91
Boruch, R. F. 269
Bozdogan, H. 67
Bradburn, N. M. 126
Breiger, R. L. 155, 156
Breitsamer, J. 319
Bridge, R. G. 270
Brier, S. S. 81
Brinser, A. 270
Brody, C. 268
Brown, M. B. 61, 162, 211, 361
Buchhofer, B. 320
Buck, S. F. 270
Burke P. J. 27, 90
Burstein, L. 358
Buss, A. R. 320
Butler, D. 270
Bye, B. V. 195

Caillez, F. 21
Camburn, D. 270
Campbell, D. T. 203, 204, 212, 213, 215, 217, 218, 220, 221, 240, 249, 250, 263, 269, 270, 272, 273, 301
Carlsson, G. 18
Carmines, E. 93
Carroll, S. S. 240
CBS – Centraal Bureau voor de Statistiek 336

385

Chaffee, S. H. 270
Chapman, D. W. 270
Chen, T. 253, 257, 259
Chuang, C. 83
Clarridge, B. R. 270
Clayton, K. N. 218, 220, 221
Clogg, C. C. 84, 88, 89, 90, 92, 94, 112, 113,
    115, 129, 145, 146, 187, 188, 190, 201,
    303, 360
Cobben, N. P. 315, 316, 321, 325, 327
Cochran, W. G. 210
Cohen, J. 65, 91
Coleman, J. S. 16, 240
Converse, P. E. 166, 277, 315, 348
Cook, T. D. 203, 212, 215, 217, 249, 250,
    263, 269, 301
Corby, C. 270
Cornelius, S. W. 328
Cox, D. R. 88
Cressie, N. A. 49
Crider, D. M. 248, 270
Croon, M. 143, 145
Cuttance, P. 93

Daniel, W. W. 250
Darroch, J. N. 91
Davis, J. A. 21, 147, 200, 208, 241, 269, 272
Davis, K. 319
Dayton, C. M. 194
De Leeuw, J. 21, 195
Deming, W. E. 270
Dempster, A. P. 145
Dillon, W. R. 114
Droege, R. C. 270
Duncan, O. D. 54, 149, 166, 169, 177, 181,
    240, 268
Dykstra, R. L. 83

Eckland, B. K. 270
Ecob, R. 93
Ehrenberg, A. S. C. 270
Eisenstadt, S. N. 319
Eliason, S. R. 88, 90, 92
Emerson, P. L. 312
Erdfelder, E. 91
Erlebacher, A. 269
Evan, W. M. 320
Everitt, B. S. 145

Evers, M. 55

Fairclough, E. H. 270
Fay, R. E. 253, 259, 263, 360
Fein, S. B. 66
Feiring, B. R. 330
Felson, M. 330
Fenech, A. P. 59
Ferber, R. 249, 263, 270
Fienberg, S. E. 21, 49, 51, 52, 74, 75, 81,
    90, 92, 253, 257, 259, 315, 327, 330,
    331, 335, 345, 347, 348, 349, 358, 359
Fiske, M. 147
Fitzgerald, R. 270
Foner, A. 315
Forthofer, R. N. 90
Freedman, D. 270
Friedrichs, J. 320
Fuchs, C. 253, 256, 257, 258, 259, 260
Fuller, L. 270
Furlong, M. J. 319

Gage, N. L. 240
Gart, J. J. 88
Gaudet, H. 147
Gifi, A. 21
Glenn, N. D. 315, 347
Glock, C. Y. 220, 221, 249, 270
Goe, S. J. 270
Gokhale, D. V. 21
Goldstein, H. 21, 218, 248
Goldthorpe, J. H. 155
Goodman, L. A. 15, 21, 23, 27, 45, 59, 61,
    65, 70, 77, 80, 81, 84, 88, 94, 97, 106,
    108, 111, 112, 113, 129, 153, 156, 162,
    163, 195, 198, 267, 279, 287, 328, 360
Gottman, J. M. 21, 278, 301, 303, 312, 313
Goudy, W. J. 270
Graybill, F. A. 50, 330, 345
Greenberg, D. F. 21, 164, 240, 248, 264,
    265, 358
Grizzle, J. E. 21
Grover, R. 105
Guttentag, M. 269

Haber, M. 162, 211, 361
Haberman, S. J. 44, 45, 49, 53, 54, 55, 59,
    62, 65, 84, 85, 87, 89, 90, 94, 104, 105,

107, 108, 111, 112, 149, 157, 161, 162, 175, 181, 211, 259, 270, 278, 279, 300, 312, 360

Hagenaars, J. A. 28, 95, 96, 106, 107, 115, 124, 126, 128, 129, 145, 155, 169, 170, 171, 178, 182, 183, 184, 193, 198, 210, 233, 234, 242, 248, 249, 253, 256, 257, 261, 263, 265, 267, 268, 270, 315, 316, 321, 325, 327, 358, 360
Hakim, C. 14
Halman, L. C. 96
Hamdan, M. A. 16, 210
Hamers, P. A. M. 95
Hannan, M. T. 240
Hansen, T. S. 270
Harackiewicz, J. M. 240
Harder, Th. 22
Hauck, W. W. 88
Hauser, R. M. 155, 156, 188, 270
Hay, R. A. 21, 303, 312, 313
Hays, W. L. 26, 61, 65, 90, 205, 206, 209, 210, 281
Heckman, J. 327
Heinen, A. G. 28, 95, 145, 263, 358
Heise, D. R. 241
Henry, N. W. 94, 113, 145, 146
Hibbs, D. A. jr. 303
Hildebrand, D. K. 21
Hocking, R. 253
Holland, P. W. 21
Hollis, M. 253
Holm, K. 24
Hope, K. 358
Hornung, C. A. 358
Hout, M. 84, 149, 163, 328, 347
Hsiao, C. 21
Huckfeldt, R. R. 22
Hum, D. P. 250
Hyman, H. H. 14, 23, 263

Inglehart, R. 100

Jacob, H. 14
Jagodzinski, W. 194, 327
Jephcott, J. S. G. 270
Jöreskog, K. G. 93, 113
Johnson, M. 315
Johnston, J. 24

Judd, C. M. 217, 219, 229, 230, 231, 269, 303

Kaase, M. 99, 128, 136, 149, 150
Kaplan, D. 63, 253
Kaufman, R. L. 36
Kawasaki, S. 89
Kendall, P. 182
Kenny, D. A. 217, 219, 229, 230, 231, 240, 269, 303
Kerlinger, F. N. 36
Kessler, R. C. 21, 164, 240, 248, 264, 265, 358
Kezouh, A. 83
Kiecolt, K. J. 14
Kiiveri, H. 80, 98, 107, 145
Kim, J. O. 89, 330
Kimberly, J. R. 248
Knoke, D. 27, 90, 347
Koch, G. C. 21, 166, 272
Koehler, K. J. 49
Kohfeld, C. W. 22
Krippendorf, K. 21
Kritzer, H. M. 166
Kruskal, W. H. 89
Kühnel, S. M. 194
Kullback, S. 21

Labouvie, E. W. 328
Laing, J. D. 21
Laird, N. M. 145
Lana, R. E. 270
Land, K. C. 330
Landis, J. R. 21, 166, 272
Langeheine, R. 94, 145
Larntz, K. 49
Laufer, R. S. 319
Lawal, H. B. 49
Lazarsfeld, P. F. 23, 94, 97, 113, 145, 146, 147, 149, 182, 183, 200, 240
Leaky III, F. J. 88
Lee, E. S. 90
Lemke, J. H. 83
Lenski, G. E. 358
Levenson, B. 200
Lew, R. A. 241
Light, R. J. 49
Likens, T. W. 22

Little, R. J. 250, 251, 258, 260
Lock, O. 248
Long, J. S. 36, 56, 58, 93
Lorimer, R. J. 90
Lösel, F. 252
Luijben, T. 63
Lüdtke, H. 320
Luijkx, R. 106, 124, 257
Lumsdaine, A. A. 269

MacCallum, R. 63
Maccoby, E. M. 152, 187
Macdonald, K. I. 155
Madden, T. J. 114
Madow, W. G. 250
Magidson, J. 51, 54, 89
Mallows, C. L. 270
Mannheim, K. 319, 320
Marascuilo, L. A. 210
Mare, R. D. 259
Margolin, B. H. 49
Marini, M. M. 253
Marsden, P. V. 156, 187, 189
Martin, E. 126, 270
Mason, K. O. 330, 345, 346, 347
Mason, W. M. 315, 327, 330, 331, 335, 345, 347, 348, 349, 358, 359
Matsueda, R. L. 91
Mayer, L. S. 240
Maynes, E. S. 248
McAllister, R. J. 270
McCain, L. J. 313
McCleary, R. 303, 312, 313
McCullagh, P. 91
McCullough, B. C. 240
McCutcheon, A. L. 94, 114
McGuire, W. J. 194
McLeod, J. M. 270
McNemar, Q. 210
McReady, G. B. 194
Medoff, D. R. 36
Meilijson, I. 105
Milligan, G. W. 49, 91
Mohr, L. B. 269
Mood, A. M. 50, 90
Mooijaart, A. 145
Mulaik, S. A. 61, 64
Muthén, B. 253

Namboodiri, N. K. 55
Nathan, L. E. 14
Nelder, J. A. 91
Nesselroade, J. R. 21, 328
Neter, J. 248, 270
Nosanchuk, T. A. 270
Nunnally, J. C. 362

O'Grady, K. E. 36
Olsen, A. R. 253
Oxspring, H. 253

Padioleau, J. G. 320
Pagès, J. P. 21
Palmore, E. 327
Parsons, T. 319
Pasanella, A. K. 200
Pedhazur, E. J. 36
Pelz, D. C. 240, 241
Pettigrew, H. M. 88
Pfeil, E. 320
Pirie, W. R. 210
Plewis, I. 21, 218
Poole, W. K. 330
Popper K. R. 61
Pressat, R. 314
Presser, S. 264
Price, D. O. 335
Pullum, T. W. 327, 329, 348

Raftery, A. E. 67
Ramanathan, R. 248
Rao, J. N. 90
Ratcliff, D. 91
Read, T. R. 49
Reese, H. W. 328
Reichardt, C. S. 269
Reiser, M. 145
Reynolds, H. T. 21, 88, 90
Riley, J. W. Jr. 315
Riley, M. W. 315, 321, 331
Rindskopf, D. 108
Ringer, D. W. C. 270
Robb, R. 327
Robinson, R. 240
Rodgers, W. L. 327, 348
Rosenberg, M. 23, 147, 200, 248
Rosenthal, H. 21

Rosow, I. 320
Rost, J. 94, 145
Rozelle, R. M. 240
Rubin, D. B. 88, 145, 250, 251, 253, 258, 260
Ryder, N. B. 315, 318

Salert, B. 22
Saris, W. E. 63, 91
Satorra, A. 63, 91
Sawyer, D. O. 145, 146
Schaie, K. W. 320, 323, 324
Schechter, E. S. 195
Schenker, N. 88
Schervish, P. G. 36
Scheuren, F. 248
Schmidt, P. 194
Schoenberg, R. 18
Schönemann, P. H. 115
Schultz, B. 88
Schuman, H. 264
Schwartz, J. E. 145
Schwarz, G. 67
Sclove, S. L. 67
Serlin, R. C. 210
Sewell, W. H. 270
Sheehy, L. L. 270
Silvia, M. 63
Simonton, D. K. 303
Sloane, D. M. 268
Smith, H. L. 331, 348, 358
Smith, K. W. 88, 89
Smith, T. W. 270
Sobel, M. E. 149, 201
Sobol, M. G. 249, 270
Sörbom, D. 63, 93, 113
Speed, T. P. 80, 98, 107, 145
Spencer, N. 248
Stanley J. S. 203, 204, 212, 215, 217, 270, 273, 301
Starmer, C. F. 21
Steeh, C. G. 270
Steiger, J. H. 115
Stephan, F. F. 270
Stokes, D. 270
Stouthard, Ph. 249
Stronkhorst, H. 91

Struening, E. L. 269
Stuart, A. 211, 312
Sutcliffe, J. P. 145, 201
Suyapa, E. 63
Swafford, M. 21
Swaminathan, H. 303
Swan, J. 51

Thielens, W. 147
Thomas, D. G. 88
Thomas, D. R. 90
Thomas, K. 13, 14
Thornton, A. 270
Tuma, N. B. 16
Turner, C. F. 126, 270

Van de Pol, F. 195
Van der Eijk, C. 274
Van der Heijden, P. G. M. 21
Visser, R. 21

Waksberg, J. 270
Warren, J. T. 240
Wauters, F. 249
Weidman, L. 88
Westfall, P. H. 59
Wheaton, B. 60, 65
Wiggins, L. M. 94, 146, 183
Williams, W. H. 270
Willits, F. K. 248, 270
Wilson, K. L. 358
Winship, C. 259
Winsborough, H. H. 317, 330, 347
Wohlwill, J. F. 325
Wüstendörfer, E. 252

Yamane, T. 32
Yee, A. H. 240

Zahn, D. A. 66
Zeisel, H. 209
Zeller, R. A. 93
Zelterman, D. 87
Zimmermann, K. F. 89
Zweifel, J. R. 88
Zwick, C. J. 270

Subject Index

Absolute difference scores 165
Additive model 33, 35, 91
Adjusted residual frequency 62, 87, 152
Adjustment for age differences 327
Adjustment for cohort differences 327
Adjustment for trend differences 327
Adjustment of posttest differences by
    pretest differences 216, 217
After only design 216
Age 17, 20, 315, 317-318
Age adjustment 327
Age-centrism 324
Age differences 17, 20
Age × cohort design 340-343
Age × period design 337-340
Age × period × cohort design 345-354
Aggregate change See Net change
AIC 67
Akaike's information criterion 67
Alternative χ^2-test statistics 87
Amount of information 61, 66
Analysis of covariance 218, 219, 221-223,
    230, 232
Analysis of variance 36, 54, 218, 219,
    223-228, 230
Apparent change 18
Approximation of chi-square distribution
    See Chi-square distribution
A priori zero cell See Structural zero cell
ARIMA 313
Arithmetic mean 32, 33
Assignment variable 217, 229, 230, 231
Association coefficient phi 65
Association model 84-87, 143, 163, 164
Attenuation of correlation 142, 214

Autocorrelation 234
Availability of data 13, 14, 19, 361
Average 32, 33
Average rate of change 32

Backward selection 62
Baseline model 63
Bayesian approach to zero cells 88 See also
    Zero cell
Bayesian information criterion BIC 67, 68
BIC 67, 68
Birth cohort 314, 318
BMDP 360
Boundary of parameter space 146
Boundary values 87, 107, 108, 110, 146

Categorical data 14-16, 21
Categorical (latent) variable 95-103
Causal analysis 25, 70-81
Causal feedback 81
Causal lag 241, 242
Causal model 77-81
Causal model with latent variables 135-143
Causal order 70, 79, 233-235, 241, 243
CDAS 91, 360
Change model 230, 231
Changes in association 176-180, 193
Changes in different periods 180-181, 193
Changes in different subgroups 166-169,
    193, 285-289
Changes in related characteristics 169-176,
    193, 289-300
Chi-square distribution 48, 49, 59, 87, 80,
    88, 146, 209
Chi-square statistics 87, 210

Classical error theory 185
Classification error 114, 183
Closed form expressions 52-54, 72, 125, 256
Cluster sampling 90
Coder error 182
Coding schemes for categories 86
Cohort 17, 19-21, 22, 314, 315
Cohort adjustment 327
Cohort-centrism 323
Cohort composition 318
Cohort design 20, 21, 314-316
Cohort fallacy 321, 324
Cohort replacement 19
Cohort-sequential design 324
Cohort table 315
Collapsed table 103
Collapsibility 25, 28, 41, 68, 69, 79, 88, 92, 230
Collinearity 327
Comparative analysis 127-135
Complete marginal homogeneity 172 See also Marginal homogeneity
Complexity of models 66
Complex sampling schemes 90
Composite hypothesis 79
Compound-interest formula 32
Conditional analysis 218, 219, 221-223, 230
Conditional association 42, 43, 91
Conditional logit 71
Conditional L^2 tests 50, 57, 58, 59, 61, 62, 88, 91, 120, 146, 258
Conditional modal latent class 114
Conditional odds 30, 31, 71
Conditional proportions 92
Conditional response probability 98, 109
Conditional test of marginal homogeneity 161, 162
Conditional two-variable effect 43
Confounding factors 204, 212, 213, 217, 273
Construct validity 263
Contextual variable 354, 355, 358
Continuation odds 92
Continuity correction 211
Continuous change 16, 22
Continuous characteristics 14

Continuous observation 15, 16
Control group 215, 217
Correction for attenuation 246 See also Attenuation of correlation
Corrections for nonresponse 250-253 See also Nonresponse
Correlated error terms 95, 148, 231, 242, 263-365
Correlated response error 126
Correspondence analysis 21
Covariance matrix 112
Cross-cultural analysis 127-135
Cross-lagged association 237, 240, 241
Cross-lagged panel correlation technique 234, 240-234, 241
Cross-product ratio α 30, 31, 43
Cross-sectional design 320, 321-323
Cross-sectional survey 17, 18, 147, 148, 271
Cross-sequential design 324
Cubic trend 303
Cumulative scale 122, 143, 145, 146
Curvilinear change 279-281

d-system 21
Data archives 14
Davis's d-system 21
Decaying effects 305
Degrees of freedom 48, 112, 113, 188, 205, 206, 258, 342
$\hat{\delta}$ Bonett and Bentler's goodness-of-fit index 66
Delta method 91
Demarcation line 64, 65
Demography 314
Dependent observations 169, 208, 335
Descriptive measures of goodness-of-fit 57-68
Design matrix 54-56, 85, 91, 279-280
DESMAT 312
Diagonal model 163, 328
Dichotomous dependent variable 24, 71-74
Difference equations 22
Difference scores 164-165, 218, 219, 232, 233
Differential equations 22
Direct effects between indicators 95, 126
Discrete change 16, 22

Discrete data 15, 16
Discrete observation 15, 16
Distance model 163
Disturbing factors *See* Confounding factors
Dummy coding 36, 56
Dummy variable 24, 36, 209
Dynamic equilibrium 18
Dynamic models 22

e — effect size 65
E-step 105, 106, 107 *See also* EM-algorithm
Ecological fallacy 355
ECTA 54, 341, 360
Effect coding 36, 37, 56
Effect component of times-series 303-305
Effect magnitudes 44, 45
Effect model 25, 70-81, 102
Effect parameter 35, 39
Effect size (e, w) 64, 65
Elaboration procedure 23, 24, 97
EM-algorithm 94, 103-108, 117, 124, 145, 146, 253-263
Empty cells *See* Zero cells
Equality restriction 108-110, 345, 348
Equivalent indicators 94, 110, 111, 145
Error type I 49, 91
Error type II 60
Error variance 185
Estimated expected frequency 27, 28, 39, 46, 48, 52, 81, 103, 161, 162 *See also* Expected frequencies
Estimated partial odds 40
Estimated sufficient statistics 104, 105, 107, 117, 253, 255 *See also* Sufficient statistics
Estimated unobserved frequency 104
Event-history analysis 15
Exogenous variable 264
Expected gross change 151
Experimental group 215, 217
Experimental stimulus 215
Extended table 157, 169, 193-200
External validity 212, 263
External variables 94, 95, 113-119, 138, 139, 360
Extreme values *See* Boundary values

F-distribution 65

F-statistic 65
Factor analysis 93, 94, 113, 145
Factor scores 94, 113, 115
Falsifiability 61
Fisher's information matrix 91
Fitting models 56-68 *See also* Model selection
Fixed marginals 50, 51, 52, 53
Fixed probabilities 110
Fixed sample size 26
Fixed sample size per stratum 26
Forward selection 61, 62
FREQ 54, 85, 278, 279, 305, 341, 360
Frequency model 25, 33-54, 70
Frequency table 23
Full table 28

Genealogical generation 319
Generalizability 213
Generalized IPF 91
Generalized least squares procedures 24
Generation 17, 19-21, 22, 319-320
Generation as actuality 320
Generation location 319
Generation replacement 19, 314, 315
Generation unit 320
Geometric mean 27, 32, 33, 39
GLIM 9, 360
Global maximum 108
Goodman and Kruskal's tau, gamma 89
Goodman's variant of EM-algorithm 106-108, 124
Goodness-of-fit statistics 57-68
Gross change 19, 147, 149-152, 271, 360
Guttman scale *See* Cumulative scale

Hauser's level model *See* Level model
Hierarchically nested hypotheses 58, 81
Hierarchical model 47, 48, 52-54, 75, 76, 91, 105, 106
Higher-order association coefficient α' 43 *See also* Cross-product ratio
Higher-order odds ratio α' 31
HILOGLINEAR 54
History 213, 215, 217, 273, 284
Homogeneity analysis 21
Hypothesis testing 46, 57-61

Identifiability 35, 36, 94, 98, 99, 108, 111-113, 115, 125, 146, 260, 326-332, 341, 345-348, 359
Identification problem *See* Identifiability
Identifying restrictions *See* Identifiability
Ignorable response mechanism 251, 258, 259, 262 *See also* Response mechanism
Impact panel 204
Impossible estimated values 24, 108
Imputing scores 253 *See also* Nonresponse, correcting for nonresponse
Independence gross change 151
Independence model 47, 48, 152, 200
Independence restrictions 98
Indicator 93, 95-103
Indicator variables *See* Gross change
Individual change *See* Latent class scores
Individual latent scores 113-115, 138, 139
Inequality restrictions 83, 143, 163, 330
Information gain 66
Information matrix 91, 105
Information theoretical approach 21, 89
Information theoretical measures of fit 66-68
Initial estimates *See* Start values
Instantaneous causal influence 239, 241
Instrumental variable 265
Internal validity 212, 263
Interocular test 278
Interpolation 335
Interrupted time-series design 272, 300-305
Interval variables 14, 15, 21, 24, 84
Interviewer effect 126, 358
Interviewer error 182
IPF 52-54, 91, 104, 105, 106, 254, 255
Isotonic regression 83, 143
Iterative proportional fitting *See* IPF

Joint (latent) variable 117, 146

Lambda λ 115
Large sample 49, 60
LAT 104, 105, 112, 145, 197, 360
Latent analysis of covariance 232
Latent change 181-183
Latent class 95-103
Latent class analysis 93-145
Latent class scores 94, 113-119, 138, 139
Latent cumulative scale 122

Latent distance model 146
Latent distribution 109
Latent extended table 193-200
Latent Guttman scale *See* Latent cumulative scale
Latent marginal homogeneity 193-200, 232
Latent modified path model 95
Latent partial symmetry 193-200
Latent probability model 183
Latent quasi-independence model 197
Latent quasi-symmetry 193-200, 232
Latent structure model 145
Latent symmetry model 193-200, 232
Latent trait model 145
Latent variable 93, 94, 95-103
LCAG 106, 124, 125, 188, 195, 244, 257, 267, 310, 311, 360
Level model 155, 159, 188
Life-course differences 17, 20
Life-course fallacy 321, 323
Likelihood ratio chi-square *See* Loglikelihood ratio chi-square
Likert scale 145
Linear change 279-281
Linear effect 56
Linear interpolation 335
Linear restriction 345
Linear time-series analysis 312
Linear trend 303, 304
Linear-by-linear interaction 163
LISREL 56, 63, 91, 93, 94, 95, 113, 125, 126, 253
Listwise deletion 252 *See also* Nonresponse
Local dependence model 126, 127, 263
Local identifiability 112
Local independence 97, 98, 126, 266
Local maximum 108
Log-bilinear association model 84, 86, 87, 143, 163, 164
Logit 70
Log-linear association model 84-86, 143, 163, 164
Log-linear model with latent variables 94, 98-145
Log-linear parameters 41
Logit model 25, 70-81
Loglikelihood ratio chi-square L^2 48-50, 74, 87, 88, 103, 106, 108, 112, 257, 258
LOGLINEAR 54

Longitudinal design 320, 323-324
Long term social change 19, 271

M-step 105, 106, 107 See also EM-
  algorithm
Magnitudes of effect 44, 45
Manifest change 181-183, 184
Manifest variable 93, 95-103
Marginal frequencies 50
Marginal homogeneity model 158, 161,
  169, 172, 174, 193-200, 211, 224-228,
  232, 295-298, 336
Marginal odds 30, 31, 33
Marginal table 50, 68, 79, 103
Markovian change 195
Mathematical programming techniques 330
Maximum effects 89
Maximum gross change 151
Maximum likelihood 21, 24, 26, 28, 39, 47,
  48, 49, 66, 90, 94, 103-108, 115, 253,
  254
Maximum negative effect 45
Maximum positive effect 45
McNemar's test 210
Mean effect parameter 89
Measurement error See Reliability
Measurement model 93, 135
Measurement unreliability See Reliability
Mediational model 229
Migration 336
Minimum chi-square procedures 21
Minimum gross change 151
Misclassification 114, 183
Mismatches 248, 249
Missing at random 251 See also Non-
  response
Missing completely at random 251 See also
  Nonresponse
Missing data 145
Missing in-between observations 16
MLLSA 106, 112, 360
Mobility table 149
Modal latent class 114
Model complexity 66
Model fitting 21, 94
Model selection 56-68, 91, 94, 113
Model testing 46, 94, 112
Modification index 63
Modified latent path model 123-126

Modified latent path model with latent vari-
  ables 135-143
Modified LISREL approach 95
Modified multiple regression 70-76
Modified path model 70, 77-82, 95,
  124-126, 195, 233-248
Monotonicity hypothesis 83
Mood 182
Mortality 273, 336 See also Nonresponse
Multicollinearity 347
Multinomial distribution 26, 90
Multiple correlation coefficient 65
Multiple regression 77
Multiplicative model 33, 34
MULTIQUAL 91
Multistage sampling 90
Multivariate frequency table 23, 24
Multivariate procedures 24

Nested design 340, 342, 343
Nested nonresponse pattern 256, 261
Net change 19, 149-152, 204, 210, 272, 360
Net change in two or more characteristics
  272, 289-300
Net changes in subgroups 285-289
NEWTON 104, 105, 112, 143, 145, 198,
  259, 310, 311, 360
Newton-Raphson procedure 52, 53-56, 76,
  104, 105
Nominal variable 15, 24
Nonequivalent control group design 203,
  215-233
Nonhierarchical model See Hierarchical
  model
Nonignorable response mechanism 251 See
  also Response mechanism
Nonrecursive causal model 81
Nonresponse 18, 19, 148, 203, 213, 217,
  249-263, 270, 273, 360
Nonresponse corrections 250-253
Normal distribution 24, 26, 45, 87
Normed parameters Q 89
Null hypothesis 61, 91

Observed change 181-183, 184
Observed frequency 27, 46, 103
Observed proportion 27, 28
Observed variable 93, 95-103
Observed variance 185

Odds 24, 27, 29-31, 35, 75, 92
Odds ratio 24, 27, 30, 31, 35
Omitted variable 126, 242
One-group pretest-posttest design 203-215, 216
One-shot survey 17, 18, 216, 271
One-variable effect 34, 35, 39, 72
Operational true score 201
Order restrictions See Inequality restrictions
Ordinal latent variable 143-145
Ordinal manifest variables 143
Ordinal variable 15, 24, 25, 82-87, 92, 95, 162-164
Orthonormal polynomials 312
Outlier 90
Overall effect 34, 35, 39, 71
Overfitting 56, 60, 62, 63

Pairwise deletion 252 See also Nonresponse
Panel attrition See Nonresponse
Panel design 18, 19, 21, 147, 148, 323
Panel mortality See nonresponse
Panel problems 148, 203, 248-269
Parabolic change 279-281
Parabolic effect 56
Parabolic trend 303
Parallel indicators 94, 110, 111
Parametrization of parameters 36, 346, 347
Parsimony 46, 61, 62, 117
Partial correlation coefficient 65
Partial cross-lagged correlations 240
Partial marginal homogeneity 172, 174
Partial odds 31-33, 39, 40, 41
Partial symmetry model 172, 173, 193-200, 224-228, 232, 296-298
Partitioning of (log)likelihood ratio chi-square L^2 50, 59
Partitioning of total association 81
Path analysis 79, 81, 203
Pearson chi-square 48-50, 74, 87, 88, 106, 112, 257, 258
Percentage difference d 89
Perfect indicators 96
Period 20, 21, 312, 315, 318 See also Time
Period analysis 314, 321
Period-centrism 321, 323

Period × cohort design 343
Period fallacy 323, 324
Phi (ϕ) 65
Platonic true score 201
Poisson distribution 90
Polytomous dependent variable 75-77
Polytomous variable 25, 44, 88, 89
Population replacement 273 See also Mortality
Posttest difference 216, 217
Posttest only design 216
Posttest scores 205
Power 59, 60, 63, 65, 87, 88, 91, 205, 206, 209, 211, 249
Practical panel problems 248
Prediction analysis 21
Pretest differences 216, 217
Pretest-posttest design 272
Pretest scores 205
Probability of misclassification See Misclassification
Processing errors 182
Product-moment correlation 89
Product-multinomial distribution 26, 90
"Psychometric" mistakes 182

Qualitative data 14, 21
Quantitative variables 14
Quasi-experimental design 148, 203
Quasi-independence model 152-154, 159, 190, 197
Quasi-latent variables 117-119
Quasi-symmetry model 158, 169, 193-200, 224-228, 232, 295-298

Raking 270
Random fluctuations around trend 301-305
Randomizing 205, 216, 272
Random nonresponse 249
Random peaks 301
Random valleys 301
Rasch scale 145
Recording errors 182
Recursive causal model 81
Regression analysis 24, 36, 54, 84, 91
Regularity conditions 146
Reinterview effect 18, 19, 148, 212, 217, 263, 265, 270

Reliability 93, 94, 95, 96, 110, 111, 142, 145, 148, 152, 181-183, 184, 213, 214, 215, 232, 233, 243, 246, 272, 306-312, 360
Repeated interviewing effect See Test-retest effect
Repeated surveys 18
Reproduction of distribution of independent variables 74
Reproduction of marginal tables 50, 51, 52, 53, 74
Residual frequency 61, 62, 87, 152, 266, 267, 268
Residual table 267
Response consistency effect 126
Response error 182 See also Reliability
Response mechanism 250-252, 258, 259, 262
Response probability 98, 109
Restricted latent class model 108-111

Sampling distribution 26, 90
Sampling fluctuations 24, 83, 277, 284
Sampling zeroes See Zero cell
SAS 360
Saturated effect model 71-73
Saturated frequency model 33-45
Saturated latent model 119-123
Saturated model 24
Scoring algorithm 104, 105
Secondary analysis 14, 273, 301
Selection 217, 229
Separate-sample pretest-posttest design 204-215, 216, 273, 300, 301
Sequential design 324-326
Simple random sampling 90
Simultaneous analysis in several groups 95, 127-135, 306
Simultaneous causal effects 81
Simultaneous latent class analyses 127-135
Small cell frequencies See Small sample, Zero cell
Small sample 23, 25, 49, 60, 87, 89, 360 See also Zero cell
Social desirability 126
Socratic effect 194, 212, 263
Software 360, 361
Sparse table See Small sample, Zero cell
SPSS-X 54, 360

Square table 148-152
Standard error of estimate 45, 46, 52, 54, 90, 105, 112, 206
Standardized effects 45, 46, 64, 72, 87
Standardized regression coefficients 46
Standardized residual frequency 62, 152, 266 See also Residual frequency
Start values 52-54, 106, 255, 256
Statistical conclusion validity 249
Statistical independence 30, 208
Status inconsistency 358
Stratified sample 26, 50
Stratified simple random sampling 90
Stratifying variable 31
Structural-functionalist generation concept 319
Structural modeling 93
Structural zero cell 90, 152, 188, 341, 342
Stuart's Q 211
Sufficient statistics 50, 52, 104, 107 117, 253, 255
Symmetric association 30, 31, 81
Symmetric measure of interaction 43
Symmetry model 156, 169, 193-200, 211, 224-228, 232, 295-298
Systematic trend component 301-305
Systematic true change 195

t-distribution 208
t-test 46, 205, 206, 209
Terminal maxima 108 See also Boundary values
Testing 212, 217
Testing models See Model selection
Testing procedures 56-68
Test-retest effects 18, 19, 95, 126, 203, 212, 249, 263, 264, 265, 360
Test statistic L^2 See Loglikelihood ratio chi-square L^2
Test statistics 74, 87, 90, 106
Theoretical variable 93, 95-103
Three-factor design 345
Three-variable effect 35, 43, 73
Time 16, 20, 21, 272, 273, 274, 312, 315 See also Period
Time lag 241, 242
Time-lag design 320, 324
Time-sequential design 324
Time series 21, 301

Transposed table 157
Trend *See* Net change
Trend adjustment 327
Trend component 301-305
Trend design 19, 20, 21, 147, 148, 204,
   271-274, 323
True change 18, 181-183, 184
True latent change 214
True model 56, 64
True score 182, 201
Turnover table 148-152, 162-164, 183-193
Turnover table with ordered categories 162-
   164
Two-factor design 340
Two-variable effect 35, 41, 43, 72
Type I error 49, 91
Type II error 60 *See also* Power
Typical panel problems *See* Panel problems
Typical panel questions 148, 165-181

Unconditional analysis 218, 219, 223-228,
   230
Unconditional tests 58, 88

Underfitting 56, 60, 62, 63
Unobserved frequency 104
Unreliability *See* Reliability
Unreliable measurements *See* Reliability
Unrestricted model 58
Unsaturated effect models 73,74
Unsaturated frequency model 46-55
Unsaturated hierarchical model 50
Unsaturated latent model 123-126
Unsaturated models 24, 25
Unstandardized regression coefficients 46

Validity 93, 94, 95
Variance of estimates *See* Standard error of
   estimate

w — square root of effect size 65
Weighted least squares procedures 21

Yea/nay-saying 126

z-distribution 208
z-test 209

About the Author

Jacques A. Hagenaars, born in 1945, is full professor of Methodology of the Social Sciences at Tilburg University, The Netherlands. He studied at Tilburg University specializing in Sociological Theory and Research Methods. He was a fellow at the Political Center of the Institute for Social Research at the University of Michigan in 1969/1970. His research interests are in causal analysis, especially with categorical data, reliability and validity of surveys, and designs, techniques, and methods for studying social change. He is the author of LCAG, a program for log-linear causal modeling with latent variables (and missing data), coeditor of a Dutch textbook on causal analysis, and has published many articles on the above topics in Dutch scholarly journals and in *Sociological Methods and Research, Social Science Research, Quality and Quantity, European Sociological Review,* and *The Netherlands Journal of Sociology.* At present, he is involved in the research projects "Missing data in log-linear analysis with latent variables," "Ordinal latent class and latent trait models," "The use of latent structure models in cross-national comparative research," and "The use and meaning of Mannheim's concept Generation in cohort analysis."